Signals, switches, regulons and cascades: control of bacterial gene expression

DNA sequencing has provided a wealth of information on the genetic material stored in bacterial genomes. The use of DNA arrays and proteomics will transform the scale of our ability to describe the patterns of gene expression as bacteria respond to their environments. However, the ability to control bacteria in a clinical context or exploit them in industrial or environmental contexts also depends on understanding the regulatory mechanisms which connect input experience to output response at the genetic level. This book deals with our current knowledge of the circuits and networks that govern bacterial gene expression – from the single gene to the whole genome – and which provide the framework for explaining the data from the post-genomics revolution.

David A. Hodgson is a Reader in Microbiology in the Department of Biological Sciences at the University of Warwick, UK.
Christopher M. Thomas is Professor of Molecular Genetics in the School of Biological Sciences at the University of Birmingham, UK.

Symposia of the Society for General Microbiology

Managing Editor: Dr Melanie Scourfield, SGM, Reading, UK
Volumes currently available:

SIXTY-FIRST SYMPOSIUM OF THE
SOCIETY FOR GENERAL MICROBIOLOGY
HELD AT WARWICK UNIVERSITY APRIL 2002

Edited by
D. A. Hodgson, C. M. Thomas

Signals, switches, regulons and cascades: control of bacterial gene expression

Published for the Society for General Microbiology

PUBLISHED BY THE PRESS SYNDICATE OF THE UNIVERSITY OF CAMBRIDGE
The Pitt Building, Trumpington Street, Cambridge, United Kingdom

CAMBRIDGE UNIVERSITY PRESS
The Edinburgh Building, Cambridge CB2 2RU, UK
40 West 20th Street, New York, NY 10011-4211, USA
477 Williamstown Road, Port Melbourne, VIC 3207, Australia
Ruiz de Alarcón 13, 28014 Madrid, Spain
Dock House, The Waterfront, Cape Town 8001, South Africa

http://www.cambridge.org

© Society for General Microbiology 2002

First published 2002

Printed in the United Kingdom at the University Press, Cambridge

Typeface Sabon (Adobe) 10/13.5pt   *System* QuarkXPress™   [SE]

A catalogue record for this book is available from the British Library

Library of Congress Cataloguing in Publication data

ISBN 0521 81388 3 hardback

Front cover illustration: Phase variant colonies of a *fimA–lacZ* fusion of a *fimE* mutant of *Escherichia coli* K-12 grown on lactose-MacConkey agar. Phase ON (pink, smaller) and phase OFF (white, larger) isolates can be distinguished as colony variants. Photograph provided by Ian Blomfield and Ray Newsam, Research School of Biosciences, University of Kent at Canterbury, UK.

CONTENTS

CONTRIBUTORS

Bibb, M. J.
Department of Molecular Microbiology, John Innes Centre, Colney, Norwich NR4 7UH, UK

Blomfield, I. C.
Department of Biosciences, University of Kent at Canterbury, Canterbury CT2 7NJ, UK

Browning, D.
School of Biosciences, The University of Birmingham, Birmingham B15 2TT, UK

Buck, M.
Department of Biological Sciences, Imperial College, London SW7 2AZ, UK

Busby, S.
School of Biosciences, The University of Birmingham, Birmingham B15 2TT, UK

Buttner, M. J.
Department of Molecular Microbiology, John Innes Centre, Colney, Norwich NR4 7UH, UK

Dixon, R.
Department of Molecular Microbiology, John Innes Centre, Colney, Norwich NR4 7UH, UK

Dorman, C. J.
Department of Microbiology, Moyne Institute of Preventive Medicine, Trinity College, Dublin 2, Republic of Ireland

Finn, R. D.
Wellcome Sanger Institute, Wellcome Trust Genome Campus, Hinxton, Cambs CB11 1SA, UK

Gerdes, K.
Department of Biochemistry and Molecular Biology, University of Southern Denmark, DK-5230 Odense M, Denmark

Green, J.
Department of Molecular Biology and Biotechnology, The University of Sheffield, Sheffield S10 2TN, UK

Harris, A. K. P.
Department of Biochemistry, University of Cambridge, Cambridge, UK

van Heel, M.
Department of Biological Sciences, Imperial College, London SW7 2AZ, UK

Henkin, T. M.
Department of Microbiology, The Ohio State University, 484 W. 12th Avenue, Columbus, OH 43210, USA

Hoch, J. A.
Division of Cellular Biology, Department of Molecular and Experimental Medicine, The Scripps Research Institute, 10550 North Torrey Pines Road, La Jolla, CA 92037, USA

Hong, H.-J.
Department of Molecular Microbiology, John Innes Centre, Colney, Norwich NR4 7UH, UK

Hood, D.
University of Oxford, Department of Paediatrics, John Radcliffe Hospital, Oxford OX3 9DU, UK

Lee, D.
School of Biosciences, The University of Birmingham, Birmingham B15 2TT, UK

Møller-Jensen, J.
Department of Biochemistry and Molecular Biology, University of Southern Denmark, DK-5230 Odense M, Denmark

Morrison, D. A.
Laboratory for Molecular Biology, Department of Biological Sciences, University of Illinois at Chicago, Chicago, IL 60607, USA

Moxon, R.
University of Oxford, Department of Paediatrics, John Radcliffe Hospital, Oxford OX3 9DU, UK

Müller-Hill, B.
Institut für Genetik der Universität zu Köln, Weyertal 121, 50931 Köln, Germany

Neidhardt, F. C.
Department of Microbiology & Immunology, University of Michigan Medical School, Ann Arbor, Michigan, USA

Orlova, E. V.
Department of Crystallography, Birkbeck College, University of London, London WC1E 7HX, UK

Paget, M. S. B.
School of Biological Sciences, University of Sussex, Brighton BN1 9QG, UK

Salmond, G. P. C.
Department of Biochemistry, University of Cambridge, Cambridge, UK

Schleif, R.
Biology Department, Johns Hopkins University, 3400 N. Charles St, Baltimore, MD 21218, USA

Whitehead, N. A.
Department of Biochemistry, University of Cambridge, Cambridge, UK

Williams, P.
Institute of Infections and Immunity, Queen's Medical Centre, University of Nottingham, Nottingham, UK

EDITORS' PREFACE

The current accumulation of complete genome sequences of bacteria is showing that, as genomes increase in size, the number of associated regulatory genes increases exponentially. That is, the more genes a bacterium has, the greater the need to regulate which genes are on or off and so tailor the bacterial properties to the prevailing circumstances, or selected niche. Thus, if we are to make sense of the growing genomic sequence resources, we need to understand not only the genetic content but also the regulatory networks controlling its expression. This symposium volume aims to help readers do just that.

The last time bacterial gene expression was the sole subject of an SGM symposium was 'Regulation of Gene Expression – 25 Years on' in 1986, which celebrated 25 years of the operon model of Jacob and Monod. The operon model has proved to be a seminal concept in our understanding of gene expression. This symposium aims to show that, in addition to transcription initiation, there are many other levels at which gene expression can be regulated in bacteria. For example, in the low-G + C ratio Gram-positives, regulated anti-termination of transcription is as common a mode of regulation as regulation of the initiation of transcription.

The title of the symposium is meant to reflect the different levels at which gene expression can be controlled in bacteria. Signals and switches activate expression of individual genes and operons. Individual genes and operons can be co-ordinately expressed to form regulons. Activation of expression of one gene can lead to activation of other genes that, in turn, activate still other genes to form cascades of gene expression.

David A. Hodgson
Christopher M. Thomas

Microbial reaction to environment: bacterial stress responses revisited in the genomic–proteomic era

Frederick C. Neidhardt

Department of Microbiology & Immunology, University of Michigan Medical School, Ann Arbor, Michigan, USA

INTRODUCTION

How microbial cells respond to their environment has always been at the core of microbial physiology. Now that genomics and proteomics are providing startling new opportunities for microbial physiologists, it is appropriate to assess the current state of this field and to hazard a guess about the direction of future studies.

It is particularly fitting that the Society for General Microbiology has provided an opportunity for this overview of bacterial adaptation, given the history of involvement of the Society with this subject. In conjunction with the third of its annual symposia, the Society for General Microbiology in 1953 published *Adaptation in Micro-Organisms* (Davies & Gale, 1953). Its 15 chapters examined different aspects of the response of microbial cells (chiefly bacterial) to changes in their environment. That volume was a testimony to the growing interest in the means by which microbial cells adjust to environmental circumstance, and was a landmark in the sorting out of mutational and physiological responses. Eight years later the same topic occupied the 11th symposium volume, *Microbial Reaction to Environment* (Meynell & Gooder, 1961). This volume proved to be the last devoted exclusively to responses to the environment before the era of molecular genetics made possible spectacular growth in understanding the molecular processes that are fundamental to microbial adaptability.

This review will begin with a discussion of some central notions of bacterial stress physiology. It will then develop the idea that stress responses, approached with the new tools of genomics and proteomics, provide a key to uncovering the central workings of the cell.

SGM symposium 61: Signals, switches, regulons and cascades: control of bacterial gene expression.
Editors D. A. Hodgson, C. M. Thomas. Cambridge University Press. ISBN 0 521 81388 3 ©SGM 2002.

STRESS RESPONSES AND BACTERIAL LIFE

Size matters. The microscopic dimensions of bacteria make possible their hallmark features: short generation time, simple body plan, enormous numbers, rapid evolution, metabolic diversity, persistence and ubiquity. Not every bacterial species displays all of these features, but the exceptions underscore the enormous variety of bacterial life.

With small size come disadvantages, the most prominent one being vulnerability to environmental forces. It is axiomatic that bacterial cells can be subject to life-threatening stresses as the chemical and physical properties of their surroundings change.

Many bacteria form macroscopic associations of billions of individual cells. The resulting *biofilms*, with their mass permeated by fluid-filled channels, enable some degree of modification of the local environment (for reviews see Costerton *et al.*, 1995; Rittman, 1999). On the whole, however, coping with environmental change is a feat accomplished by individual cells, whether free-floating (planktonic) or associated in biofilms. The structure and molecular processes of the prokaryotes are uniquely well suited to this end.

As will be seen later, genomic studies confirm the intuitive guess that capacity to resist stress is generally correlated with the nature of the environmental niche occupied by a particular bacterial species. A few bacterial species (few, relative to the total number) have opted for the comfort and stability of the *milieu interieur* of larger organisms by becoming obligatory parasites, even choosing in some cases a permanent intracellular existence. Evolution gradually deprives such bacteria of many of the coping mechanisms that are necessary in the wider world. One need only point to the mitochondria and plastids of eukaryotic cells to see this specialization and simplification carried to extreme.

Definitions of stress and the response to stress

Any change in the environment can evoke a response in bacterial cells. Commonly one thinks of *stress* as an environmental force that if unchecked decreases the growth rate of the organism, or causes it to cease growth entirely or to die. But even environmental changes that *increase* growth rate cause adaptive responses, and many regulatory systems respond to both beneficial and deleterious changes in the environment. Only the latter will be considered here.

The response a microbial cell makes to stress is generally called *adaptation*. At the population level, this term has also (regrettably) been applied to the selection of genetic variants better able to cope with the particular stress.

Table 1. Common environmental forces producing stress

Physical forces
Temperature

Hydrostatic pressure

Osmotic pressure

Radiation – UV, visible light; ionizing

Chemical forces
Presence of deleterious agents

   Inorganic compounds and ions

      Heavy metals

      Oxygen and its derivatives

      Protons

      Hydroxyl ions

   Organic compounds

      Antibiotics

      Acids, phenols, alcohols, etc.

Nutrient depletion or restriction

   Carbon and energy sources

   Nitrogen, sulfur, phosphorous sources

   Metal ions (Fe, Zn, Mg, Mo, Se, etc.)

   Oxygen or other electron acceptors

Endogenously produced compounds

   Protons

   Organic acids, alcohols, etc.

Variety of stresses

Stress can be caused by physical as well as by chemical forces (Table 1). Temperature and radiation (visible light, UV light and ionizing radiation) are the most intensively studied of the physical forces, but there is still much to learn about the responses to each of these, including further characterization of the heat- and cold-shock response systems. The ability of most bacteria to withstand a wide range of osmolarity is the result of both structural and metabolic adjustments called into play by changes in solute concentration. (Of course, the availability of water is one of the major factors – chemical or physical – in providing an hospitable environment; desiccation is a special stress, and one not easily examined.) Hydrostatic pressure is less studied in model organisms such as *Escherichia coli* than among barophilic marine species.

Chemical stresses are extremely numerous and can be grouped in a number of useful ways. First, there are adventitious components of the environment. Encounter with toxic substances is what customarily comes to mind when one thinks of bacterial stress responses; the number of inorganic and organic substances that damage cells is very great indeed. Inorganic agents include heavy metals and other toxic ions, oxygen and its derivatives (hydrogen peroxide and superoxide), and protons and hydroxyl ions. Organic substances, including antibiotics, inimical to growth of bacteria are likewise numerous. On the other hand, the chemical environment includes also organic and inorganic substances (nutrients) that are needed for cell growth, as well as substances that can be utilized as supplements to assist biosynthesis for growth. The absence of these nutrients, or their restricted concentration, is a common cause of nutritional stress.

Cells affect their environment, and compounds that are endogenously produced by the cells comprise an important chemical aspect of the environment. These substances can be quite toxic. Changes in the pH of the medium accompanying the production of short-chain organic acids and related by-products are prominent among the stresses produced by metabolism.

Biological stresses are not really separate, but rather provide special cases of chemical stress; nonetheless it is useful to note the special circumstances encountered by bacteria when a metazoan organism constitutes the environment. The general and specific defences of animals to infection are formidable, including the entire array of agents and cells constituting the immune response. It is not surprising that bacteria that have solved the problem of living intimately with animal hosts possess special systems triggered by the stresses encountered upon entering a potential host (for reviews see Mahan *et al.*, 1996; Dorman & Smith, 2001).

Given the panoply of potential hardships, it seems miraculous that a bacterial cell can cope with the real world. Indeed, informatic analysis of sequenced microbial genomes reveals that a considerable fraction of the genome may be devoted to regulatory genes, including the two-component regulatory systems that are prime cellular elements in sensing the environment and responding to change. As already pointed out, however, not every bacterial species is equally adept in responding to stress. Analysis (Stover *et al.*, 2000) of recently sequenced microbial genomes has revealed a close relationship between the lifestyle of an organism and the size and nature of its genome. The more environmental stresses an organism faces, the larger is its genome, and the greater is the fraction of its genome devoted to regulatory genes, including two-component response regulatory systems (B. Roizman, personal communication). As Stover *et al.* (2000) point out, *Pseudomonas aeruginosa*, noted for both its metabolic versatility and its ubiquity in the varying environment of soil, has a genome of approximately 5570 open reading frames (ORFs) of which nearly 470 (8.4 %) appear to encode regulatory proteins. Not as versatile, but still able to withstand a multitude of environments in its life cycle, is *E. coli*, which spends most of its life in the gut of animals, but which must survive transit from one host to another; it has a genome size of 4289 ORFs, of which nearly 250 are putative regulators (5.8 %). On the other hand, *Helicobacter pylori*, an organism restricted to the stomach as a habitat, has a genome of 1553 ORFs with only 17 putative regulatory genes (1.1 %). The *Mycoplasma genitalium* genome reveals that this obligate intracellular parasite possesses only 480 genes, and far fewer than 1 % exhibit the motifs of regulatory genes.

The problem of surviving a myriad of stresses is made more tractable by the fact that a cell need not mount a totally different response for each stress. Many toxic stresses share target structures, which can be protected and repaired by common mechanisms. Several deleterious conditions can be diminished in severity by the same modification of the cell envelope. Starvation triggers a general stress response independent of the specific nature of the missing nutrient (in addition to nutrient-specific alterations in metabolic fluxes). Similarly, the global stress response systems – catabolite repression (Saier *et al.*, 1996), stringency (Cashel *et al.*, 1996), the heat-shock (Yura *et al.*, 2000) and cold-shock (Phadtare *et al.*, 2000) responses, the SOS response (Walker, 1996) and the systems geared for oxidative stress (Storz & Zheng, 2000) – are effective against many hundreds of different specific stress factors. Finally, when all else fails, the cells can activate an emergency stress response system that prepares the cell for the broadly resistant non-growth state (Hengge-Aronis, 1996, 2000; Huisman *et al.*, 1996).

Response circuits

A greatly simplified diagram of a stress response system is shown in Fig. 1. Imbedded in it are all the major topics of this symposium: signals, switches, regulons and cascades. A *stress*, brought about by some change in the environment, is sensed by the bacterium. Commonly the *sensor* is a protein within the cell envelope. Deformation or modification of the sensor generates a *signal* (frequently a phosphorylation of the sensor) which is transmitted, in some cases through a train of phosphorylated proteins, to a *regulator* (commonly a DNA transcription modulator). The regulator modulates the expression of a set of *effector operons*, collectively termed a *regulon*, to produce the *responding proteins*, which bring about the cellular response. Feedback controls modulate the cellular response to match the extent of the deformation caused by the stress.

This flow chart is inadequate to represent the true complexity of the bacterial response to even simple stress stimuli. Fig. 2 more accurately depicts the real situation, and suggests the complexity of the regulatory circuit of a stress response. In reality, a stress

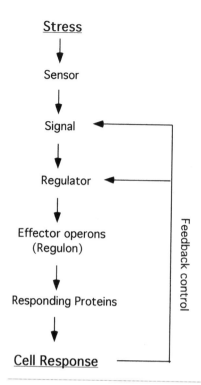

Fig. 1. Diagram of a simplified stress response circuit, including its feedback control.

usually triggers more than one signal, activating more than one regulatory cascade. Multiple regulons become involved in the response, some of them being activated secondarily as consequences of the activation of more primary regulons. One way to visualize this situation is by recognizing that some of the effector proteins involved in the primary response of a regulon to a stress signal may in turn be regulatory proteins involved in the subsequent activation of other regulons. The ensemble of responding proteins has been given the name *stimulon*. It is the stimulon of a stress that is usually observed; the task of sorting the responders of each stimulon into regulons requires the skilful application of the major tools of molecular genetics and physiology.

The challenge to contemporary molecular physiologists is formidable: to discover all the elements of the cell that respond to a stress stimulus, and then to account for the regulatory devices that achieve the overall response. The techniques of transcriptome and proteome monitoring to be discussed below should make easy work of the former task – which in any event is largely descriptive in its initial phases. On the other hand, the latter task will require all the approaches of traditional molecular genetics and physiology.

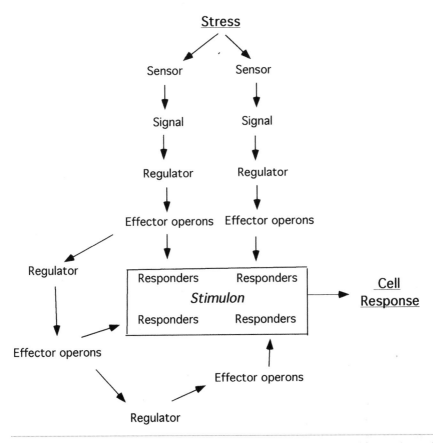

Fig. 2. Diagram of an idealized bacterial stress response circuit. An environmental force is depicted as triggering two signals from two separate protein sensors. Each signal is transmitted to a regulator protein, which results in the activation of several effector operons (a regulon). The regulon on the left includes a gene encoding an additional regulator, which in turn controls a regulon encoding still another regulator. From the four regulons of effector operons are produced protein responders that constitute the stimulon for this particular stress. For clarity, the numerous feedback control circuits have been omitted.

Outcomes of adaptation

The outcome of a bacterial stress response may fall anywhere along the spectrum from complete restoration of the pre-stress growth rate, to formation of a differentiated non-growth state that assures survival in the face of an insurmountable barrier to cell growth.

This situation may be visualized in Fig. 3, which depicts the range of responses observed upon initiation of a stress in a growing bacterial culture. Immediately or

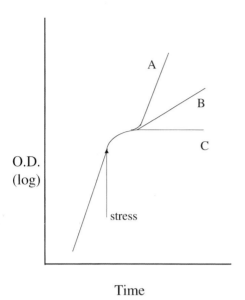

Fig. 3. Idealized bacterial growth curves depicting possible outcomes of response to an environmental stress. The optical density (OD) of a liquid culture is plotted on a semi-logarithmic scale against time. See text for discussion of curves A, B and C.

shortly after the onset of stress in the example given, there is a perceptible slowing of growth, resulting in a time during which the regulatory response is being invoked. As the effector molecules accumulate, the ability of the cell to cope with the stress is developed. The extent of success determines the subsequent growth rate. If the response of the cell completely overcomes the stress (e.g. the elimination of hydrogen peroxide by the action of catalase), resumption of the pre-stress growth rate would be expected (curve A, Fig. 3). If the cellular response ameliorates the stress, but cannot eliminate its effect on growth, a new steady-state is achieved at diminished growth rate (curve B, Fig. 3). If the cellular response is totally inadequate (e.g. complete depletion of the sole source of carbon and energy), growth never resumes, and the cells manage as best they will to differentiate into stationary phase (curve C, Fig. 3).

Adaptation physiology includes growth physiology

Growth studies lead to an important recognition. The successful outcome of a stress response can, and usually does, include establishment of a new state of balanced growth. Consider, for example, a population of *E. coli* cells shifted from rich to a leaner medium; the outcome, after a suitable adaptive response involving the formation of biosynthetic enzymes, is resumption of growth at a diminished rate. One can consider that *all states of balanced growth at rates lower than the maximum for that organism involve adaptation to stress.* This viewpoint may seem to stretch the issue somewhat, but indeed is quite on target because it generates the corollary that one can expect *every stress response of cells to include adjustments that optimize metabolism and maximize the growth rate,* in addition to components that are unique to the particular stress. Thus curtailment of synthesis of rRNA and ribosomal protein, for example, is to be expected during the transition to slow growth after imposition of a stress. The myriad adjustments that are made to optimize growth (Neidhardt *et al.,* 1990) should be kept in mind when analysing the results of global monitoring of adaptive responses in cells exposed to stress. Needless to say, the very large set of genes (for a summary see Hengge-Aronis, 1996) that respond to the switches and regulatory cascades preparing the cell for non-growth (stationary phase) can be expected to come into play during severe stress.

PROTEOMICS AND GENOMICS IN ADAPTATION PHYSIOLOGY

Role of global monitoring in studying stress responses

The early 1970s were a watershed in bacterial physiology. A phase change in research was brought about by the introduction by Patrick O'Farrell (1975) of two-dimensional polyacrylamide gel electrophoresis to separate the individual proteins of cells in a manner permitting their global observation and quantitation. The result was a fundamental change in the way physiologists could approach the cell. For the first time it became possible to phrase research questions about the response of a cell to its environment without prejudicing the answer. Instead of asking about the synthesis of this or that protein, or this or that RNA molecule of personal interest, the investigator could ask what proteins and genes *the cell* considered interesting during a given stress. Therein lies one of the landmark advances in bacterial physiology. Two-dimensional gels facilitate many kinds of studies on individual proteins (including making reverse genetics of expanded applicability), but their supreme contribution is lifting the veil from the eyes of physiologists. Recognition of the large number of proteins responding to an environmental change, introduction of the term *stimulon* to specify these responders (Fig. 2), and discovery of the widespread existence of stress *regulons* (sets of genes controlled by the same regulator) quickly followed the adoption of two-dimensional gel technology.

Though many insights were gained into the nature of the *proteome* (the cell's complement of proteins) and the workings of the bacterial cell by this technique, the transformation of bacterial physiology was slow (25 years slow!) in coming. The reasons are easily discerned. First, the introduction of facile recombinant DNA techniques for locating, isolating, splicing and cloning genes along with improved methods of *in vivo* genetic analysis reinforced the one-gene-at-a-time approach that had been so successful for decades. Second, the drawbacks to two-dimensional gel technology – chiefly the difficulties of obtaining high-quality resolution on the gels and of matching pattern images – discouraged its widespread adoption. Also, the task of establishing the genetic and biochemical identity of the two-dimensional gel 'spots' was formidable, and only a handful of investigators recognized the value of studying changes in the protein profile of the cell in the absence of spot identifications. Nevertheless, the proteomic gel approach flourished in selected laboratories and led to many advances in cell stress physiology (reviewed by Neidhardt & VanBogelen, 2000).

Completion of the genome sequences of many bacteria and other organisms in the latter half of the 1990s stimulated the development of informatics to interpret the vast amounts of information being revealed in these nucleotide sequences. At the same time, tools were developed from knowledge of nucleic acid chemistry, polymerase chain reaction and reverse transcriptase that enabled simulation of the complete genome of a bacterial species on a single chip. Using this tool, the *transcriptome* profile (the array of mRNA molecules present at any one time in a cell population) could be displayed in a manner inviting quantitative evaluation.

By the end of the millennium, bacterial physiologists had the ability to monitor the transcriptome, at least semi-quantitatively, and to assign relative expression levels to individual, identified genes. The subsequent studies provided the first glimpse into the transcriptional response (however semi-quantitative the results) of the cells to stress. The early results have conditioned a generation of microbial physiologists to appreciate the imperative of global monitoring of gene expression from cistron to mature protein product.

Comparison of proteomic and genomic monitoring

The importance of global monitoring is now thoroughly recognized. Perhaps less appreciated is the necessity of employing *both* proteomic and genomic techniques. Some reflection on the uses of each approach might be appropriate at this point (and might stimulate advances in these two vital technologies).

Genomic (transcriptional) monitoring. Completion of the sequencing of many microbial genomes opened the door to *transcriptional monitoring* (sometimes referred to as *genomic monitoring* because of its use of known fragments of the total genome), and brought about the current hectic pace of experiments that are generating thousands of pieces of information about the presence and amount of individual mRNA transcripts. The set of related methods employed need no description here (see, for example, the study on yeast by Shalon *et al.*, 1996). The first transcriptional profiling of *E. coli* (Richmond *et al.*, 1999; Tao *et al.*, 1999) described the central methods of the approach and the impressive display of information it can reveal about the cell. Since then, variations of the methods have been introduced – key differences lie in the nature of the DNA molecules used to capture the cDNA produced from the cells' mRNA.

The ready identification of every detected transcript – an incredible achievement – is perhaps the supreme success of genomic monitoring, and next in importance must be the completeness of the survey. Since there are no extraction or solubility problems (as plague protein preparations), a truly universal monitoring is afforded. Finally, the sensitivity of detection of minor transcripts is also quite impressive.

There are shortcomings, to be sure (perhaps the most prominent is the difficulty of obtaining absolute vs relative values for the individual mRNA species), but these will surely be overcome eventually. The technical triumphs of genome-based monitoring take on special significance when one recalls the degree to which regulation of gene expression in bacteria occurs by controls on transcription, particularly initiation. Knowledge of the transcript profile of the cell is of inestimable value in discerning the operation of transcriptional control circuits, quite independent of the correspondence of mRNA levels with the levels of their protein products.

Proteomic monitoring. While it is true that O'Farrell's (1975) development of two-dimensional gel electrophoresis for resolving total cell protein was a landmark event, this technique has frustrated scores of investigators for nearly a quarter of a century. In fact, the prominent characteristics of transcriptional monitoring underscore what have been the weakest aspects of proteomic monitoring. One-fifth of the proteins of *E. coli* are integral membrane proteins, and therefore possess characteristics that are inimical to ready solubilization; their hydrophobicity favours aggregation and precipitation from aqueous solutions. Many of the ORFs revealed in the *E. coli* genome by informatics encode quite small polypeptides – small enough to be lost during manipulation of samples in gel electrophoresis. Likewise, there are abundant basic proteins in *E. coli* with pI values above the range for which ampholytes can maintain a stable pH gradient. Finally, the range of abundance of cellular proteins extends over many orders of magnitude, creating an inherent difficulty of detection of minor species by a two-dimensional matrix. Of course, the very small, the very basic, the membrane-bound and the few in number are the proteins of great interest to the bacterial physiologist for among them are the sensors of the environment, the regulators of gene expression, and, more broadly, the keepers of the homeostatic potential of the cell.

Since monitoring mRNA avoids the difficulties of size, solubility, composition and abundance that plague monitoring proteins, why struggle with recalcitrant proteomics? The answer is that monitoring the proteome, that is, determining the *protein expression profile*, which is the quantitative catalogue of the proteins made by a cell under a partic-ular circumstance, cannot be avoided in any serious attempt to understand and to model the living cell. This is the case because of two paramount facts – almost trite in their obviousness. Half the cell's biomass is protein; and proteins, not mRNA, are the functional products of gene expression. It is their activities as signallers, carriers, regula-tors and metabolic catalysts that give life to the cell. Knowing the global pattern of transcription is important, but it cannot, by a long shot, provide the information physiologists need.

Physiologists need to have an account of the proteins of the cell – their kind and individ-ual number, how fast each is made and destroyed, what post-translational modifica-tions are made to each, and where they are placed in the structure of the cell (or outside it). These are minimal requirements…beyond which lay the matters of assessing for each protein its state of activity and the functional contacts it makes with other proteins and with nucleic acids.

For all its power, transcriptional (genomic) monitoring cannot provide any of this infor-mation. On the other hand, despite its weaknesses, proteomic monitoring by means of two-dimensional gel electrophoresis makes a surprisingly good pass at providing all six of the minimal parameters for those proteins that fall within its purview. The use of radioactive labels, singly or in pairs, and fed to the cells in steady-state or in pulse-labelling protocols created a powerful approach to ascertaining with reasonable preci-sion, sensitivity and accuracy the rates of synthesis and degradation of the nearly one thousand readily measured proteins in standard two-dimensional gels. The results can in most instances be expressed either relative to a standard (such as the rate of synthesis or cellular amount in a control culture), as is most frequently desired, or in absolute terms by reference to a protein with known cellular level in the same cell sample, or, less precisely, to the total protein content of the sample. Prior fractionation of the cell culture sample into envelope, outer and inner membrane, periplasm and cytosol frac-tions can be used in conjunction with isotopic labelling to ascertain the protein compo-sition of these cell compartments and monitor the appearance of new proteins in response to stress. Some protein modifications (including phosphorylations, acylations, methylations) can be discerned by the change in gel location of their products.

Nevertheless, despite projects that have led to genetic or biochemical identification of some 700–800 gel spots for *E. coli* samples, the effort has not come close to the total success of DNA arrays. Furthermore, use of these identifications in each experiment relies totally on accurate gel image matching. While image matching within sets of related gels and between gel sets has been a weak and labour-intensive step in the analytical process, recent advances (Compugen's Z3 software in particular) have greatly changed the picture.

But are technical advances in two-dimensional gel electrophoresis going to be the answer to proteomics problems? Many investigators are doubtful, and because the problems reviewed above (particularly protein solubility) seem to be intractable, attention has shifted to the application of mass spectrometry to proteomics. Various ways to utilize the exquisite mass discriminating capability of spectrometry have proved to be ideal for detecting and identifying proteins, either as intact polypeptides or as proteolytic fragments, and these techniques overcome or bypass the problems inherent in polyacrylamide gel electrophoresis. These several applications of mass spectrometry have been discussed and compared by Aebersold *et al.* (2000).

The major drawback to current mass spectrometric analytical techniques is the absence of ready quantitation. The signals generated by a polypeptide or polypeptide fragment are not precisely related to the quantity of the material detected, and the use of radioactive labelling is of course proscribed. A novel approach to achieve quantitation has been introduced by Aebersold and his colleagues (Gygi *et al.*, 1999; Aebersold *et al.*, 2000), in which isotope coded affinity tags (ICAT) are used to label proteins from two samples, one with an isotopically light and one with an isotopically heavy (deuterium-containing) form of the tag. The ICAT reagents are thiol-specific compounds that react with cysteine residues. The tags contain biotin to permit isolation of tagged proteolytic fragments by avidin–biotin affinity chromatography. The ratio of heavy to light forms of the derivatized peptides for individual proteins reveals the relative abundance of the proteins in the two samples.

While there is much to be done to advance the current technology, there seems to be no theoretical barrier to developing a mass spectrometric based technology that will achieve the goal of quantitative monitoring of the expressed proteome of the cell. If this happens (and if the process becomes 'democratized', i.e. becomes financially affordable for more than a handful of laboratories), two-dimensional gel electrophoresis will be relegated to simpler tasks than being the central technique in stress response studies.

Results from global monitoring

The first large-scale monitoring of the proteome of *E. coli* during a stress response was that of VanBogelen *et al.* (1996). A total of 816 proteins, representing nearly 3/4 of the cell's protein mass, were monitored while cells exhausted the phosphate in their medium and entered a non-growth state. The response was dramatic; 208 proteins were induced (differential rates of synthesis >twofold elevated) and 205 were repressed (differential rates of synthesis <0.5 of the pre-shift rate); the magnitude of change was as much as several hundred fold for many of these proteins. From the abundance of these proteins it could be determined that the 413 responders accounted for nearly 40 % of the cell's total protein-synthesizing capacity. In the same study, the proteome was monitored in a culture in steady-state growth at 42 % the normal rate, with phosphonate rather than phosphate serving as the source of phosphorous. Comparison of the responders in the two situations made it possible to tell which might be specifically related to the phosphate restriction and which to the shift into the non-growth state. In short, the adaptive response to phosphate restriction involved most of the induced proteins (i.e. 118 induced proteins were shared in the two conditions, while only 19 repressed proteins responded to both conditions). Adaptation to stationary phase seems to account for the large number of repressed proteins during phosphate exhaustion. Only 100 or so of the 800 proteins monitored in the study had been identified, underscoring the weakness of proteome monitoring by two-dimensional gels. Contrast this situation with the first global transcriptional monitoring of *E. coli* (Tao *et al.*, 1999), in which transcripts of all 4290 genes and ORFs were compared in cultures growing in minimal and rich media and the response of the various functional classes of genes (e.g. those encoding proteins of biosynthesis, macromolecule synthesis, energy metabolism, etc.) could be evaluated.

By now there have been dozens of global studies of stress responses, in a variety of organisms, and all point to one generalization: exactly as presaged by the results of the phosphate restriction study, stress responses usually involve many genes, as presaged in the phosphate limitation experiment. Part of the explanation has already been mentioned – stress usually results in a change in growth rate, and hence triggers all the adjustments that cells make to optimize metabolism at different rates of growth. These adjustments can easily include up to 75 protein-encoding genes, plus those for stable RNA (for reviews, see relevant chapters in Neidhardt *et al.*, 1996). At least four other functional classes of stress-responding proteins may be recognized: (a) proteins that deal directly with the given stress (e.g. catalase when hydrogen peroxide is the stress); (b) proteins that assist in coping with many stresses (e.g. the chaperones and proteases of the heat-shock response); (c) proteins that prepare the cell for survival during non-growth (e.g. proteins of the regulatory cascade headed by the stationary-phase sigma factor, sigma-s); and (d) proteins involved in repair of stress-damaged structures (e.g. RecA in the case of DNA damage). In addition, because the cell is a closed system, any large change in the differential rates of synthesis of major proteins can have indirect, passive effects on the synthesis of other proteins; this possibility is not easily tested, but proteins that show small, consistent increases or decreases in synthesis no matter what the stress are candidates for passive control effects.

A second discovery has emerged from these early global studies: *proteomic signatures –* tell-tale protein patterns that signify a dysfunction or a unique physiological state of some cellular process (VanBogelen *et al.*, 1999). Hundreds of stress experiments have been conducted with *E. coli* using proteomic monitoring by means of two-dimensional gels over the past 20 years. The accumulated data enable one to tabulate signatures of the cellular state of different systems within the cell. Thus high temperature, low temperature, different sorts of ribosome dysfunction, blockage of protein secretion, restriction for phosphorus or nitrogen source, oxidative toxicity, chromosome damage or malfunction – all these states exhibit characteristic protein profiles. Tabulation of these signatures, besides providing a means to diagnose the physiological state of a population of cells, is of great value in assessing the mode of action of new antibacterial agents.

RECONSTRUCTION AND MODELLING IN STRESS STUDIES

When a bacterial cell encounters a challenge in its environment, it responds. The physiologist wants to learn how the response is brought about. An initial step is to record what genes respond and learn how their responses change the activities of the cell; this involves both qualitative and quantitative challenges: identifying the responders, and measuring their response. The next step is to discern the control circuitry of the response network, or, stated in molecular genetic terms, to discover the regulatory elements and their functions. This task involves application of the classical techniques of mutant analysis and related biochemical follow-up using the information gained in the first step as to the nature of the players.

Up to this point the exploration involves the standard reductionist approach of molecular biology. There is growing support for the view that this approach simply will not suffice to achieve the end goal of a real understanding of the workings of the cell, for one wishes eventually to have knowledge sufficient to model the cell's activities and accurately predict its behaviour during steady-state growth and during newly imposed stress.

What is needed is a radical addition to microbial physiology. To understand the bacterial cell as a living entity will require tools traditionally associated more with engineering than biology. No regulon functions in isolation. Until one studies how genes act in context one cannot begin to sort out the rationale for the myriad patterns of molecular regulation in the cell. Biochemical systems analysis must be called upon to provide approaches to a *reconstruction*, as Savageau (1996) has called it, of the cell from the information gained by reductionist molecular biology. He has described (Savageau, 1976, 1996; Neidhardt & Savageau, 1996) one modelling tool, the power-law formalism, for dealing with non-linear systems such as the intact cell, but further development of modelling methods is greatly needed.

Quantitative information – of high precision – is needed to perform a systems analysis of the behaviour of the individual genes and proteins in intact stress response networks. This analysis is needed to discover the design principles of these networks and to construct a model that eventually will be powerful enough to predict correctly cell behaviour during stress. With the advent of genomic and proteomic techniques for global monitoring of gene function in the intact cell during stress, the way is open to this new physiology.

ACKNOWLEDGEMENTS

The author is grateful for helpful discussions with and assistance from many colleagues, particularly Ruth A. VanBogelen, Michael A. Savageau, Eduardo A. Roisman and Samuel I. Miller. Work in the author's laboratory was supported by the US Public Health Service (grant GM17892) and the National Science Foundation (grants DMB8903787 and MCB9417897).

REFERENCES

Aebersold, R., Rist, B. & Gygi, S. P. (2000). Quantitative proteome analysis; methods and applications. *Ann NY Acad Sci* **919**, 33–47.

Cashel, M., Gentry, D. R., Hernandez, V. J. & Vinella, D. (1996). The stringent response. In *Escherichia coli and Salmonella: Cellular and Molecular Biology*, 2nd edn, vol. 1, pp. 1458–1496. Edited by F. C. Neidhardt, R. Curtiss, III, J. L. Ingraham, E. C. C. Lin, K. B. Low, B. Magasanik, W. S. Reznikoff, M. Riley, M. Schaechter & H. E. Umbarger. Washington, DC: American Society for Microbiology.

Costerton, J. W., Lewandowski, Z., Caldwell, D. E., Korber, D. R. & Lappin-Scott, H. M. (1995). Microbial biofilms. *Annu Rev Microbiol* **49**, 711–745.

Davies, R. & Gale, E. F. (editors) (1953). *Adaptation in Micro-Organisms*. Third symposium of the Society for General Microbiology. Cambridge: Cambridge University Press.

Dorman, C. J. & Smith, S. G. J. (2001). Regulation of virulence gene expression in bacterial pathogens. In *Principles of Bacterial Pathogenesis*, pp. 76–133. Edited by E. A. Groisman. San Diego, CA: Academic Press.

Gygi, S. P., Rist, B., Gerber, S. A., Turecek, F., Gelb, M. H. & Aebersold, R. (1999). Quantitative analysis of complex protein mixtures using isotope-coded affinity tags. *Nat Biotechnol* **17**, 994–999.

Hengge-Aronis, R. (1996). Regulation of gene expression during entry into stationary phase. In *Escherichia coli and Salmonella: Cellular and Molecular Biology*, 2nd edn, vol. 2, pp. 1497–1512. Edited by F. C. Neidhardt, R. Curtiss, III, J. L. Ingraham, E. C. C. Lin, K. B. Low, B. Magasanik, W. S. Reznikoff, M. Riley, M. Schaechter & H. E. Umbarger. Washington, DC: American Society for Microbiology.

Hengge-Aronis, R. (2000). The general stress response in *Escherichia coli*. In *Bacterial Stress Responses*, pp. 161–178. Edited by G. Storz & R. Hengge-Aronis. Washington, DC: American Society for Microbiology.

Huisman, G. W., Siegele, D. A., Zambrano, M. M. & Kolter, R. (1996). Morphological and physiological changes during stationary phase. In *Escherichia coli and Salmonella: Cellular and Molecular Biology*, 2nd edn, vol. 2, pp. 1672–1682. Edited by F. C. Neidhardt, R. Curtiss, III, J. L. Ingraham, E. C. C. Lin, K. B. Low, B. Magasanik,

W. S. Reznikoff, M. Riley, M. Schaechter & H. E. Umbarger. Washington, DC: American Society for Microbiology.

Mahan, M. J., Slauch, J. M. & Mekalanos, J. J. (1996). Environmental regulation of virulence gene expression in *Escherichia, Salmonella*, and *Shigella spp.* In *Escherichia coli and Salmonella: Cellular and Molecular Biology*, 2nd edn, vol. 2, pp. 2803–2815. Edited by F. C. Neidhardt, R. Curtiss, III, J. L. Ingraham, E. C. C. Lin, K. B. Low, B. Magasanik, W. S. Reznikoff, M. Riley, M. Schaechter & H. E. Umbarger. Washington, DC: American Society for Microbiology.

Meynell, G. G. & Gooder, H. (editors) (1961). *Microbial Reaction to Environment.* Eleventh symposium of the Society for General Microbiology. Cambridge: Cambridge University Press.

Neidhardt, F. C. & Savageau, M. A. (1996). Regulation beyond the operon. In *Escherichia coli and Salmonella: Cellular and Molecular Biology*, 2nd edn, vol. 1, pp. 1310–1324. Edited by F. C. Neidhardt, R. Curtiss, III, J. L. Ingraham, E. C. C. Lin, K. B. Low, B. Magasanik, W. S. Reznikoff, M. Riley, M. Schaechter & H. E. Umbarger. Washington, DC: American Society for Microbiology.

Neidhardt, F. C. & VanBogelen, R. A. (2000). Proteomic analysis of bacterial stress responses. In *Bacterial Stress Responses*, pp. 445–452. Edited by G. Storz & R. Hengge-Aronis. Washington, DC: American Society for Microbiology.

Neidhardt, F. C., Ingraham, J. L. & Schaechter, M. (1990). *Physiology of the Bacterial Cell: a Molecular Approach.* Sunderland, MA: Sinauer Associates.

Neidhardt, F. C., Curtiss, R., III, Ingraham, J. L., Lin, E. C. C., Low, K. B., Magasanik, B., Reznikoff, W. S., Riley, M., Schaechter, M. & Umbarger, H. E. (editors) (1996). *Escherichia coli and Salmonella: Cellular and Molecular Biology*, 2nd edn. Washington, DC: American Society for Microbiology.

O'Farrell, P. H. (1975). High resolution two-dimensional electrophoresis of proteins. *J Biol Chem* **250**, 4007–4021

Phadtare, S., Yamanaka, K. & Inouye, M. (2000). The cold shock response. In *Bacterial Stress Responses*, pp. 33–46. Edited by G. Storz & R. Hengge-Aronis. Washington, DC: American Society for Microbiology.

Richmond, C. S., Glasner, J. D., Mau, R., Jin, H. & Blattner, F. R. (1999). Genome-wide expression profiling in *Escherichia coli* K-12. *Nucleic Acids Res* **27**, 3821–3835.

Rittman, B. (editor) (1999). *Microbial Ecology of Biofilms.* New York: Pergamon Press.

Saier, M. H., Jr, Ramseier, T. M. & Reizer, J. (1996). Regulation of carbon utilization. In *Escherichia coli and Salmonella: Cellular and Molecular Biology*, 2nd edn, vol. 1, pp. 1325–1343. Edited by F. C. Neidhardt, R. Curtiss, III, J. L. Ingraham, E. C. C. Lin, K. B. Low, B. Magasanik, W. S. Reznikoff, M. Riley, M. Schaechter & H. E. Umbarger. Washington, DC: American Society for Microbiology.

Savageau, M. (1976). *Biochemical Systems Analysis: a Study of Function and Design in Molecular Biology.* Reading, MA: Addison-Wesley.

Savageau, M. A. (1996). A kinetic formalism for integrative molecular biology: manifestation in biochemical systems theory and use in elucidating design principles for gene circuits. In *Integrative Approaches to Molecular Biology*, pp. 115–146. Edited by J. Collado-Vides, B. Magasanik & T. F. Smith. Cambridge, MA: The MIT Press.

Shalon, D., Smith, S. J. & Brown, P. O. (1996). A DNA microarray system for analyzing complex DNA samples using two-color fluorescent probe hybridization. *Genome Res* **6**, 639–645.

Storz, G. & Zheng, M. (2000). Oxidative stress. In *Bacterial Stress Responses*, pp. 47–60.

Edited by G. Storz & R. Hengge-Aronis. Washington, DC: American Society for Microbiology.

Stover, C. K., Pham, X.-Q. T., Erwin, A. L. & 28 other authors (2000). Complete genome sequence of *Pseudomonas aeruginosa* PA01, an opportunistic pathogen. *Nature* **406**, 959–964.

Tao, H., Bausch, C., Richmond, C., Blattner, F. R. & Conway, T. (1999). Functional genomics expression analysis of *Escherichia coli* growing on minimal and rich media. *J Bacteriol* **181**, 6425–6440.

VanBogelen, R. A., Olson, E. R., Wanner, B. L. & Neidhardt, F. C. (1996). Global analysis of proteins synthesized during phosphorus restriction in *Escherichia coli*. *J Bacteriol* **178**, 4344–4366.

VanBogelen, R. A., Schiller, E. E., Thomas, J. D. & Neidhardt, F. C. (1999). Diagnosis of cellular states of microbial organisms using proteomics. *Electrophoresis* **20**, 2149–2159.

Walker, G. C. (1996). The SOS response in *Escherichia coli*. In *Escherichia coli and Salmonella: Cellular and Molecular Biology*, 2nd edn, vol. 1, pp. 1400–1416. Edited by F. C. Neidhardt, R. Curtiss, III, J. L. Ingraham, E. C. C. Lin, K. B. Low, B. Magasanik, W. S. Reznikoff, M. Riley, M. Schaechter & H. E. Umbarger. Washington, DC: American Society for Microbiology.

Yura, T., Kanemori, M. & Morita, M. T. (2000). The heat shock response: function and regulation. In *Bacterial Stress Responses*, pp. 3–18. Edited by G. Storz & R. Hengge-Aronis. Washington, DC: American Society for Microbiology.

Gene variation and gene regulation in bacterial pathogenesis

Derek Hood and Richard Moxon

University of Oxford, Department of Paediatrics, John Radcliffe Hospital, Oxford OX3 9DU, UK

INTRODUCTION

Within each of the 12 major evolutionary branches of the bacteria, there are organisms capable of causing disease (pathogens). These pathogenic microbes, although a tiny minority of the total, are of immense importance because infection is a major cause of morbidity and mortality in animals and plants. In human terms, the global toll of pathogenic bacteria accounts directly for about 40 % of all deaths from infections, examples of which are tuberculosis, pneumonia, dysentery and meningitis. Thus, despite improved hygiene (especially clean water), antibiotics and vaccines, pathogenic bacteria pose formidable challenges. The application of molecular genetics, cell biology and genomics has revolutionized our knowledge of the epidemiology, diagnosis, therapy and prevention of infection. Looking to the future, it can be plausibly argued that future successes lie in coming to terms with the molecular minutiae of pathogenic microbes. Our current techniques and methodologies have taken us from the classical concepts of Koch to a reductionist mindset of defining host–microbial interactions (pathogenesis) at the molecular and even the atomic level. Virulence – classically defined through experimental infections in terms of, say, LD_{50} – can be detailed through defining the role of specific molecules that mediate adherence/penetration of host cells, evasion of host defences (innate and acquired), scavenging of nutrients and damage to host tissue. Characterizing these interactions at the molecular level and verifying their biological (*in vivo*) relevance currently represents one of the most exciting areas of applied science. Critical to these concepts is the iterative, dynamic nature of host–microbial interactions. Hence the particular relevance of this topic to this symposium on signals, switches, regulons and cascades. This mutuality has been captured in

SGM symposium 61: Signals, switches, regulons and cascades: control of bacterial gene expression.
Editors D. A. Hodgson, C. M. Thomas. Cambridge University Press. ISBN 0 521 81388 3 ©SGM 2002.

the metaphor of the 'gene for gene arms race' (Dawkins & Krebs, 1979) and the paradigm of the Red Queen (Van Valen, 1973). To survive and propagate, all organisms must maintain their fitness in diverse and changing environments and cells have therefore evolved mechanisms for responding to such changes. Pathogenic bacteria face particularly exacting tests of their adaptive potential due to the immense potential for generating diversity that is characteristic of their hosts, embodied in the repertoire of the B and T cells of adaptive immunity. Typically, bacterial infection occurs within a matter of days or even hours and involves a sequence of events that includes transmission (between genetically distinct hosts), colonization, dissemination and – albeit rarely – disease. The capacity of bacterial pathogens to run the gauntlet of the differing environments of its hosts is remarkable, especially since infections may involve clonal expansion of a single bacterial cell (Moxon & Murphy, 1978). The molecular basis of this adaptive potential ultimately depends on mechanisms that promote gene regulation and gene variation.

SIGNALS, REGULONS AND CASCADES

Host environmental factors, such as osmolarity, pH and available nutrients, influence bacterial pathogenicity by affecting the production of virulence determinants and by controlling the growth rate (Miller *et al.*, 1989). For example, when the animal pathogen *Salmonella typhimurium* is present in the intestinal lumen, several environmental and regulatory conditions modulate the expression of factors required for bacterial entry into host cells. An excellent example of the complex and co-ordinate regulation of virulence genes *in vivo* is expression of six different invasion genes, located on the pathogenicity island SPI-1, that are co-ordinately regulated by oxygen, osmolarity and pH (Bajaj *et al.*, 1996). An important point is that the activation of these virulence genes is obligatorily dependent on a multiplicity of environmental cues. Specific, combinatorial responsiveness through regulons provides a mechanism whereby bacteria behave in a very organized manner, subject to subtle variations in the different extracellular or intracellular environments encountered in the host. Another fascinating issue is that, as a general rule, physical contact between bacteria and host cells is required for the expression of virulence genes in animal hosts. Upon contact with a host cell, *Yersinia pseudotuberculosis* increases the rate of transcription of certain virulence genes in sequential fashion. The microbial–host interaction trigger for this 'cascade' is the export of LcrQ, a component of a type III secretion system and a negative regulator of the expression of a family of secreted proteins required for virulence (Yops). A decrease in the intracellular concentration of LcrQ mediates increased expression of Yops. Thus the type III secretion system plays a key role in the co-ordinate elaboration of virulence factors after physical contact with the target cell (Petterson *et al.*, 1996). Expression of virulence factors may depend on the microbial population attaining a critical density to trigger the elaboration of bacterial cell-to-cell signalling molecules (e.g. *N*-acetyl-L-homoserine lactone in *Pseudomonas aeruginosa*) (Winson *et al.*, 1995). *Staphylococcus aureus* uses a global regulator, *agr*, which is activated by secreted auto-inducing peptides, to control the expression of its major virulence genes. The auto-inducing peptides show sequence variation that affects their specificity. These peptides are thiolactones that either activate or inhibit depending on the conformation of the ligand–receptor interaction, the fine structural details of which are now understood in detail (Balaban *et al.*, 1998). This has made it possible to use synthetic variants of these peptides for *in vitro* and *in vivo* assays to establish their relevance and to open the door to novel strategies of infection control (Mayville *et al.*, 1999).

SWITCHES

Classical gene regulation provides only one example of the genetic mechanisms that have evolved to facilitate bacterial acclimation to their host. As emphasized already, bacterial infections typically occur within a matter of hours in which bacteria, disseminating within and translocating between hosts, encounter landscapes of such diversity that the prescriptive mechanisms of gene regulation may be inadequate to encompass the plethora of host factors that have evolved to eliminate them. How then do the relatively small numbers of bacteria that make up a potentially infectious inoculum generate the necessary diversity to adapt and thereby to evade these multifarious host clearance mechanisms? One strategy is through increasing the mutation frequency of those genes that are involved in critical interactions with their hosts. In many pathogenic bacteria, this capacity for variation is mediated through specialized DNA sequences, for example transposons or runs of repetitive DNA (microsatellites) that are hypermutable. DNA rearrangements through the translocation, loss or gain of DNA cause high-frequency and reversible on/off switching of associated genes. Given that there are often several such genetic loci in a single pathogen genome and that mutation occurs at random, the combinatorial effect on the phenotypic diversity of the pathogen population resulting from the independent switching of just a few such loci can be substantial (Moxon & Thaler, 1997). These genetic elements have been called contingency loci to emphasize their potential to enable at least a few bacteria in a given population to adapt to unpredictable and precipitous changes in the host environment (Moxon *et al.*, 1994; Barry & McCulloch, 2001; Deitsch *et al.*, 1997). Traits encoded by contingency genes include those governing recognition by the immune system, motility, attachment to and invasion of host cells, and acquisition of nutrients. In this review, we have sought to show how the interplay of gene variation and gene regulation impacts on the commensal and virulence behaviour of *Haemophilus influenzae*, an important pathogen of humans.

PHENOTYPIC SWITCHING OF CELL SURFACE ANTIGENS IS CHARACTERISTIC OF *HAEMOPHILUS INFLUENZAE*

H. influenzae (Hi), an obligate commensal of humans, resides predominantly in the upper respiratory tract. Encapsulated, mainly type b, strains are associated with occasional serious invasive diseases such as bacteraemic infections and meningitis. Non-typable strains (NTHi) lacking a capsule are less virulent but are a common cause of otitis media, sinusitis and lower respiratory tract infections, including life-threatening pneumonias. A feature of *Hi* is its propensity for switching the phenotypes of surface molecules, often referred to as phase variation. This reversible on/off switching occurs at high frequency, typically 1 in 1000 cells per generation, and involves surface molecules that interact with host structures (Weiser *et al.*, 1990b).

The major genetic mechanism mediating phase variation in *Hi* involves changes in the number of nucleotides in simple repeat elements, tandem iterations of short (1–7) nucleotides. In eukaryotes, these simple repeat tracts are called microsatellites, but until quite recently microsatellites were not thought to be a conspicuous feature of prokaryotes. Simple nucleotide repeats are relatively unstable and are hypermutable, allegedly through polymerase slippage (slipped-strand mispairing) during nucleic acid replication (Levinson & Gutman, 1987). This is a *recA*-independent mechanism that results in loss or gain of usually one, but sometimes more, repeat units. These tracts of repetitive DNA provide a mechanism for varying genes in which nucleotide repeats are prone to mispair as the complementary strands, annealed by classical Watson–Crick base-pairing, are separated spontaneously as DNA 'breathes' during transcription or replication. As the strands separate and the complementary base-pairing is temporarily disrupted, slippage may occur so that one strand is displaced relative to the other in either the 5′ or the 3′ direction. This results in one or more repeat units being mispaired when the strands realign (Fig. 1a). On the next round of replication, mismatch repair mechanisms correct the incompatibility, such that one or more repeats is gained or lost depending on the direction of slippage. However, loss or gain of a repeat unit (say one copy of a tetranucleotide, 5′-CAAT-3′, to take one example) will cause a frame shift, for translation depends upon having the correct triplet code. Thus, if there are n repeat units and this number of repeats is in-frame for translation, then $n-1$ or $n+1$ repeats, resulting from slippage, would cause a frame shift and either a truncated or altered peptide depending on whether the frame shift introduced a stop, or merely altered the triplet code. Thus, in this example, repetitive DNA promotes phase-variable expression of genes through its effect on translation of the encoded molecule. Phase variation mediated by simple nucleotide repeats can also occur in the promoter region of genes through its effect on RNA polymerase binding. An example is the transcriptional variation of *Hi* fimbriae, an adhesin mediating attachment to respiratory epithelia (van Ham *et al.*, 1993), resulting from slipped-strand mispairing of TA dinucleotide repeats associated with two divergently transcribed genes with overlapping promoters. The variable expression of outer-membrane proteins (HMW) is also subject to a switching mechanism upstream of the reading frame where tandem repeats of a heptanucleotide repeat are located.

Given the availability of complete genome sequences of many of the major bacterial pathogens, the presence of long tracts of simple nucleotide repeats can be used as a short cut to identify genes involved with bacterial virulence. A search of the complete genome sequence of *Hi* strain Rd for repetitive DNA sequences identified several genes possessing tetranucleotide repeats. These genes were predicted to encode proteins involved in commensal or virulence functions (Hood *et al.*, 1996a) and included genes required for sequestration of iron-containing compounds, lipopolysaccharide (LPS) biosynthesis, a putative adhesin and, perhaps surprisingly, components of restriction modification systems. A similar pattern of genes, but predominantly involving pentanucleotide and mononucleotide repeats, has been described in *Neisseria meningitidis* and *Neisseria gonorrhoeae* (Stern *et al.*, 1986; Yang & Gotschlich, 1996; Sarkari *et al.*, 1994; Saunders *et al.*, 2000). With the availability of complete genome sequences for many of the major bacterial pathogens, sensitive and specific software programs have been developed that can identify potential phase-variable genes. Some further recent examples include *Helicobacter pylori* (Saunders *et al.*, 1998) and *Campylobacter jejuni* (Parkhill *et al.*, 2000).

To put the above into context, simple repeats that mediate phase variation through slippage of repetitive DNA is but one of a number of distinct and well-described mechanisms of molecular switching that have been identified in pathogenic bacteria. Antigenic variation of the flagellar antigens of *Salmonella* was recognized by Andrews (1922) although its molecular basis was elucidated much later (Lederberg & Iino, 1956; Simon *et al.*, 1980). Variation of pilin genes of *Neisseria* occurs through recombination between a functional and many silent copies of the major structural gene (Stern & Meyer, 1987). Surface proteins of *Streptococcus pyogenes* have large repetitive tracts within the genes that are altered by recombinational events (Gravekamp *et al.*, 1996). In parasites, classical examples are the variant surface antigens of malaria parasites (Roberts *et al.*, 1992) and of trypanosomes (Borst & Rudenko, 1994).

What has been proposed for many years is that the capacity of pathogens to generate these repertoires of variant antigens is one of the critical mechanisms by which these microbes adapt to the differing microenvironments of the host and evade immune responses. In this chapter, we review the phase-variable genes that are involved in the biosynthesis of the LPS of *Hi*, and how this diversity is biologically relevant to its commensal and virulence behaviour.

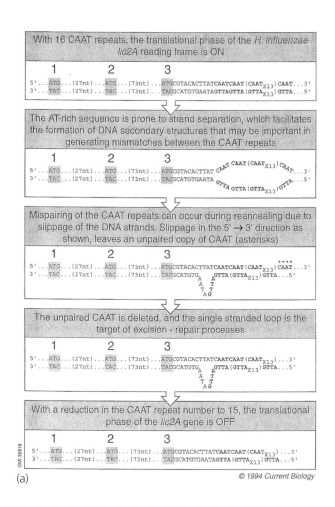

With 16 CAAT repeats, the translational phase of the *H. influenzae* *lic2A* reading frame is ON

```
        1              2              3
5'...ATG...(27nt)...ATG...(73nt)...ATGCGTACACTTATCAATCAAT(CAATx13)CAAT...3'
3'...TAC...(27nt)...TAC...(73nt)...TACGCATGTGAATAGTTAGTTA(GTTAx13)GTTA...5'
```

The AT-rich sequence is prone to strand separation, which facilitates the formation of DNA secondary structures that may be important in generating mismatches between the CAAT repeats

```
        1              2              3
5'...ATG...(27nt)...ATG...(73nt)...ATGCGTACACTTAT CAAT CAAT(CAATx13)CAAT...3'
3'...TAC...(27nt)...TAC...(73nt)...TACGCATGTGAATA GTTA GTTA(GTTAx13)GTTA...5'
```

Mispairing of the CAAT repeats can occur during reannealing due to slippage of the DNA strands. Slippage in the 5' → 3' direction as shown, leaves an unpaired copy of CAAT (asterisks)

```
        1              2              3                              ****
5'...ATG...(27nt)...ATG...(73nt)...ATGCGTACACTTATCAATCAAT(CAATx13)CAAT...3'
3'...TAC...(27nt)...TAC...(73nt)...TACGCATGTG      GTTA(GTTAx13)GTTA...5'
                                            A   T
                                            T   T
                                             AG
```

The unpaired CAAT is deleted, and the single stranded loop is the target of excision - repair processes

```
        1              2              3
5'...ATG...(27nt)...ATG...(73nt)...ATGCGTACACTTATCAATCAAT(CAATx13)...3'
3'...TAC...(27nt)...TAC...(73nt)...TACGCATGTG      GTTA(GTTAx13)GTTA...5'
                                            A   T
                                            T   T
                                             AG
```

With a reduction in the CAAT repeat number to 15, the translational phase of the *lic2A* gene is OFF

```
        1              2              3
5'...ATG...(27nt)...ATG...(73nt)...ATGCGTACACTTATCAATCAAT(CAATx13)...3'
3'...TAC...(27nt)...TAC...(73nt)...TACGCATGTGAATAGTTA(GTTAx13)GTTA...5'
```

(a)

© 1994 Current Biology

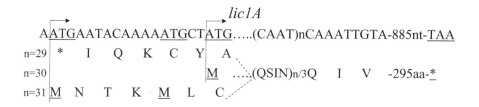

lic1A

AATGAATACAAAAATGCTATG.....(CAAT)nCAAATTGTA-885nt-TAA

n=29 * I   Q   K   C   Y   A

n=30   M   .....(QSIN)n/3Q   I   V   -295aa-*

n=31 M   N   T   K   M   L   C

lic2A

ATG-27nt-ATG-72nt-ATG-11nt-(CAAT)n-698nt-TGA

lic3A

(b)   ATGAACGGTACAATATGT-(CAAT)n-737nt-TAG

THE GENES FOR SEVERAL TRANSFERASES OF *H. INFLUENZAE* LPS BIOSYNTHESIS MEDIATE PHASE VARIATION

LPS is a critical, if not essential, component of *Hi*. LPS largely defines the permeability barrier around the cell and has an important role in mediating adherence to host cells, dissemination of bacteria by contiguous or systemic routes and in mediating inflammation and injury to host tissues. The membrane-anchoring lipid A portion (endotoxin) is responsible for much of the pathological damage associated with disease whereas the inner- and outer-core oligosaccharides promote adherence, invasion and extracellular survival. *Hi* LPS lacks the polymerized side chains (O-antigens) distal to the core in many pathogenic Gram-negative species such as the *Enterobacteriaceae*. These more truncated glycolipids of *Hi* have been designated lipo-oligosaccharides (LOS) by some investigators. However, we prefer to retain the designation LPS. Given the absence in *Hi* of LPS outer-core extensions similar to O-side chains, it could be concluded that its LPS is a simpler molecule than that from other Gram-negative bacteria. This is not so since it exhibits a significant degree of inter- and intra-strain heterogeneity (Fig. 2) as detected by reactivity with monoclonal antibodies (mAbs) (Kimura & Hansen, 1986) and staining of LPS by gel electrophoresis. Some LPS variation in *Hi* will be as a result of the complex biosynthetic processes that link the sugars and substituents (phosphate, phosphoethanolamine, pyrophosphoethanolamine, *O*-acetyl, phosphorylcholine or sialic acid) to complete the tertiary structure of the molecule (microheterogeneity). This structural variation will be a function of whether or not the biosynthetic processes go to completion since all possible sugar units and substituents may be non-stoichiometric, in the absence of any known proof-reading function. However, the major determinants of LPS heterogeneity in *Hi* are the specific genes that promote phase variation.

Fig. 1. (a) Possible mutational mechanism responsible for LPS phase variation in *H. influenzae*, based on slipped-strand mispairing (Levinson & Gutman, 1987). The DNA sequence of the 5′ end of the *H. influenzae lic2* gene, required for the biosynthesis of a Galα(1–4)Galβ disaccharide in the LPS core, is shown. Phase variation of the digalactoside is mediated through changes in the number of copies of tandem repeats of CAAT (bold); 16 copies of the tetranucleotide sequence are shown. Potential translational start codons (boxed ATG) for a long open reading frame are shown and are positioned such that ATG 1 and 2 are in-frame when *lic2* has 16 copies of CAAT, ATG 3 is in-frame if *lic2* has 17 copies of CAAT and there is no ATG in-frame if there are 15 copies of CAAT. Reprinted from *Current Biology*, **4**, Moxon, E. R., Rainey, P. B., Nowak, M. A. & Lenski, R. E., Adaptive evolution of highly mutable loci in pathogenic bacteria, 24–33, Copyright 1994, with permission from Elsevier Science. (b) Multiple initiation codons in the *lic* genes of *H. influenzae*. Initiation (ATG) and termination (TAA/TGA/TAG) codons are underlined. Shown in greater detail is the relationship between the number (*n*) of repeats of the tetranucleotide 5′-CAAT-3′ and the reading frame of the *lic1A* gene. The nucleotide sequence and the alternative initiation codons are shown with the amino acid sequence below. When 30 or 31 copies of the repeat are present, correct translation of the reading frame can take place from alternative ATGs in two of the three reading frames (shown by arrows). With 29 copies of 5′-CAAT-3′ present, no initiation codon is available for correct translation. The relative spacing of the alternative translational start codons for both *lic2A* and *lic3A* are also illustrated.

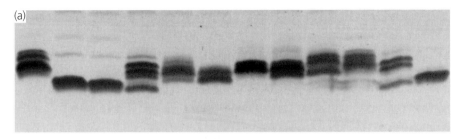

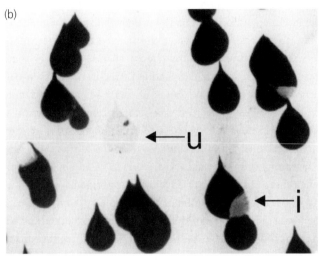

Fig. 2. (a) Fractionation profiles of LPS purified from 12 representative strains of *H. influenzae* after polyacrylamide gel electrophoresis and staining with silver. Each band within a given profile is considered to represent at least one glycoform of the LPS. (b) Colony immunoblots of *H. influenzae* staining with LPS-specific monoclonal antibody. The majority of colonies stain strongly. Arrows indicate intermediate (i) and unreactive (u) colonies or colony segments.

Phase variation results in multiple LPS structures within a population of cells (Fig. 2) and has confounded efforts both to study the biology of the molecule and to develop its potential as a common antigen for development of a vaccine to control disease. mAbs specific to LPS provide a necessary tool for detecting variations in LPS structure. They have been used to characterize the extensive variability and cross-reactivity of LPS structural epitopes from different organisms (Kimura & Hansen, 1986; Gulig *et al.*, 1987) and to study phase variation of LPS epitopes in individual strains of *Hi*.

A total of five LPS biosynthetic loci relevant to the phase-variable expression of LPS epitopes have currently been identified and characterized in *Hi*. The genes and some of the relevant features are listed (Table 1). Four of these phase-variable genes were iden-

Table 1. Phase-variable LPS biosynthetic loci identified in *H. influenzae*

Gene	HI number*	Repeat	Function	Reference
lic1A	HI1537	CAAT	Addition of phosphorylcholine	Weiser *et al.* (1989b, 1997)
lic2A	HI0550	CAAT	β-Galactosyltransferase	High *et al.* (1993)
lic3A	HI0352	CAAT	α2,3-Sialyltransferase	Maskell *et al.* (1991); Hood *et al.* (2001b)
lgtC	HI0259	GACA	α-Galactosyltransferase	Hood *et al.* (1996a)
lex2	–	GCAA	β-Glucosyltransferase	Jarosik & Hansen (1994)

*HI number as described for the strain Rd genome sequence (www.tigr.org).

tified by classical genetic approaches using LPS-specific mAbs to monitor LPS epitope expression patterns. The fifth gene, *lgtC*, was identified from the complete genome sequence of strain Rd (Fleischmann *et al.*, 1995; Hood *et al.*, 1996a). In all cases, the phase-variable genes are involved in synthesis of structures that are part of the outer core (Fig. 3) of the LPS molecule. Structural details are now available for the LPS from several *Hi* strains [RM153 (Eagan), RM7004, A2, RM118 (Rd), 486, 375, 1003, 2019, 9274]. However, it was not until the complete genome sequence of *Hi* strain Rd became available that a comprehensive study of LPS biosynthesis could be carried out. Over 30 genes were identified by homology comparisons with known LPS genes from other organisms (Hood *et al.*, 1996b) and gene function predictions for many have been confirmed by structural analysis of LPS from the relevant mutant strains. All of the major biosynthetic steps for the saccharide portion of the LPS in the *Hi* type d strain, RM118 (Hood *et al.*, 2001a), and the majority for the *Hi* type b strain RM153 (Hood *et al.*, 1996b) have been assigned.

The molecular basis of LPS phase variation in *Hi* was first characterized by Weiser and colleagues (Weiser *et al.*, 1989a, b). A genomic library from a type b strain, RM7004, was screened for clones that would allow a recipient strain, Rd, to express novel LPS epitopes recognized by murine mAbs to outer-core structures of *Hi* LPS (Weiser *et al.*, 1989b). Strain Rd does not naturally react with these mAbs, whereas strain RM7004 reacts in a phase-variable manner with each (Patrick *et al.*, 1989). A DNA clone from a genomic library made from RM7004 was identified which conferred some strong reactivity with two out of the four mAbs. This locus was designated *lic1* (lipopolysaccharide core epitope 1).

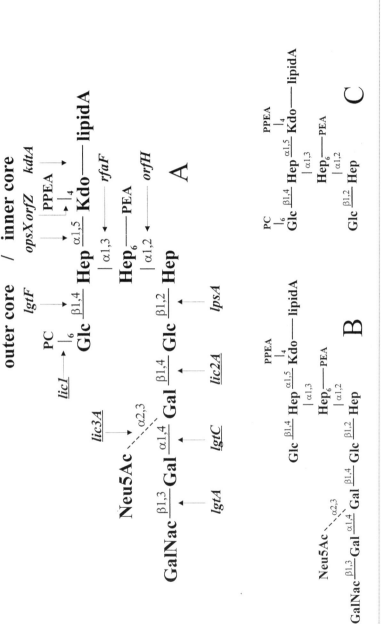

Fig. 3. Schematic representation of the structure of LPS from *H. influenzae* strain RM118 based on the results of the analysis of Risberg *et al.* (1999). Structure A, the full-length glycoform synthesized when all phase-variable genes are expressed. The proposed site of action of products of LPS biosynthesis loci is shown, linked by arrows to the relevant saccharide linkage (Hood *et al.*, 2001a). Phase-variable loci are underlined. The *lic3A* gene product can add an alternative sugar residue to the *lgtC* gene product; the latter predominates in LPS synthesis under *in vitro* growth conditions. Structure B results when *lic1* is not expressed, and structure C when *lic2A* is not expressed. Represented in the LPS structure: Kdo, 2-keto-3-deoxyoctulosonic acid; Hep, L-glycero-D-manno-heptose; Glc, D-glucose; Gal, D-galactose; GalNAc, *N*-acetylgalactosamine; Neu5Ac, 5-acetylneuraminic acid; PEA, phosphoethanolamine; P, phosphate; PC, phosphorylcholine. For the heptose residues, listed top to bottom are heptose I, heptose II then heptose III.

TETRANUCLEOTIDE REPEATS IN *lic1A* MEDIATE PHASE VARIATION OF PHOSPHORYLCHOLINE AND FACILITATE ADAPTATION TO DIFFERENT HOST ENVIRONMENTS

The nucleotide sequence of the *lic1* locus comprises four open reading frames (ORFs) encompassing 3.4 kb of DNA. The ORFs, designated *lic1A*, *lic1B*, *lic1C* and *lic1D*, are transcribed in the same direction, each adjacent ORF having overlapping stop and start codons (Weiser *et al.*, 1989a). This suggests that the *lic1* locus is an operon co-transcribing four ORFs. The *lic1A* gene contains the tetranucleotide 5′-CAAT-3′ repeat tract located within the ORF at its 5′ end (Fig. 1b). Loss or gain of copies of this repeat by slipped-strand mispairing would cause frameshifting with respect to the alternative start codons found in two of the three possible reading frames, resulting in variable gene expression. Weiser has shown that the *lic1* locus encodes genes for the incorporation of phosphorylcholine (PC) into *Hi* LPS, and mutant strains containing deletions encompassing all four ORFs of the *lic* locus have confirmed their essential role in the expression of PC. Lic1A has similarity to eukaryotic choline kinases, Lic1B to choline permease, Lic1C to a pyrophosphorylase and Lic1D is the predicted diphosphonucleoside choline transferase (Weiser *et al.*, 1997). The function of Lic1D, the phosphorylcholine transferase, is dependent upon an exogenous source of choline, a major component of host lipids, and it is proposed that choline is acquired from the host and assimilated into LPS. There are three levels of expression of PC using colony immunoblotting: strong ($++++$), weak ($+$) and undetectable ($-$). Oligonucleotides flanking the tetranucleotide region were used to amplify a portion of *lic1A* from chromosomal DNA obtained from colonies exhibiting each of the three levels of PC expression. The most prevalent size of DNA repeat (5′-CAAT-3′) found in non-reactive colonies was 184 bp equivalent to 29 repeats, whereas the size of the repeat for the $++++$ reactive colonies was 192 bp (corresponding to 31 repeats) and for $+$ colonies, 188 bp equivalent to 30 (5′-CAAT-3′) repeats. These results show that the degree of expression of PC in the LPS correlates with translation of *lic1A* from two alternative closely sited ATGs, or no translation (Fig. 1b). The predominant state appears to be 30 repeats under the experimental conditions, but loss or gain of one or more CAAT repeats results in strong or undetectable mAb binding, respectively. Further analysis of *lic1* gene expression using a *lacZ* reporter fused in-frame with *lic1D* in *Hi* (Moxon & Maskell, 1992) showed reversible high-frequency variation between three levels of β-galactosidase expression. The results were not entirely consistent with the patterns of PC expression described above, indicating that the number of repeats in *lic1A* influences the expression of PC but that there are other critical factors involved.

Although the genetic basis of the transcription and translation of the *lic1* locus is complex, the phenotypic consequences of the phase variation mediated by *lic1A* are that the population of bacteria includes a proportion of individual cells that either express or lack PC on LPS glycoforms. It is now apparent from experimental studies in animals that the presence or absence of PC may be advantageous in certain host compartments and under certain conditions, but not others. Weiser and colleagues have shown that expression of PC was favoured during nasopharyngeal colonization and carriage in the infant rat and that *Hi* expressing PC is prevalent in organisms cultured from human respiratory secretions (Weiser & Pan, 1998). In a different model of *Hi* infection, the induction of otitis media in chinchillas, it has been shown that expression of PC greatly favours survival of NT*Hi* strains in the nasopharynx and subsequent infection of the middle ear (Tong *et al.*, 2000). This raises the important issue of how the presence of PC on *Hi* LPS glycoforms mediates these biological functions and the significance of phase variation in the commensal and virulence behaviour of *Hi*.

Studies on *Streptococcus pneumoniae* were the first to implicate PC in mediating attachment to host epithelial cells through interactions with the platelet adherence factor receptor (PAF-R) (Cundell *et al.*, 1995). PAF-R is found on both human epithelial and endothelial cells and can result in a series of host signalling events. Subsequent investigations have shown that strains of *Hi* adhere to and invade human bronchial epithelial cells through an interaction of *Hi* LPS with PAF-R (Swords *et al.*, 2000). A further role stems from work showing that PC-expressing bacteria are more susceptible to the bactericidal activity of human serum. This increase in susceptibility occurs through the binding of C-reactive protein (CRP) (Weiser *et al.*, 1998) and subsequent activation of the classical pathway of complement. However, CRP (and complement) can be synthesized in the upper respiratory tract and might sterically interfere with the interaction of PC with PAF-R. The complexity of these potential interactions may have been a decisive factor in the evolution of phase variation of PC. Presumably, it has the potential to afford an adaptive mechanism through the natural selection of variants best suited to the differing host microniches and provide flexibility over the time course of colonization or infection (i.e. in the presence of inflammation). An additional proposed mechanism through which PC may affect bacterial survival is to confer relative resistance to antimicrobial peptides (Lysenko *et al.*, 2000a).

Distinct from the phase variation of PC within a clonal population, the precise location of PC on LPS glycoforms of genetically distinct strains can be different. The altered molecular environment of PC seems to be a factor in determining its accessibility, for example for binding of CRP and its subsequent effector role in mediating susceptibility to complement-mediated killing (Lysenko *et al.*, 2000b). Furthermore, allelic differences in Lic1D (the PC transferase) apparently preferentially direct the addition of PC to different glycoforms. An analysis of LPS structure from different strains of *Hi* has shown that, depending on the strain, each of the three oligosaccharide extensions of the triheptosyl backbone of the inner core can be substituted with PC (Lysenko *et al.*, 2000b; Mansson *et al.*, 2002).

H. INFLUENZAE SYNTHESIZES A PHASE-VARIABLE GALABIOSE EPITOPE ON ITS LPS THAT IS A MOLECULAR MIMIC OF HOST STRUCTURES AND IMPLICATED IN VIRULENCE

When digested chromosomal DNA from strain RM7004 was hybridized against the tetranucleotide repeats of the *lic1A* locus, it resulted in the discovery of two additional, distinct genetic loci, designated *lic2* and *lic3*. Deletion and insertion mutagenesis of the *lic2* locus indicated that the ORF (designated *lic2A*) contained multiple copies of 5'-CAAT-3' within its 5' end and had a function in LPS biosynthesis, based on loss of reactivity with LPS-specific mAbs (Weiser *et al.*, 1990b), related to formation of the digalactoside αGal(1–4)βGal epitope. This structure is found on the LPS of other bacteria (Virji *et al.*, 1990), is a component of the globoseries glycolipids expressed by some human epithelial cells and can act as receptor for the P fimbriae expressed by *Escherichia coli* and the B subunit of shiga toxin. A potential biological role for the αGal(1–4)βGal structure as a virulence determinant was hypothesized based upon its potential to mimic host structures. *lic2A* contains two possible start codons (ATGx and ATGy) upstream of the repeats in one frame and a third potential start codon (ATGz) closer to the repeats in another frame (Fig. 1b). By PCR amplification and direct sequencing of the repeat region, some correlation was again found between the number of 5'-CAAT-3' repeats in *lic2A* and the expression of the LPS epitopes recognized by an mAb specific for the digalactoside (High *et al.*, 1993). Three levels of mAb reactivity ('on', 'intermediate' and 'off') could be detected upon colony immunoblotting of *Hi* type b strains. These occurred at different relative frequencies in the strains. Direct sequencing of DNA amplified by PCR from immunostained colonies of each strain showed some correlation between the number of repeats in *lic2A* and the observed mAb reactivity. However, no absolute correlation of colony reactivity and the number of repeats in *lic2A* has been made. Much of this incomplete correlation of *lic2A* expression, digalactoside incorporation and mAb reactivity was made apparent with the discovery of a further LPS phase-variable gene, *lgtC*. Analysis of the complete genome sequence of strain Rd (Hood *et al.*, 1996a) revealed nine novel loci with multiple tandem tetranucleotide repeats. One of these was a homologue of a gene encoding a glycosyltransferase, *lgtC*, implicated in LPS biosynthesis of *Neisseria*. *lgtC* (the third of three contiguous ORFs) has 22 copies of the tetranucleotide 5'-GACA-3' located just within its 5' end and just 3' to a single ATG start codon. The function of *lgtC* was investigated by mutation of the gene in strain RM7004, producing a mutant with a very similar phenotype to that of strain RM7004*lic2A*. Deletion of *lgtC* in strain RM7004 showed loss of reactivity to mAbs specific for the digalactoside and the isolated LPS had a simpler electrophoretic profile when compared to wild-type. Mass spectrometric analysis of the LPS isolated from a number of *lgtC* mutant strains has indicated in each case that the terminal residue of the αGal(1–4)βGal epitope was absent (Fig. 2) (Hood *et al.*, 1996b, 2001a). Biochemical analysis of LgtC has confirmed that it is an α-galactosyltransferase (Hood *et al.*, 2001a). The importance of the digalactoside epitope to LPS biology is underlined by the evidence that two apparently independent phase-variable genes, *lgtC* and *lic2A*, contribute to its biosynthesis. Indeed, experiments have inferred that the digalactoside that is an extension from HepI of *Hi* type b LPS is added to a glucose residue which is itself added in a phase-variable manner by the *lex2* gene product. The apparent contradictions observed in experiments to monitor LPS gene repeat numbers and phenotype (as described above for *lic2*) are likely to be due to multiple phase-variable genes whose contributions to LPS phenotype are interdependent and capable of producing multiple reactive epitopes within the LPS molecule.

Phase variation of the digalactoside, a molecular mimic of a host structure, is likely to have a role in evasion of host immune defences. Weiser and colleagues demonstrated in human patients that greater than 99 % of cells isolated from the CSF of patients newly diagnosed with meningitis reacted with the appropriately specific mAbs (Weiser *et al.*, 1989a; Weiser & Pan, 1998). In contrast, among *Hi* cells isolated from the respiratory tract of people colonized with the organism, <0.1 % were reactive. This is consistent with the occurrence of phase variation *in vivo*. Expression of a digalactoside favours the survival of the organism against the killing effect of human serum (Cope *et al.*, 1991; Kimura & Hansen, 1986; Hood *et al.*, 1996a; Weiser *et al.*, 1990a; Weiser & Pan, 1998). An impairment of virulence has also been noted for a *lic1/lic2* double mutant strain that constitutively lacks the digalactoside epitope, when compared to the wild-type strain. This mutant is comparable to the wild-type in its ability to colonize the nasopharynx and survive in the bloodstream of the infant rat, but shows a reduced incidence and magnitude of bacteraemia after intra-nasal colonization (Weiser *et al.*, 1990a). The combined results from these experiments would indicate that the digalactoside structure is likely to play a role in translocation from the respiratory tract to the bloodstream and subsequent survival in the blood. The role of the digalactoside in serum resistance is presumed to be by presentation of a self-antigen (molecular mimic) to the host immune system. It is reasonable to propose that expression of this epitope is disadvantageous in alternative niches, such as the nasopharynx.

The fifth LPS-associated phase-variable gene characterized for *Hi* is *lex2*. A genomic library was constructed from the *Hi* type b strain DL42, which reacts with the LPS-specific mAb 5G8. Clones from this library were used to transform another type b strain, DL180, which is non-reactive with mAb 5G8 (Jarosik & Hansen, 1994). A clone was identified which conferred mAb 5G8 reactivity and this was sequenced to

reveal two contiguous ORFs. The first ORF, designated *lex2A*, contained the tetranucleotide 5′-GCAA-3′ repeated 18 times within its 5′ end (Jarosik & Hansen, 1994).
Transposon insertion mutagenesis indicated that the second ORF, *lex2B*, is required for
expression of the mAb 5G8-reactive LPS epitope. The *lex2B* gene has recently been
shown in our laboratory to encode a β-glucosyltransferase in type b strains (R. Aubrey,
D. Hood & E. R. Moxon, unpublished). This transferase permits further oligosaccharide chain extension to occur in the LPS of these strains. Thus three loci, each containing a gene with tetranucleotide repeats, have been identified that are relevant to
galabiose expression. Two of these loci, *lic2A* and *lgtC*, are transferases involved in the
synthesis of the same glycoform extension.

To study the impact on virulence of interdependent genes involved in phase-variable epitopes that may switch spontaneously during an experimental infection poses some
problem. To simplify interpretation, the tetranucleotide repeats can be deleted so as to
result in constitutive translation in the 'ON' phase. A mutant of *lic2A*, lacking the 5′-
CAAT-3′ repeats, was constructed in strain RM7004 (High *et al.*, 1996). The LPS from
this strain was identical to the wild-type strain as assessed by colony immunoblotting
and gel fractionation of purified LPS. Deletion of the 5′-CAAT-3′ repeats reduced the
rate of phase variation, but did not abolish it, indicating once more the involvement of
multiple genes in the synthesis of the relevant epitope. Using phase locked 'ON' mutants,
there was enhanced survival in infant rats of the *Hi* strain expressing two, as compared
to one, digalactosides in its LPS (R. Aubrey, D. Hood & E. R. Moxon, unpublished).

In conclusion, it must be biologically significant that so much has been invested in the
molecular switching of galabiose. It seems reasonable that this mimicry may be relevant
to avoidance of human host immune responses, but more evidence is required to
confirm this possibility.

A SIALYLTRANSFERASE MEDIATING SERUM RESISTANCE IS PHASE VARIABLE

The third *lic* locus identified, *lic3*, also contained the 5′-CAAT-3′ repeats in *Hi* type b
strains. DNA sequence analysis of cloned DNA containing the *lic3* locus identified four
ORFs of potential relevance to the 5′-CAAT-3′ repeats, designated *lic3* ORFs 1–4
(Maskell *et al.*, 1991). The first ORF, *lic3A*, contains 5′-CAAT-3′ repeats just within its
5′ end. Two possible start codons are located 1 and 15 bp upstream of the repeats (Fig.
1b). In the original experiments, mutation of *lic3A* had no detectable effect on LPS
structure and so gene expression could not be correlated to a phase-varying phenotype.
Recently, based on its homology to a sialyltransferase in *C. jejuni* (Gilbert *et al.*, 2000),
lic3A has been shown to encode an α-2,3-sialyltransferase required for the phase-variable addition of Neu5Ac to a lactose structure in *Hi* LPS (Hood *et al.*, 2001b).

Based on *in vitro* studies, sialic acid is likely to be critical in conferring resistance of *Hi* to complement-mediated killing and phagocytic ingestion. Incorporation of sialic acid as a terminal epitope in *Hi* LPS has been shown to be particularly relevant for resistance of capsule-deficient or NT*Hi* strains in the pathogenesis of otitis media, sinusitis and lower respiratory tract infections (Hood *et al.*, 1999, 2001b). Cell surface sialylation is a characteristic of many host cells and prevents complement activation by binding the serum regulatory protein, factor H (Varki, 1993). A majority, if not all, NT*Hi* strains display LPS sialylation and analysis of the strain Rd genome (a strain known to be capable of synthesizing sialylated glycoforms) and detailed metabolic studies (Vimr *et al.*, 2000) indicate that *Hi* must obtain sialic acid from an external source. Since *Hi* is an obligate commensal/pathogen of man, the source of sialic acid must be from the human host environment. However, sialic acid metabolism is evidently a balance between catabolic pathways and activation for incorporation into macromolecules. Furthermore, in addition to Lic3A that adds Neu5Ac in an $\alpha2,3$-linkage to a lactosyl residue, there are other enzymes involved in the incorporation of sialic acid into LPS in *Hi* (Hood *et al.*, 2001b). Recent experiments in a chinchilla model of acute otitis media suggest that LPS sialylation may be crucial in the pathogenesis of disease (S. Pelton and others, unpublished results). As shown for the phase-variable expression of PC, *Hi* can modulate its surface sialylation and this may be important for optimizing survival in different host compartments such as the nasopharynx, the Eustachian tube and the middle ear. For example, in another mucosal pathogen, *N. gonorrhoeae*, LPS sialylation down-regulates interactions with host epithelial cells, thus modulating microbial–host interactions.

Based on the complete genome sequence of strain Rd, *Hi* utilizes a limited number of exogenous sugars. Thus the dual role of sialic acid as both a major carbon and energy source and also as a structural component of LPS is intriguing. It has been shown that synthesis of the acceptor for LPS sialylation is altered in cells grown without exogenous sialic acid (Vimr *et al.*, 2000). Relatively little is known about how carbohydrates such as glucose and galactose from the host environment are utilized, presumably through substrate level phosphorylation, and how this influences other metabolic pathways and the supply of precursors for synthesis of macromolecules. The challenge is now to integrate details of the regulation of metabolism of sugars such as sialic acid with that of LPS biosynthesis, phase variation and the biology of commensal/virulence behaviour *in vivo* using microarrays. Some clues to the coupling of sialylation of LPS to the general metabolic activities and energy content of the cell may come from considering the location of the *lic3A* gene. DNA sequence analysis of the clone containing the *lic3* locus identified three potentially relevant additional ORFs downstream of *lic3A*. Immediately downstream of *lic3A* is an ORF predicted to make a protein that has 56 % identity with UDP-galactose 4-epimerase, designated *galE*. Mutation of the *galE* gene would perturb the balance of activated galactose and its availability for incorporation into LPS synthesis. Downstream of *galE*, the third and fourth ORFs of a possible *lic3* transcriptional unit are proteins with homology to enterobacterial AmpG and adenylate kinase (Adk). We have considered the possibility that the clustering of these genes in the *lic3* locus may involve a regulatory mechanism linking the biosynthesis of LPS (*lic3A* and *galE*) with the energy status of the cell (Adk) through a transducing mechanism involving AmpG. However, there is to date little evidence to support this hypothesis.

CONCLUDING REMARKS

Every organism is a special case and one of the challenges of biology is to identify unifying themes within a complex tapestry of detail. An overarching theme in biology is the power of diversity that, coupled with natural selection, provides the raw material for adaptation. This capacity to evolve fitness is counterintuitive in that evolution is not deterministic. It is worth emphasizing that *Hi* is predominantly a commensal and, for all intents and purposes, the forces driving its survival are those that promote persistence in the human respiratory tract. Thus its pathogenic potential is accidental or at least incidental to the maintenance of its basic reproductive rate, the number of new infections. The microbial determinants contributing to the basic reproductive rate include the capacity to regulate gene activity and to diversify genes. In response to the environment, there is synergy of gene regulation and gene variation at both the population (allelic) and individual (antigenic) level. Together, these are the mechanisms that provide the phenotypic substrate for the natural selection of fitness. In this context, LPS is a fascinating model macromolecule. It is the most prevalent component of the outer leaflet of the asymmetric Gram-negative bacterial cell envelope and confers important properties on the outer membrane facilitating the influx of essential nutrients while acting as a barrier to other substances. It is required for the efficient surface assembly of proteins that control, for example, cation or anion channels. However, LPS is also a key determinant of colonization, dissemination and disease. Thus the functions of LPS are essential to the commensal and pathological lifestyle of the organism. This mutuality and its complexity is reflected in the properties of gene regulation and gene switching and can be exemplified by considering sialic acid, one of the component sugars of LPS. Sialic acid is not essential to cell viability, but it does contribute to overall fitness. Sialic acid cannot be synthesized *de novo* but must be scavenged from the host. Furthermore, at least *in vitro*, given an exogenous source of sialic acid *Hi* can 'choose' whether to use it as a carbon and energy source through an anabolic pathway, or whether to incorporate sialic acid as a terminal sugar unit of its LPS. Here gene regulation plays a key role, although the details are poorly understood and, in particular, there are huge gaps in our knowledge concerning the relevance of sialic acid expression *in vivo* in both commensal and virulence behaviour. However, recent studies in an animal model of otitis media provide compelling evidence for the essential role of sialic acid in pathogenesis. In addition to the importance of gene regulation in the sialylation phenotype, a gene encoding a sialyltransferase, *lic3A*, is one of the several phase-variable LPS biosynthetic genes. Thus expression of sialic acid is subject to high-frequency, reversible on/off switching. A consideration of the role of sialic acid in microbial–host interactions suggests a plausible basis for the evolution of this gene switching. Sialic acid confers resistance to killing by complement, one of the key components of innate host immunity. This is achieved in part by its ability to bind factor H, one of the regulatory proteins of the complement cascade inhibiting the formation of C3b convertase. However, since sialic acid is an acidic negatively charged molecule, it can also down-regulate interactions of *Hi* with host cells. In the case of phagocytes, this may facilitate bacterial survival. On the other hand, this same property of sialic acid can decrease interactions with epithelial cells and thereby interfere with the efficiency of colonization. The capacity to generate variants with or without sialic acid could therefore represent a mechanism for altering phenotype contingent upon the circumstances of the host environment. In the case of the contingency (sic) gene *lic1A*, there is compelling evidence that phase variation is adaptive. As pointed out elsewhere, the evolution of multiple contingency genes provides a powerful stochastic process whereby a population of bacteria uses localized hypermutation to generate population diversity. If the proportion of hypermutable contingency genes is small compared to the total number of genes of the genome, this nonetheless confers the capacity for substantial phenotypic diversity within a clonal population, while minimizing the deleterious effects on fitness that accrue from the global hypermutation of a mutator genotype. As a concluding comment, we find it fascinating to analyse the relative extents to which different species of pathogenic and non-pathogenic bacteria have invested in gene regulation and gene variation for facilitating their adaptive potential. Life is about trade-offs. The phenotypic flexibility and fitness potential conferred by gene variation must be balanced against the increased probability of evolving virulence through the generation of variant phenotypes which, by chance or coincidence, have enhanced pathogenic potential.

REFERENCES

Andrews, F. W. (1922). Studies on group agglutination. *J Pathol Bacteriol* **25**, 515–521.

Bajaj, V., Lucas, R. L., Hwang, C. & Lee, C. A. (1996). Coordinate regulation of *Salmonella typhimurium* invasion genes by environmental and regulatory factors is mediated by control of *hilA* expression. *Mol Microbiol* **22**, 703–714.

Balaban, N., Goldkorn, T., Nhan, R. T. & 8 other authors (1998). Autoinducer of virulence as a target for vaccine and therapy against *Staphylococcus aureus*. *Science* **280**, 438–440.

Barry, J. D. & McCulloch, R. (2001). Antigenic variation in trypanosomes: enhanced phenotypic variation in a eukaryotic parasite. *Adv Parasitol* **49**, 1–55.

Borst, P. & Rudenko, G. (1994). Antigenic variation in African trypanosomes. *Science* **264**, 1872–1873.

Cope, L. D., Yogev, R., Mertsola, J., Latimer, L. J., Hanson, M. S., McCracken, G. H. & Hansen, E. J. (1991). Molecular cloning of a gene involved in lipooligosaccharide biosynthesis and virulence expression by *Haemophilus influenzae* type b. *Mol Microbiol* **5**, 1113–1124.

Cundell, D. R., Gerard, N. P., Gerard, C., Idanpaan-Heikkila, I. & Toumanen, E. I. (1995). *Streptococcus pneumoniae* anchor to activated human cells by the receptor for platelet activating factor. *Nature* **377**, 435–438.

Dawkins, R. & Krebs, J. R. (1979). Arms races within and between species. *Proc R Soc Lond* **205**, 489–511.

Deitsch, K. W., Moxon, E. R. & Wellems, T. E. (1997). Shared themes of antigenic variation and virulence in bacterial, protozoal and fungal infections. *Microbiol Mol Biol Rev* **61**, 281–293.

Fleischmann, R. D., Adams, M. D., White, O. & 37 other authors (1995). The genome of *Haemophilus influenzae* Rd. *Science* **269**, 496–512.

Gilbert, M., Brisson, J.-R., Karwaski, M.-F., Michniewicz, J. J., Cunningham, A. M., Wu, Y., Young, N. M. & Wakarchuk, W. W. (2000). Biosynthesis of ganglioside mimics in *Campylobacter jejuni* OH4384. Identification of the glycosyltransferase genes, enzymatic synthesis of model compounds, and characterization of nanomole amounts by 600 MHz ^{1}H and ^{13}C NMR. *J Biol Chem* **275**, 3896–3906.

Gravekamp, C., Horensky, D. S., Michel, J. L. & Modoff, L. C. (1996). Variation in repeat number within the alpha C protein of group B streptococci alters antigenicity and protective epitopes. *Infect Immun* **64**, 3576–3583.

Gulig, P. A., Patrick, C. C., Hermanstorfer, L., McCracken, G. H., Jr & Hansen, E. J. (1987). Conservation of epitopes in the oligosaccharide portion of the lipooligosac-charide of *Haemophilus influenzae* type b. *Infect Immun* **55**, 513–520.

van Ham, S. M., van Alphen, L., Mool, F. R. & van Putten, J. P. M. (1993). Phase variation of *H. influenzae* fimbriae: transcriptional control of two divergent genes through a variable combined promoter region. *Cell* **73**, 1187–1196.

High, N. J., Deadman, M. E. & Moxon, E. R. (1993). The role of the repetitive DNA motif (5′-CAAT-3′) in the variable expression of the *Haemophilus influenzae* lipopolysac-charide epitope Galα(1-4)βGal. *Mol Microbiol* **9**, 1275–1282.

High, N. J., Jennings, M. P. & Moxon, E. R. (1996). Tandem repeats of the tetramer 5′-CAAT-3′ present in lic2A are required for phase variation but not LPS biosynthesis in *Haemophilus influenzae*. *Mol Microbiol* **20**, 165–174.

Hood, D. W., Deadman, M. E., Jennings, M. P., Bisceric, M., Fleischmann, R. D., Venter, J. C. & Moxon, E. R. (1996a). DNA repeats identify novel virulence genes in *Haemophilus influenzae*. *Proc Natl Acad Sci U S A* **93**, 11121–11125.

Hood, D. W., Deadman, M. E., Allen, T., Masoud, H., Martin, A., Brisson, J. R., Fleischmann, R., Venter, J. C., Richards, J. C. & Moxon E. R. (1996b). Use of the complete genome sequence information of *Haemophilus influenzae* strain Rd to investigate lipopolysaccharide biosynthesis. *Mol Microbiol* **22**, 951–965.

Hood, D. W., Makepeace, K., Deadman, M. E., Rest, R. F., Thibault, P., Martin, A., Richards, J. C. & Moxon, E. R. (1999). Sialic acid in the lipopolysaccharide of *Haemophilus influenzae*: strain distribution, influence on serum resistance and struc-tural characterisation. *Mol Microbiol* **33**, 679–692.

Hood, D. W., Cox, A. D., Schweda, E. K. H., Walsh, S., Deadman, M. E., Martin, A.,

Moxon, E. R. & Richards, J. C. (2001a). Genetic basis for expression of the major globotetraose containing lipopolysaccharide from *Haemophilus influenzae* strain Rd (RM118). *Glycobiology* **11**, 957–967.

Hood, D. W., Cox, A. D., Gilbert, M. & 8 other authors (2001b). Identification of a lipopolysaccharide α-2,3-sialyltransferase from *Haemophilus influenzae*. *Mol Microbiol* **39**, 341–351.

Jarosik, G. P. & Hansen, E. J. (1994). Identification of a new locus involved in the expression of *Haemophilus influenzae* type b lipooligosaccharide. *Infect Immun* **62**, 4861–4867.

Kimura, A. & Hansen, E. J. (1986). Antigenic and phenotypic variants of *Haemophilus influenzae* type b lipopolysaccharide and their relationship in virulence. *Infect Immun* **51**, 60–79.

Lederberg, J. & Iino, T. (1956). Phase variation in *Salmonella*. *Genetics* **41**, 744–757.

Levinson, G. & Gutman, G. A. (1987). Slipped strand mispairing: a major mechanism for DNA sequence evolution. *Mol Biol Evol* **4**, 203–221.

Lysenko, E. S., Gould, J., Bals, R., Wilson, J. M. & Weiser, J. N. (2000a). Bacterial phosphorylcholine decreases susceptibility to the antimicrobial peptide LL-37/hCAP18 expressed in the upper respiratory tract. *Infect Immun* **68**, 1664–1671.

Lysenko, E., Richards, J. C., Cox, A. D., Stewart, A., Martin, A., Kapoor, M. & Weiser, J. N. (2000b). The position of phosphorylcholine on the lipopolysaccharide of *Haemophilus influenzae* affects binding and sensitivity to C-reactive protein-mediated killing. *Mol Microbiol* **35**, 234–245.

Mansson, M., Hood, D. W., Richards, J. C., Moxon, E. R. & Schweda, E. K. H. (2002). Structural analysis of the lipopolysaccharide from non-typeable *Haemophilus influenzae* strain 1003. *Eur J Biochem* (in press).

Maskell, D. J., Szabo, M. J., Butler, P. D., Williams, A. E. & Moxon, E. R. (1991). Molecular analysis of a complex locus from *Haemophilus influenzae* involved in phase-variable lipopolysaccharide biosynthesis. *Mol Microbiol* **5**, 1013–1022.

Mayville, P., Ji, G., Beavis, R., Yang, H., Goger, M., Novick, R. P. & Muir, T. M. (1999). Structure-activity analysis of synthetic autoinducing thiolactone peptides from *Staphylococcus aureus* responsible for virulence. *Proc Natl Acad Sci U S A* **96**, 1218–1223.

Miller, J. F., Mekalanos, J. J. & Falkow, S. (1989). Coordinate regulation and sensory transduction in the control of bacterial virulence. *Science* **243**, 916–922.

Moxon, E. R. & Maskell, D. J. (1992). *Haemophilus influenzae* lipopolysaccharide: the biochemistry and biology of a virulence factor. In *Molecular Biology of Bacterial Infection, Current Status and Future Perspectives*, pp. 75–96. Society for General Microbiology Symposium 49. Edited by C. E. Hormaeche, C. W. Penn & C. J. Smyth. Cambridge: Cambridge University Press.

Moxon, E. R. & Murphy, P. A. (1978). *Haemophilus influenzae* bacteraemia and meningitis resulting from survival of a single organism. *Proc Natl Acad Sci U S A* **75**, 1534–1536.

Moxon, E. R. & Thaler, D. S. (1997). The tinkerer's evolving toolbox. *Nature* **387**, 659–662.

Moxon, E. R., Rainey, P. B., Nowak, M. A. & Lenski, R. E. (1994). Adaptive evolution of highly mutable loci in pathogenic bacteria. *Curr Biol* **4**, 24–33.

Parkhill, J., Wren, B. W., Mungall, K. & 18 other authors (2000). The genome sequence of the food-borne pathogen *Campylobacter jejuni* reveals hypervariable sequences. *Nature* **403**, 665–668.

Patrick, C. C., Pelzel, S. E., Miller, E. E., Haanes-Fritz, E., Ruolf, J. D., Gulig, P. A., McCracken, G. H. & Hansen, E. J. (1989). Antigenic evidence for simultaneous expression of two different lipooligosaccharides by some strains of *Haemophilus influenzae* type b. *Infect Immun* **57**, 1971–1978.

Petterson, J., Northfelth, R., Dubinina, E., Bergman, T., Gustafsson, M., Magnusson, K. E. & Wolf-Watz, H. (1996). Modulations of virulence factor expression by pathogen target cell contact. *Science* **273**, 1231–1233.

Risberg, A., Masoud, H., Martin, A., Richards, J. C., Moxon, E. R. & Schweda, E. K. H. (1999). Structural analysis of the lipopolysaccharide oligosaccharide epitopes expressed by a capsule-deficient strain of *Haemophilus influenzae* Rd. *Eur J Biochem* **261**, 171–180.

Roberts, D. J., Craig, A. G., Berendt, A. R., Pinches, R., Nash, G., Marsh, K. & Newbold, C. I. (1992). Rapid switching to multiple antigenic and adhesive phenotypes in malaria. *Nature* **357**, 689–692.

Sarkari, J., Pantid, N., Moxon, E. R. & Achtman, M. (1994). Variable expression of the Opc outer membrane protein in *Neisseria meningitidis* is caused by size variation of a promoter containing poly-cytidine. *Mol Microbiol* **13**, 207–217.

Saunders, N. J., Peden, J. F., Hood, D. W. & Moxon, E. R. (1998). Simple sequence repeats in the *Helicobacter pylori* genome. *Mol Microbiol* **27**, 1091–1098.

Saunders, N. J., Jeffries, A. C., Peden, J. F., Hood, D. W., Tettelin, H., Rappuoli, R. & Moxon, E. R. (2000). Repeat-associated phase variable genes in the complete genome sequence of *Neisseria meningitidis* strain MC58. *Mol Microbiol* **37**, 207–215.

Simon, M., Zieg, J., Silverman, M., Mandel, G. & Doolittle, R. (1980). Phase variation in the evolution of a controlling element. *Science* **209**, 1370–1374.

Stern, A. & Meyer, T. F. (1987). Common mechanisms controlling phase and antigenic variation in pathogenic neisseriae. *Mol Microbiol* **1**, 5–12.

Stern, A., Brown, M., Nickel, P. & Meyer, T. F. (1986). Opacity genes in *Neisseria gonorrhoeae*: control of phase and antigenic variation. *Cell* **47**, 61–67.

Swords, W. E., Buscher, B. A., Ver Steeg, I. K., Preston, A., Nichols, W. A. & Weiser, J. N. (2000). Non-typeable *Haemophilus influenzae* adhere to and invade human bronchial epithelial cells via an interaction of lipooligosaccharide with the PAF receptor. *Mol Microbiol* **37**, 13–27.

Tong, H. H., Blue, L. E., James, M. A., Chen, Y. P. & DeMaria, T. F. (2000). Evaluation of phase variation of nontypeable *Haemophilus influenzae* lipooligosaccharide during nasopharyngeal colonization and development of otitis media in the chinchilla model. *Infect Immun* **68**, 4593–4597.

Van Valen, L. (1973). A new evolutionary law. *Evol Theory* **1**, 1–30.

Varki, A. (1993). Biological roles of oligosaccharides: all of the theories are correct. *Glycobiology* **3**, 97–130.

Vimr, E., Lichtensteiger, C. & Steenbergen, S. (2000). Sialic acid metabolism's dual function in *Haemophilus influenzae*. *Mol Microbiol* **36**, 1113–1123.

Virji, M., Weiser, J. N., Lindberg, A. A. & Moxon, E. R. (1990). Antigenic similarities in lipopolysaccharides of *Haemophilus influenzae* and *Neisseria* and expression of a digalactoside structure also present on human cells. *Microb Pathog* **9**, 441–450.

Weiser, J. N. & Pan, N. (1998). Adaptation of *Haemophilus influenzae* to acquired and innate immunity based on phase variation of lipopolysaccharide. *Mol Microbiol* **30**, 767–775.

Weiser, J. N., Lindberg, A. A., Manning, E. J., Hansen, E. J. & Moxon, E. R. (1989a). Identification of a chromosomal locus for expression of lipopolysaccharide epitopes in *Haemophilus influenzae*. *Infect Immun* **57**, 3045–3052.

Weiser, J. N., Love, J. M. & Moxon, E. R. (1989b). The molecular mechanism of *Haemophilus influenzae* lipopolysaccharide epitopes. *Cell* **59**, 657–665.

Weiser, J. N., Williams, A. & Moxon, E. R. (1990a). Phase-variable lipopolysaccharide structures enhance the invasive capacity of *Haemophilus influenzae*. *Infect Immun* **58**, 3455–3457.

Weiser, J. N., Maskell, D. J., Butler, P. D., Lindberg, A. A. & Moxon, E. R. (1990b). Characterisation of repetitive sequences controlling phase variation of *Haemophilus influenzae* lipopolysaccharide. *J Bacteriol* **172**, 3304–3309.

Weiser, J. N., Shchepetov, M. & Chong, S. T. H. (1997). Decoration of lipopolysaccharide with phosphorylcholine: a phase variable characteristic of *Haemophilus influenzae*. *Infect Immun* **65**, 943–950.

Weiser, J. N., Pan, N., McGowan, K. L., Musher, D., Martin, A. & Richards, J. (1998). Phosphorylcholine on the lipopolysaccharide of *Haemophilus influenzae* contributes to persistence in the respiratory tract and sensitivity to serum killing mediated by C-reactive protein. *J Exp Med* **187**, 631–640.

Winson, M. K., Camara, M., Latifi, A. & 10 other authors (1995). Multiple *N*-acyl-L-homoserine lactone signal molecules regulate production of virulence determinants and secondary metabolites in *Pseudomonas aeruginosa*. *Proc Natl Acad Sci U S A* **92**, 9427–9431.

Yang, Q. L. & Gotschlich, E. C. (1996). Variation of gonococcal lipooligosaccharide structure is due to alterations in poly-G tracts in *lgt* genes encoding glycosyl transferases. *J Exp Med* **183**, 323–327.

DNA topology and regulation of bacterial gene expression

Charles J. Dorman

Department of Microbiology, Moyne Institute of Preventive Medicine, Trinity College, Dublin 2, Republic of Ireland

INTRODUCTION

DNA supercoiling is perhaps the best characterized aspect of bacterial chromosomal DNA topology and this feature will form the focus of this chapter. Supercoiling arises due to overwinding or underwinding of the DNA helix (Drlica *et al.*, 1999). These create torsional stress in the molecule, which in turn seeks to adopt a conformation of minimal energy (i.e. compensate for the stress) by writhing about the helical axis. Supercoiling produced by overwinding the double helix is termed positive while that resulting from underwinding is termed negative. In nature, negative supercoiling is the usual state of DNA, although positive supercoils can be created transiently by the common transactions of DNA, such as transcription and replication (see below). In some thermophiles, especially those adapted to very high temperature environments, positively supercoiled DNA is the norm (Kikuchi & Asai, 1984), at least for DNA free from nucleosomes (Musgrave *et al.*, 2000). Since positive supercoils tighten the helix, these may lend stability in an environment where DNA melting is strongly favoured (Kikuchi, 1990).

Negative supercoiling can facilitate melting of portions of the double helix whereas positive supercoiling can inhibit it. This permits the state of supercoiling to influence those reactions that require DNA strand separation if they are to proceed. From the point of view of this discussion, the key reaction is transcription initiation, specifically the isomerization of a closed to an open transcription complex. Naturally, other aspects of DNA topology such as DNA looping and intrinsic or induced DNA curvature can also influence this process. All of these influences must be integrated with what one

SGM symposium 61: Signals, switches, regulons and cascades: control of bacterial gene expression.
Editors D. A. Hodgson, C. M. Thomas. Cambridge University Press. ISBN 0 521 81388 3 ©SGM 2002.

might regard as the 'conventional' components of the gene expression and regulation apparatus, i.e. RNA polymerase and protein transcription factors.

GENERATION AND REGULATION OF DNA SUPERCOILING

The level of supercoiling in DNA is managed by enzymes called topoisomerases (Drlica, 1992; Wang, 1996). *Escherichia coli* and related eubacteria have four topoisomerases and their properties are summarized in Table 1. DNA gyrase (topoisomerase II) has the unique ability to introduce negative supercoils into DNA (it can also remove supercoils). Supercoils are also created by processes in which macromolecular complexes unwind the duplex locally and translocate along one of the DNA strands, as happens in transcription with RNA polymerase and in DNA replication with DNA polymerase. If the moving complex is unable to rotate round the DNA axis and the DNA is itself unable to rotate, the consequences are the generation of positive supercoils ahead of the moving complex and negative ones behind (Koo *et al.*, 1990; Liu & Wang, 1987; Wu *et al.*, 1988). If this situation is not resolved, the protein–DNA complex soon locks up and translocation ceases (Fig. 1). Resolution of this crisis requires the intervention of topoisomerases. Positive supercoils are a substrate for DNA gyrase while negative ones are relaxed by DNA topoisomerase I (Massé & Drolet, 1999a). Negative supercoils can also be relaxed by topoisomerases III and IV (Table 1). In the case of DNA replication, topoisomerase IV plays a critical role in relaxing positive supercoils in front of the moving replication fork (Khodursky *et al.*, 2000).

The levels of gyrase and topoisomerase I proteins in the cell are thought to be set by a homeostatic mechanism acting at the level of transcription (Jensen *et al.*, 1999). The *topA* gene encoding topoisomerase I is activated by increases in negative supercoiling (Tse-Dinh, 1985; Tse-Dinh & Beran, 1988) while the promoters of the genes *gyrA* and *gyrB*, encoding the A and B subunits of gyrase, respectively, are inhibited by increases in negative supercoiling (Menzel & Gellert, 1983, 1987). In mutants deficient in topoisomerase I, DNA is found to have increased levels of negative supercoiling, and this situation is compensated by second site mutations that map to the *gyr* genes (DiNardo *et al.*, 1981; Pruss *et al.*, 1982) or by DNA amplifications that result in overexpression of topoisomerase IV, an alternative source of DNA relaxation activity (Dorman *et al.*, 1989; Free & Dorman, 1994; Kato *et al.*, 1990; McNairn *et al.*, 1995; Raji *et al.*, 1985). In the homeostatic model, feedback by supercoiling onto the *topA* and *gyr* gene promoters helps keep the levels of these enzymes balanced appropriately.

An important consequence of the loss of topoisomerase I activity in a *topA* mutant is an accumulation of R loops, in which RNA becomes hybridized with the DNA template strand (Phoenix *et al.*, 1997). These lead to an inhibition of growth and it has been suggested that it is R loop formation that represents the major deleterious property of *topA*

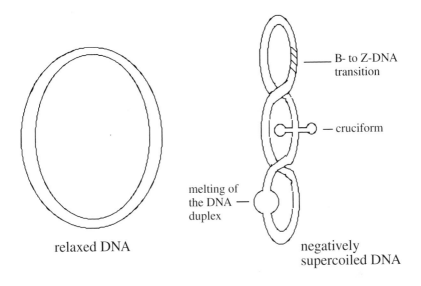

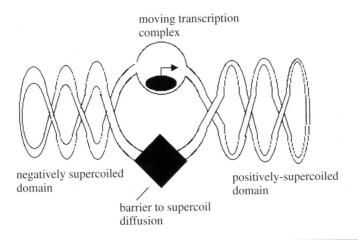

Fig. 1. DNA supercoiling and structural transitions in DNA. A relaxed closed circular DNA molecule is shown at top left. When this is negatively supercoiled (top right), the free energy imparted to the molecule causes it to adopt a writhed conformation and promotes structural transitions such as local melting of the DNA duplex, the extrusion of a cruciform composed of regions of base-paired DNA and single-stranded loops extending from the bulk DNA at a fourway junction, and the conversion of a patch of B-DNA to Z-DNA. At the bottom is shown the effect of transcription on the topology of a closed circular DNA template. The complex is moving from left to right and is unable to rotate about the DNA molecule. Differentially supercoiled domains are created ahead of and behind the moving complex. A barrier to rotational diffusion of the domains preserves them until topoisomerases intervene to remove them. In the absence of this intervention, the transcription complex will be forced to halt.

Table 1. Topoisomerases of *Escherichia coli*

Name*	Molecular mass (kDa)	Gene	Comments
Type I enzymes			
Topoisomerase I	97	*topA*	DNA swivelase; makes a transient cut in one strand of the DNA duplex, binds to cut site via a 5′-phosphotyrosine bond; relaxes negatively supercoiled DNA; requires Mg^{2+}
Topoisomerase III	73.2	*topB*	Can relax negatively supercoiled DNA; has decatenase activity; catenase activity in association with RecQ; requires Mg^{2+}
Type II enzymes			
DNA gyrase (two subunits) (topoisomerase II)	105 (A) 95 (B)	*gyrA* *gyrB*	Has ATP-dependent negative supercoiling activity; relaxes negative supercoils in an ATP-independent manner; relaxes positively supercoiled DNA; binds DNA transiently via a 5′-phosphotyrosine bond; requires Mg^{2+}
Topoisomerase IV (two subunits)	75 (ParC)† 70 (ParE)†	*parC*† *parE*†	Decatenase activity; can relax negative supercoils; cannot introduce negative supercoils; requires ATP, Mg^{2+}

*Type I enzymes alter the linking number of DNA in steps of 1 ($\Delta Lk = 1$); type II enzymes do this in steps of 2 ($\Delta Lk = 2$).

†The name Par/*par* refers to the phenotype of such mutants that exhibit defects in distribution (partitioning) of chromosomal DNA to either side of the septation plane prior to cell division.

mutants (Brocolli *et al.*, 2000). Formation of R loops is promoted by local negative supercoiling and their appearance is normally repressed by topoisomerase I as it relaxes the negative supercoils (Massé & Drolet, 1999b). Provision of an alternative DNA-relaxing activity, such as by overexpression of topoisomerase III, offsets the deleterious effect of a *topA* mutation on growth and concomitantly suppresses the formation of R loops (Brocolli *et al.*, 2000). Similar effects can be achieved by overexpressing RNase H, which initiates nucleolytic degradation of the RNA component of R loops (Drolet *et al.*, 1995).

In addition to differential regulation by DNA supercoiling, the promoters of the *gyr* and *topA* genes are also under reciprocal control by the DNA-binding protein Fis. Identified originally as an accessory factor in the site-specific recombination event that controls phase-variable expression of flagellin protein in *Salmonella typhimurium*, Fis (the Factor for inversion stimulation) is now known also to be a pleiotropic transcription factor influencing the expression of many genes, including those encoding stable RNA (Finkel & Johnson, 1992). Under laboratory growth conditions, Fis expression is confined mainly to the earliest stages of exponential growth and this has been interpreted as a mechanism for boosting expression of genes, such as tRNA and rRNA genes, required when the bacterium encounters a fresh environment replete with resources.

In the case of *topA*, Fis appears to be required for the response of the gene to oxidative stress. Three binding sites for Fis have been identified immediately upstream of the P1 promoter and the effect of Fis on *topA* transcription is positive. Environmental insults such as exposure to hydrogen peroxide activate *topA* and this activation is contingent on the presence of the Fis protein (Weinstein-Fischer *et al.*, 2000). In contrast to its role at *topA*, Fis acts as a repressor of the promoters of the *gyrA* and *gyrB* genes (Schneider *et al.*, 1999). In the case of *gyrA*, Fis does this by occluding the promoter; in the case of *gyrB*, Fis interferes with the ability of RNA polymerase to clear the promoter and to begin transcript elongation (Schneider *et al.*, 1999). In addition to its role as a topoisomerase gene regulator, Fis also functions directly to influence DNA topology by binding to and stabilizing the supercoiling of DNA molecules within a particular range of superhelical densities. In this way, the protein can modulate genome structure and dynamics as the cell grows and adapts to its environment (Schneider *et al.*, 1997). Given its contributions to the control of DNA supercoiling in bacteria, it is perhaps not surprising to discover that the gene encoding Fis is itself regulated strongly by changes in DNA supercoiling. Expression of the *fis* gene is maximal at high levels of supercoiling and any departure from its optimum superhelical density inactivates its promoter. It has been suggested that this allows Fis to function in 'fine-tuning' the homeostatic mechanism of DNA supercoiling (Schneider *et al.*, 2000).

DNA gyrase is linked to cell physiology through its dependence on ATP (Drlica, 1992; Hsieh *et al.*, 1991a, b; van Workum *et al.*, 1996). Its activity is stimulated by ATP and inhibited by ADP, so the ratio of [ATP] to [ADP] has a strong influence on its intracellular activity. This ratio is modulated by changes in cellular metabolism; it is high in bacteria in exponential growth but low in stationary phase. It is also high in bacteria experiencing osmotic stress (Hsieh *et al.*, 1991b) or anaerobic conditions (Hsieh *et al.*, 1991a). Consequently, the level of negative supercoiling correlates directly with the [ATP] to [ADP] ratio. This provides the bacterium with a mechanism for adjusting its DNA topology (and hence its transcription profile) in response to environmental conditions.

DNA SUPERCOILING AND ENVIRONMENTAL STRESS

Early work using plasmids as reporters of DNA supercoiling revealed that bacteria undergoing osmotic stress or growing under anaerobic conditions had elevated levels of negative supercoiling (Dorman *et al.*, 1988; Higgins *et al.*, 1988; Yamamoto &

Droffner, 1985), something that was explained subsequently by the sensitivity of gyrase to changes in the [ATP] to [ADP] ratio described in the previous section. These observations suggest that bacteria might have evolved mechanisms for exploiting the impact of environmental stress on DNA topology as part of their stress response machinery (Dorman & Ní Bhriain, 1992). Consistent with this model is the finding that the promoters of several genes involved in the osmotic and anaerobic responses have supercoiling-sensitive promoters (Dixon *et al.*, 1988). Moreover, some of these promoters seemed to respond to both forms of stress, for example the *ompC* promoter of *E. coli* and *S. typhimurium* (Ní Bhriain *et al.*, 1989). No general rules have emerged concerning how supercoiling sensitivity is integrated with more conventional forms of gene regulation. It is assumed that negative supercoiling facilitates the formation of an open transcription complex at the promoter, although other effects cannot be ruled out. These might include effects on the binding of transcription factors, on the oligomerization of these proteins once bound to DNA, on the interaction of the bound factors with RNA polymerase, and on the ability of RNA polymerase to interact with separate components of the promoter (Whitehall *et al.*, 1993). Supercoiling might also be capable of influencing long-range interactions through plectonemic winding of DNA to bring together sites that lie at a distance when arranged linearly (Huang *et al.*, 2001). Certainly, formation of open complexes in DNA is not the only type of DNA structural transition that can be driven by environmental stress via supercoiling; for example, the extrusion of a cruciform structure can also be facilitated in this way (McClellan *et al.*, 1990).

The supercoiling response to environmental signals is not confined to osmotic stress and anaerobiosis. DNA supercoiling also varies in response to thermal signals, to pH, and to depletion of carbon sources. Mutants of *E. coli* that have acquired thermotolerance have been found to have DNA that is more relaxed than that in wild-type parental strains (Friedman *et al.*, 1995), and wild-type strains vary the linking number of reporter plasmids as a function of the growth temperature (Goldstein & Drlica, 1984). In *S. typhimurium*, a shift from acid to alkaline pH under laboratory growth conditions is accompanied by an increase in negative supercoiling of DNA (Karem & Foster, 1993). When *S. typhimurium* cells are engulfed by macrophage, reporter plasmid DNA is found to become relaxed, and this is accompanied by an induction of gyrase gene transcription (Marshall *et al.*, 2000). Exhaustion of carbon sources in defined growth media also results in relaxation of DNA (Balke & Gralla, 1987).

THE IMPORTANCE OF CONTEXT

Several examples of context-dependent variations in the response of genes to changes in supercoiling may be found in the literature. One concerns the tRNA$_1^{tyr}$ gene, *tyrT*, of *E. coli*, a gene whose expression is modulated directly by the Fis protein. When studied in an *in vitro* transcription system, the *tyrT* promoter showed exquisite sensitivity to changes in DNA supercoiling, being inhibited by DNA relaxation and activated by negative supercoiling; however, when this promoter was studied in *E. coli* cells in the presence and absence of topoisomerase I or DNA gyrase, no significant effect of supercoiling on its activity was seen (Lamond, 1985). These *in vivo* studies were carried out in plasmids where the *tyrT* promoter was used to drive expression of a *galK* (galactokinase) reporter gene. When the *tyrT* promoter was studied *in situ* using quantitative Northern blotting, it was found to respond positively to increases in negative supercoiling and to be inhibited by DNA relaxation (Free & Dorman, 1994). Similarly, when the plasmid-based expression system was modified so that local supercoiling was focused at *tyrT*, a strong dependency on negative supercoiling for activation was revealed (Bowater *et al.*, 1994). This case serves to illustrate the importance of context when studying the response of promoters to DNA supercoiling and the desirability of performing the investigation *in situ* whenever possible.

TRANSCRIPTION COUPLING: PROMOTER–PROMOTER COMMUNICATION VIA DNA LOCAL SUPERCOILING

It has been clear for some time that the DNA supercoiling generated at a local level by transcription has the ability to influence nearby events in the same DNA molecule such as promoter activity (Fang & Wu, 1998; Tobe *et al.*, 1995), cruciform extrusion (Krasilnikov *et al.*, 1999), site-specific recombination (Dove & Dorman, 1994; Wang & Dröge, 1996) and B-to-Z DNA structural transitions (Rahmouni & Wells, 1989, 1992; Sheridan *et al.*, 1999) (Fig. 1).

The *leu-500* promoter mutation in *S. typhimurium* has played an important role in studies on local effects of supercoiling on transcription (Lilley & Higgins, 1991). The mutant promoter has an additional G:C base pair in place of an A:T pair found in the Pribnow box of the wild-type sequence. This impedes the transition to an open complex at transcription initiation in the *leuABCD* operon, resulting in leucine auxotrophy. The mutant phenotype is suppressed by mutations in the *topA* gene. This is not due to a general effect on DNA supercoiling throughout the genome resulting from loss of topoisomerase I, but to a context-specific accumulation of negative supercoils at the mutant promoter when topoisomerase I is not there to relax them (Richardson *et al.*, 1988). The source of these supercoils is the activity of an adjacent promoter. The effect can be reproduced artificially in multicopy plasmids by placing the *leu-500* mutant sequence upstream of a divergently transcribing promoter and preventing the resulting supercoils from dissipating by introduction of a topological barrier close to the *leu-500* promoter sequence (Chen *et al.*, 1992).

On the chromosome, the source of the local supercoiling is a promoter relay involving genes located upstream of *leu*, the L-leucine biosynthetic operon (Fang & Wu, 1998;

Hanafi & Bossi, 2000). The promoter of the divergently transcribed *ilvIH* operon, located 1.9 kb upstream of *leuABCD*, appears to activate the mutant *leu-500* promoter via the intervening *leuO* gene. This gene encodes a LysR-like DNA-binding protein and both the LeuO protein and the activity of the *leuO* promoter are required for *leu-500* suppression in a *topA* mutant background. Significantly, the promoter relay mechanism is not confined to suppression of the mutant promoter; the wild-type *leu* promoter is also modulated in this way, showing that the effect is of physiological relevance (Fang & Wu, 1998).

The nature of topological barriers to supercoil movement along DNA remains unclear. In the case of the *leu-500* studies referred to above, the topological barrier used in the plasmid studies of Lilley and coworkers was absolute since it was generated by the attachment of the plasmid DNA to the cytoplasmic membrane via the coupled transcription and translation of the tetracycline-resistance gene on the episome (Chen *et al.*, 1992). In natural systems, barriers might be transient, as in cases where they are generated by protein binding. Interaction of the DNA-bending protein IHF with the *E. coli* chromosome has been suggested as a source of topological barrier to transcriptionally generated supercoils at the *fim* locus (Dove & Dorman, 1994), although *in vitro* studies with purified IHF indicated that this protein could not prevent supercoils created in one part of a plasmid by transcription from modulating a site-specific recombination event in another part of the same plasmid (Wang & Dröge, 1996). However, other work has suggested that distortions in DNA, such as bends, could block supercoil movement and that the importance of such effects *in vivo* has largely been underestimated (Nelson, 1999).

The genes encoding many LysR-like transcription factors are arranged such that they are transcribed divergently from other genes (Opel *et al.*, 2001). For example, the promoter for the *leuABCD* operon is arranged divergently with respect to that of the *leuO* gene, encoding a LysR-like protein, as described earlier in this chapter. The *ilvY* gene also encodes a LysR-like transcription factor, and this gene is transcribed divergently from a structural gene, *ilvC*, involved in L-isoleucine and L-valine biosynthesis. IlvY binds in the intergenic region and negatively autoregulates the *ilvY* gene. Activation of *ilvC* requires IlvY to bind a coinducer (α-acetolactate or α-acetohydroxybutyrate). Evidence from *in vivo* and *in vitro* studies shows that the *ilvC* and *ilvY* promoters are transcriptionally coupled through DNA supercoiling, and it has been suggested that this may be the case for other divergently arranged regulatory units controlled by LysR-like proteins (Rhee *et al.*, 1999). In the case of the *ilvC* and *ilvY* promoters, the magnitude of the negative supercoiling created was found to be proportional to the sum of the supercoils generated locally by promoter activity and global levels of supercoiling, and there was a correlation with promoter strength and transcript length. Moreover, no anchorage of the DNA template (of the sort described above for the *leu-500* promoter) was required to generate a domain of local supercoiling (Opel & Hatfield, 2001).

DNA SUPERCOILING AND GENE REGULATION IN A BACTERIAL PATHOGEN

Members of the genus *Shigella* are facultative intracellular pathogens and cause bacillary dysentery in humans (Philpott *et al.*, 2000). Invasion of host cells is a complex process, and requires the products of many genes located on a 230 kb virulence plasmid. These invasion genes are subject to control at the level of transcription through a regulatory cascade in which a transcription factor, VirF, activates transcription of a regulatory gene, *virB*, whose product in turn is required to activate the structural genes of the virulence regulon (Dorman *et al.*, 2001). Key environmental signals for gene activation are temperature (37 °C is the permissive temperature; lower temperatures result in transcription repression), osmolarity (a value equivalent to that of physiological saline is required for optimal expression of the virulence genes *in vitro*) and pH (pH 7.4 is optimal). Activation of the *virB* promoter is a key event in the expression of the virulence regulon, and this promoter has a strong dependence on DNA supercoiling. Its activation is dependent absolutely on the VirF protein (a member of the AraC-like family of transcription factors) but can be achieved at the non-permissive temperature of 30 °C if the negative superhelical density at the promoter is raised artificially. This has been done through coupling of the *virB* promoter to a divergently arranged inducible promoter. When the upstream promoter is activated, *virB* is also activated, providing the VirF protein is present (Tobe *et al.*, 1995). It seems likely that a change in the supercoiling of *virB* DNA caused by an increase in temperature to 37 °C is part of the activation mechanism in the natural system. This change in supercoiling is not required for VirF binding, but might promote the interaction of VirF with RNA polymerase, with itself through DNA-dependent VirF dimer formation, or both.

DNA SUPERCOILING AND SELECTION OF RNA POLYMERASE SIGMA FACTORS

The RpoS sigma factor of RNA polymerase has been the subject of intensive investigations for many years (Hengge-Aronis, 2000; Kolter, 1999). Identified originally as a protein required for adaptation to the stationary phase of growth, it is now recognized as a requirement for successful adaptation to many stresses which result in a growth arrest. A large regulon of genes has been described as having a requirement for RpoS. The promoters of genes from the stationary phase regulon do not have a consensus promoter sequence, raising the question of what is the functional specificity of the stress sigma factor RpoS compared to RpoD, the major sigma factor used during exponential growth?

Studies with the osmotically regulated promoters of the genes *osmB* and *osmY* in *E. coli* revealed that RpoS and RpoD have different requirements in terms of the super-helicity of the promoters. RpoD can function only when the promoters are negatively supercoiled whereas RpoS can also function when the DNA template is relaxed. Since this is its condition in energy-depleted cells, such as those undergoing stress or in stationary phase, a correlation is seen between the topological state of the promoters and the selectivity of the sigma factors (Kusano *et al.*, 1996). Sigma factor selectivity by RNA polymerase is not determined by supercoiling status alone, however. In the case of the *osmY* promoter, there are also roles for the global regulatory proteins CRP (cAMP receptor protein), Lrp (leucine-responsive regulatory protein) and IHF (Colland *et al.*, 2000).

DNA SUPERCOILING AND THE EVOLUTION OF GENE REGULATORY MECHANISMS IN BACTERIA

Supercoiling may have been selected originally as a means to assist with the packaging of DNA within the cell, but with its influence on promoter function and its sensitivity to several environmental signals, DNA supercoiling is also well-placed to function as a primordial regulator of transcription. It is possible that early bacterial promoters responded to signals such as changes in pH, temperature, osmolarity, the presence or absence of oxygen, etc., simply through a change in the facility of open complex formation as determined by the level of local or global DNA supercoiling. This rather coarse control could be modulated by the action of the nucleoid-associated DNA-binding proteins with their ability to influence promoter function in addition to their roles in determination of nucleoid structure. Later, more specific regulators may have emerged in the form of the conventional transcription activator and repressor proteins, allowing much more refined regulation of gene expression. These also impart a much greater degree of specificity to the management of the gene expression profile of the cell, as each regulatory protein can specialize in transmitting information about just one, or a very restricted number of environmental signals. The hierarchical theme of these evolutionary considerations finds echoes in the make up of the gene regulatory circuits found in modern bacterial cells (see below).

THE PLACE OF DNA SUPERCOILING IN GLOBAL REGULATION

The availability of full genome sequences for an increasing number of bacteria, and the development of techniques for analysis of gene expression across entire genomes, make studies on global regulation both feasible and necessary. It has been appreciated for many years that bacterial genes are often grouped into regulatory hierarchies called operons, regulons and stimulons for purposes of coordinate control. Now one may use DNA array technology to examine the response of the entire complement of genes to a change in a single environmental parameter in order to identify stimulon members, or to study the effects of removing individual pleiotropic regulators in order to elucidate the membership of a regulon.

The potential of DNA topology to influence gene expression is considerable, at both a local and global level. Transcription is just one of the DNA transactions that requires the double helix to open, and such events are assisted by negative supercoiling (underwinding) of the DNA. The candidacy of supercoiling as a global regulator is strengthened when one considers that every promoter in the cell has the potential to respond to fluctuations in negative supercoiling and that this parameter of DNA topology varies with the energy charge of the cell.

Like the genes they control, the gene regulatory components of the cell may be regarded as being arranged in a hierarchy. Some transcription regulators are highly specific and affect just one or a very few genes. These are usually conventional protein transcription factors that bind to specific sequences in the DNA in response to a signal or alter the nature of their contact with DNA upon receipt of the signal. They then exert either a positive or a negative influence on transcription. If the recognition sequences proliferate in the genome at sites appropriate for promoter modulation, then more genes come under the command of the regulatory proteins that recognize those sequences. This is the basis of the regulon. Proteins with a very wide range of effects in the cell may have a double life, acting as both transcription factors and architectural elements in the genome. The integration host factor (IHF), a sequence-specific DNA binding and bending protein, is like this. It alters the trajectory of the DNA in the cell by introducing bends of up to 180° and can influence transcription either indirectly (action at a distance) or directly (Goosen & van de Putte, 1995; Rice *et al.*, 1996). Another example is the Fis protein described above; this can modulate DNA supercoiling directly by binding and stabilizing molecules of a particular superhelical density, and indirectly through its role as a regulator of topoisomerase gene transcription. Proteins like these, and other members of the so-called nucleoid-associated family, can have very wide-ranging effects indeed on transcription.

Changes to the global level of supercoiling have the potential to be even more wide ranging in their effects. It is clear that the DNA in a nutritionally replete bacterium in the exponential phase of growth has DNA that is more negatively supercoiled than a genetically identical cell in a depleted medium in stationary phase. It is also clear that the profile of transcription in these bacteria is very different. Given the central role of RpoS in establishing the gene expression pattern of the stationary phase cell, and the contribution of DNA relaxation to the selection of sigma factors by RNA polymerase, it can be seen that the potential of DNA supercoiling to act globally may be realized, but it is not acting in isolation. It seems to be characteristic of the contribution of supercoiling to the control of transcription that it plays mainly a role 'behind the scenes', with other, more specific, regulatory influences superimposed. In this way, its contribution is both wide-ranging and subtle. The effect of supercoiling on transcription might seem like a very tempting subject for whole-genome analysis methods, such as microarrays, in which one compares wild-type bacteria with mutants deficient in DNA topoisomerase I or cultures treated or untreated with DNA-gyrase-inhibiting antibiotics. Indeed, studies of this type are now under way, including work with pathogens such as *Haemophilus influenzae* (Gmuender *et al.*, 2001). Such surveys are very much to be welcomed; however, great caution must be exercised in interpreting the results as the effects of perturbing supercoiling may be subtle. For example, local supercoiling has the potential to generate artefacts in whole-genome analyses. It has been pointed out that supercoiling-sensitive promoters may be activated or repressed by the topological disturbance associated with the passage of a DNA replication fork (Humphery-Smith, 1999). Thus gene expression may vary for reasons that may not be immediately obvious from the perspective of conventional thinking about gene regulation.

It is also important to appreciate the importance of local DNA supercoiling in promoter coupling. Here, the activity of one promoter may alter that of another, either positively or negatively, through a DNA topological transmission that is strongly context dependent, and which may vary as a result of protein binding (Sheridan *et al.*, 1999) or as a result of induced or intrinsic distortions in DNA structure (Nelson, 1999). Transmission in a single step of topological signals over distances of up to 800 bp is detectable *in vivo* (Krasilnikov *et al.*, 1999), and over much greater distances when 'relay' mechanisms are used (Fang & Wu, 1998; Hanafi & Bossi, 2000). These local topology-dependent regulatory relationships may be among the most subtle of all. Consequently, the challenges that lie ahead in considering global regulation of gene expression are not merely technical, but intellectual as well.

ACKNOWLEDGEMENTS

Research in the author's laboratory is supported by the Wellcome Trust, the European Union, Enterprise Ireland and the Irish Health Research Board.

REFERENCES

Balke, V. L. & Gralla, J. D. (1987). Changes in linking number of supercoiled DNA accompany growth transitions in *Escherichia coli. J Bacteriol* **169**, 4499–4506.

Bowater, R. P., Chen, D.-R. & Lilley, D. M. J. (1994). Modulation of *tyrT* promoter activity by template supercoiling *in vivo. EMBO J* **13**, 5647–5655.

Brocolli, S., Phoenix, P. & Drolet, M. (2000). Isolation of the *topB* gene encoding DNA topoisomerase III as a multicopy suppressor of *topA* null mutations in *Escherichia coli. Mol Microbiol* **35**, 58–68.

Chen, D., Bowater, R., Dorman, C. J. & Lilley, D. M. J. (1992). Activity of the *leu-500* promoter depends on the transcription and translation of an adjacent gene. *Proc Natl Acad Sci U S A* **89**, 8784–8788.

Colland, F., Barth, M., Hengge-Aronis, R. & Kolb, A. (2000). σ factor selectivity of *Escherichia coli* RNA polymerase: role for CRP, IHF and Lrp transcription factors. *EMBO J* **19**, 3028–3037.

DiNardo, S., Voelkel, K. A., Sternglanz, R., Reynolds, A. E. & Wright, A. (1981). *Escherichia coli* DNA topoisomerase I mutants have compensatory mutations in DNA gyrase genes. *Cell* **31**, 43–51.

Dixon, R. A., Henderson, N. C. & Austin, S. (1988). DNA supercoiling and aerobic regulation of transcription from the *Klebsiella pneumoniae nifLA* promoter. *Nucleic Acids Res* **16**, 9933–9946.

Dorman, C. J. & Ní Bhriain, N. (1992). Global regulation of gene expression during environmental adaptation: implications for bacterial pathogens. In *Molecular Biology of Bacterial Infection*, pp. 193–230. Edited by C. E. Hormaeche, C. W. Penn & C. J. Smyth. Cambridge: Cambridge University Press.

Dorman, C. J., Barr, G. C., Ní Bhriain, N. & Higgins, C. F. (1988). DNA supercoiling and the anaerobic and growth phase regulation of *tonB* gene expression. *J Bacteriol* **170**, 2816–2826.

Dorman, C. J., Lynch, A. S., Ní Bhriain, N. & Higgins, C. F. (1989). DNA supercoiling in *Escherichia coli*: *topA* mutations can be suppressed by DNA amplifications involving the *tolC* locus. *Mol Microbiol* **3**, 531–540.

Dorman, C. J., McKenna, S. & Beloin, C. (2001). Regulation of virulence gene expression in *Shigella flexneri*, a facultative intracellular pathogen. *Int J Med Microbiol* **291**, 89–96.

Dove, S. L. & Dorman, C. J. (1994). The site-specific recombination system regulating expression of the type 1 fimbrial subunit gene of *Escherichia coli* is sensitive to changes in DNA supercoiling. *Mol Microbiol* **14**, 975–988.

Drlica, K. (1992). Control of bacterial DNA supercoiling. *Mol Microbiol* **6**, 425–433.

Drlica, K., Wu, E.-D., Chen, C.-R., Wang, J.-Y., Zhao, X., Qiu, L., Malik, M., Kayman, S. & Friedman, S. M. (1999). Prokaryotic DNA topology and gene expression. In *Prokaryotic Gene Expression*, pp. 141–168. Edited by S. Baumberg. Oxford: Oxford University Press.

Drolet, M., Phoenix, P., Menzel, R., Massé, E., Liu, L. F. & Crouch, R. J. (1995). Overexpression of RNase H partially complements the growth defect of an *Escherichia coli* Δ*topA* mutant: R-loop formation is a major problem in the absence of DNA topoisomerase I. *Proc Natl Acad Sci U S A* **92**, 3526–3530.

Fang, M. & Wu, H.-Y. (1998). A promoter relay mechanism for sequential gene activation. *J Bacteriol* **180**, 626–633.

Finkel, S. E. & Johnson, R. C. (1992). The FIS protein: it's not just for DNA inversion anymore. *Mol Microbiol* **6**, 3257–3265.

Free, A. & Dorman, C. J. (1994). *Escherichia coli tyrT* gene transcription is sensitive to DNA supercoiling in its native chromosomal context: effect of DNA topoisomerase IV overexpression on *tyrT* promoter function. *Mol Microbiol* **14**, 151–161.

Friedman, S. M., Malik, M. & Drlica, K. (1995). DNA supercoiling in a thermotolerant mutant of *Escherichia coli*. *Mol Gen Genet* **248**, 417–422.

Gmuender, H., Kuratli, K., Di Padova, K., Gray, C. P., Keck, W. & Evers, S. (2001). Gene expression changes triggered by exposure of *Haemophilus influenzae* to novobiocin or ciprofloxacin: combined transcription and translation analysis. *Genome Res* **11**, 28–42.

Goldstein, E. & Drlica, K. (1984). Regulation of bacterial DNA supercoiling: plasmid linking numbers vary with growth temperature. *Proc Natl Acad Sci U S A* **81**, 4046–4050.

Goosen, N. & van de Putte, P. (1995). The regulation of transcription initiation by integration host factor. *Mol Microbiol* **16**, 1–7.

Hanafi, E. E. & Bossi, L. (2000). Activation and silencing of *leu-500* promoter by transcription-induced DNA supercoiling in the *Salmonella* chromosome. *Mol Microbiol* **37**, 583–594.

Hengge-Aronis, R. (2000). The general stress response in *Escherichia coli*. In *Bacterial Stress Responses*, pp. 161–178. Edited by G. Storz & R. Hengge-Aronis. Washington, DC: American Society for Microbiology.

Higgins, C. F., Dorman, C. J., Stirling, D. A., Waddell, L., Booth, I. R., May, G. & Bremer, E. (1988). A physiological role for DNA supercoiling in the osmotic regulation of gene expression in *S. typhimurium* and *E. coli*. *Cell* **52**, 569–584.

Hsieh, L.-S., Burger, R. M. & Drlica, K. (1991a). Bacterial DNA supercoiling and [ATP]/[ADP]: changes associated with a transition to anaerobic growth. *J Mol Biol* **219**, 443–450.

Hsieh, L.-S., Rouviere-Yaniv, J. & Drlica, K. (1991b). Bacterial DNA supercoiling and [ATP]/[ADP] ratio: changes associated with salt shock. *J Bacteriol* **173**, 3914–3917.

Huang, J., Schlick, T. & Vologodskii, A. (2001). Dynamics of site juxaposition in supercoiled DNA. *Proc Natl Acad Sci U S A* **98**, 968–973.

Humphery-Smith, I. (1999). Replication-induced protein synthesis and its importance to proteomics. *Electrophoresis* **20**, 653–659.

Jensen, P. R., van der Weijden, C. C., Jensen, L. B., Westerhoff, H. V. & Snoep, J. L. (1999). Extensive regulation compromises the extent to which DNA gyrase controls DNA supercoiling and growth rate of *Escherichia coli*. *Eur J Biochem* **266**, 865–877.

Karem, K. & Foster, J. W. (1993). The influence of DNA topology on the environmental regulation of a pH-regulated locus in *Salmonella typhimurium*. *Mol Microbiol* **10**, 75–86.

Kato, J.-I., Nishimura, Y., Imamura, R., Niki, H., Hiraga, S. & Suzuki, H. (1990). New topoisomerase essential for chromosome segregation in *E. coli*. *Cell* **63**, 393–404.

Khodursky, A. B., Peter, B. J., Schmid, M. B., DeRisi, J., Bothstein, D., Brown, P. O. & Cozzarelli, N. R. (2000). Analysis of topoisomerase function in bacterial replication fork movement: use of DNA microarrays. *Proc Natl Acad Sci U S A* **97**, 9419–9424.

Kikuchi, A. (1990). Reverse gyrase. In *DNA Topology and its Biological Effects*, pp. 285–298. Edited by N. Cozzarelli & J. C. Wang. Cold Spring Harbor, NY: Cold Spring Harbor Laboratory.

Kikuchi, A. & Asai, K. (1984). Reverse gyrase: a topoisomerase which introduces positive superhelical turns into DNA. *Nature* **309**, 677–681.

Kolter, R. (1999). Growth in studying the cessation of growth. *J Bacteriol* **181**, 697–699.

Koo, H. S., Wu, H. Y. & Liu, L. F. (1990). Effects of transcription and translation on gyrase-mediated DNA cleavage in *Escherichia coli*. *J Biol Chem* **265**, 12300–12305.

Krasilnikov, A. S., Podtelezhnikov, A., Vologodskii, A. & Mirkin, S. M. (1999). Large-scale effects of transcriptional DNA supercoiling *in vivo*. *J Mol Biol* **292**, 1149–1160.

Kusano, S., Ding, Q., Jujita, N. & Ishihama, A. (1996). Promoter selectivity of *Escherichia coli* RNA polymerase Eσ70 and Eσ38 holoenzymes: effect of DNA supercoiling. *J Biol Chem* **271**, 1998–2004.

Lamond, A. I. (1985). Supercoiling response of a bacterial tRNA gene. *EMBO J* **4**, 501–507.

Lilley, D. M. J. & Higgins, C. F. (1991). Local DNA topology and gene expression: the case of the *leu-500* promoter. *Mol Microbiol* **5**, 779–783.

Liu, L. F. & Wang, J. C. (1987). Supercoiling of the DNA template during transcription. *Proc Natl Acad Sci U S A* **84**, 7024–7027.

McClellan, J. A., Boublikova, P., Palecek, E. & Lilley, D. M. J. (1990). Superhelical torsion in cellular DNA responds directly to environmental and genetic factors. *Proc Natl Acad Sci U S A* **87**, 8373–8377.

McNairn, E., Ní Bhriain, N. & Dorman, C. J. (1995). Overexpression of the *Shigella flexneri* genes coding for DNA topoisomerase IV compensates for loss of DNA topoisomerase I: effect on virulence gene expression. *Mol Microbiol* **15**, 507–517.

Marshall, D. G., Bowe, F., Hale, C., Dougan, G. & Dorman, C. J. (2000). DNA topology and adaptation of *Salmonella typhimurium* to an intracellular environment. *Philos Trans R Soc Lond Ser B* **355**, 565–574.

Massé, E. & Drolet, M. (1999a). Relaxation of transcription-induced negative supercoiling is an essential function of *Escherichia coli* DNA topoisomerase I. *J Biol Chem* **274**, 16654–16658.

Massé, E. & Drolet, M. (1999b). *Escherichia coli* DNA topoisomerase I inhibits R-loop formation by relaxing transcription-induced negative supercoiling. *J Biol Chem* **274**, 16659–16664.

Menzel, R. & Gellert, M. (1983). Regulation of the genes for *E. coli* DNA gyrase: homeostatic control of DNA supercoiling. *Cell* **34**, 105–113.

Menzel, R. & Gellert, M. (1987). Fusions of the *Escherichia coli gyrA* and *gyrB* control regions to the galactokinase gene are inducible by coumermycin treatment. *J Bacteriol* **169**, 1272–1278.

Musgrave, D., Forterre, P. & Slesarev, A. (2000). Negative constrained DNA supercoiling in archael nucleosomes. *Mol Microbiol* **35**, 341–349.

Nelson, P. (1999). Transport of torsional stress in DNA. *Proc Natl Acad Sci U S A* **96**, 14342–14347.

Ní Bhriain, N., Dorman, C. J. & Higgins, C. F. (1989). An overlap between osmotic and anaerobic stress responses: a potential role for DNA supercoiling. *Mol Microbiol* **3**, 933–942.

Opel, M. L. & Hatfield, G. W. (2001). DNA supercoiling-dependent transcriptional coupling between the divergently transcribed promoters of the *ilvYC* operon of *Escherichia coli* is proportional to promoter strengths and transcript lengths. *Mol Microbiol* **39**, 191–198.

Opel, M. L., Arfin, S. M. & Hatfield, G. W. (2001). The effects of DNA supercoiling on the expression of operons of the *ilv* regulon of *Escherichia coli* suggest a physiological rationale for divergently transcribed operons. *Mol Microbiol* **39**, 1109–1115.

Philpott, D. J., Edgeworth, J. D. & Sansonetti, P. J. (2000). The pathogenesis of *Shigella flexneri* infection: lessons from *in vitro* and *in vivo* studies. *Philos Trans R Soc Lond Ser B* **355**, 575–586.

Phoenix, P., Raymond, M.-A., Massé, E. & Drolet, M. (1997). Roles of DNA topoisomerases in the regulation of R-loop formation *in vitro*. *J Biol Chem* **272**, 1473–1479.

Pruss, G. J., Manes, S. H. & Drlica, K. (1982). *Escherichia coli* DNA topoisomerase mutants: increased supercoiling is corrected by mutations near gyrase genes. *Cell* **31**, 35–42.

Rahmouni, A. R. & Wells, R. D. (1989). Stabilization of Z DNA *in vitro* by localized supercoiling. *Science* **246**, 358–363.

Rahmouni, A. R. & Wells, R. D. (1992). Direct evidence for the effect of transcription on local DNA supercoiling *in vivo*. *J Mol Biol* **223**, 131–144.

Raji, A., Zabel, D. J., Laufer, C. S. & Depew, R. E. (1985). Genetic analysis of mutations that compensate for loss of *Escherichia coli* DNA topoisomerase I. *J Bacteriol* **162**, 1173–1179.

Rhee, K. Y., Opel, M., Ito, E., Hung, S.-P., Arfin, S. M. & Hatfield, G. W. (1999). Transcriptional coupling between the divergent promoters of a prototypic LysR-type regulatory system, the *ilvYC* operon of *Escherichia coli*. *Proc Natl Acad Sci U S A* **96**, 14294–14299.

Rice, P. A., Yang, S.-W., Mizuuchi, K. & Nash, H. A. (1996). Crystal structure of an IHF-DNA complex: a protein-induced DNA U-turn. *Cell* **87**, 1295–1306.

Richardson, S. M. H., Higgins, C. F. & Lilley, D. M. J. (1988). DNA supercoiling and the *leu500* promoter mutation of *Salmonella typhimurium*. *EMBO J* **7**, 1863–1869.

Schneider, R., Travers, A. & Muskhelishvili, G. (1997). FIS modulates growth phase-dependent topological transitions of DNA in *Escherichia coli*. *Mol Microbiol* **26**, 519–530.

Schneider, R., Travers, A., Kutateladze, T. & Muskhelishvili, G. (1999). A DNA architectural protein couples cellular physiology and DNA topology in *Escherichia coli*. *Mol Microbiol* **34**, 953–964.

Schneider, R., Travers, A. & Muskhelishvili, G. (2000). The expression of the *Escherichia coli fis* gene is strongly dependent on the superhelical density of DNA. *Mol Microbiol* **38**, 167–175.

Sheridan, S. D., Benham, C. J. & Hatfield, G. W. (1999). Inhibition of DNA supercoiling-dependent transcriptional activation by a distant B-DNA to Z-DNA transition. *J Biol Chem* **274**, 8169–8174.

Tobe, T., Yoshikawa, M. & Sasakawa, C. (1995). Thermoregulation of *virB* in *Shigella flexneri* by sensing changes in local DNA superhelicity. *J Bacteriol* **177**, 1094–1097.

Tse-Dinh, Y.-C. (1985). Regulation of the *Escherichia coli* topoisomerase I gene by DNA supercoiling. *Nucleic Acids Res* **13**, 4751–4763.

Tse-Dinh, Y.-C. & Beran, R. K. (1988). Multiple promoters for transcription of DNA topoisomerase I gene and their regulation by DNA supercoiling. *J Mol Biol* **202**, 735–742.

Wang, J. C. (1996). DNA topoisomerases. *Annu Rev Biochem* **65**, 635–692.

Wang, Z. & Dröge, P. (1996). Differential control of transcription-induced and overall DNA supercoiling by eukaryotic topoisomerases *in vitro*. *EMBO J* **15**, 581–589.

Weinstein-Fischer, D., Elgrably-Weiss, M. & Altuvia, S. (2000). *Escherichia coli* response to hydrogen peroxide: a role for DNA supercoiling, topoisomerase I and FIS. *Mol Microbiol* **35**, 1413–1420.

Whitehall, S., Austin, S. & Dixon, R. (1993). The function of the upstream region of the sigma 54-dependent *Klebsiella pneumoniae nifL* promoter is sensitive to DNA supercoiling. *Mol Microbiol* **9**, 1107–1117.

van Workum, M., van Dooren, S. J., Oldenburg, N., Molenaar, D., Jensen, P. R., Snoep, J. L. & Westerhoff, H. V. (1996). DNA supercoiling depends on the phosphorylation potential in *Escherichia coli*. *Mol Microbiol* **20**, 351–360.

Wu, H. Y., Shyy, S. H., Wang, J. C. & Liu, L. F. (1988). Transcription generates positively and negatively supercoiled domains in the template. *Cell* **53**, 433–440.

Yamamoto, N. & Droffner, M. (1985). Mechanisms determining aerobic or anaerobic growth in the facultative anaerobe *Salmonella typhimurium*. *Proc Natl Acad Sci U S A* **82**, 2077–2081.

DNA rearrangements and regulation of gene expression

Ian C. Blomfield

Department of Biosciences, University of Kent at Canterbury, Canterbury CT2 7NJ, UK

INTRODUCTION

Alterations in genome composition, via random mutation (base changes, duplications, deletions or inversions), or through the acquisition of new sequences that originate from other bacteria (horizontal gene transfer), provide the means by which new combinations of genes arise. Such changes, if advantageous (or at least neutral), can become relatively stable elements of the evolving genome. The acquisition of blocks of genes by horizontal transfer, termed genome islands, is recognized as a pivotal factor in bacterial evolution. For example, the development of pathogenic traits in otherwise non-pathogenic bacteria is often associated with the insertion of large groups of genes encoding virulence factors (pathogenicity islands) into tRNA genes.

In contrast to the events described above, bacterial genomes also have the capacity to undergo more specific, predictable and usually reversible, changes in genome organization. Such programmed DNA rearrangements, which include simple insertions, deletions or inversions, as well as more complex changes, play a key role in the adaptation of bacteria to their environment, turning genes on or off (phase variation), or producing recombinant proteins with altered antigenic (antigenic variation) or other properties. It seems that both phase variation and antigenic variation are strategies to enhance bacterial survival in unpredictable environments (Dybvig, 1993; Henderson *et al.*, 1999; Norris & Baumler, 1999). Thus bacteria appear unable always to foresee whether expression of specific factors, often present on the cell surface, will be beneficial or not. In the face of this uncertainty, a bacterial population producing both phase ON and phase OFF cells (phase variation), or altering the properties of proteins produced (antigenic variation) will be better prepared to survive in a capricious environment.

SGM symposium 61: Signals, switches, regulons and cascades: control of bacterial gene expression.
Editors D. A. Hodgson, C. M. Thomas. Cambridge University Press. ISBN 0 521 81388 3 ©SGM 2002.

Although programmed DNA rearrangements control phase variation in many cases, changes in DNA methylation at 5′GATC elements also play an important role in this mode of gene regulation in *Escherichia coli* and *Salmonellae*. In this latter situation, the control of gene expression is essentially identical to the familiar regulation that operates via the action of *trans*-acting factors (protein or RNA) that interact with specific *cis*-active sites, but involves the additional moderating influence of DNA methylation on protein–DNA interactions (for a review, see van der Woude *et al.*, 1996).

Programmed DNA rearrangements are mediated by a variety of different molecular mechanisms that include conservative site-specific recombination, the activity of IS element-encoded transposases, homologous recombination acting preferentially at specific short target sites and slipped-strand mispairing during replication or DNA repair (Fig. 1). They also serve diverse biological functions. Phase and antigenic variation allow differential expression of genes within the population, and require the slow (less than once per cell cycle) conversion from one expression state to another. Relatively inefficient processes involving DNA rearrangements such as recombination appear to be well suited for this purpose.

OUTLINE OF THE MOLECULAR PROCESSES INVOLVED IN PROGRAMMED GENOMIC REARRANGEMENTS

Recombinogenic processes involve two (conservative site-specific recombination and homologous recombination) or three (transposition) *cis*-active sequences upon which specific protein factors act to bring about the cutting and rejoining of DNA molecules in new combinations that is the hallmark of recombination. For both conservative site-specific recombination, and for homologous recombination, DNA cleavage and rejoining occurs within DNA sequences with at least limited identity. Moreover, the situation of these sites (whether on the same DNA molecule or different ones), and their orientation relative to each other as direct or inverted repeats, governs the outcome (integration, excision/deletion or inversion, respectively) of recombination reactions (Fig. 2).

Conservative site-specific recombination requires specific recombinase proteins that bind adjacent to the region of strand cleavage and exchange, and which form (as far as has been determined) covalent protein–DNA adducts as reaction intermediates. Three different families of recombinase proteins that carry out conservative site-specific recombination have been identified on the basis of sequence identity and conserved reaction mechanisms. For the most part, these reactions involve neither the loss of, nor synthesis of, new DNA and are hence termed 'conservative'. In contrast to conservative site-specific recombination, transposition involves a series of related processes in which an element either moves from a donor site to a target site ('cut and paste' transposition), or is duplicated when a donor element is copied into a target site ('replicative transposi-

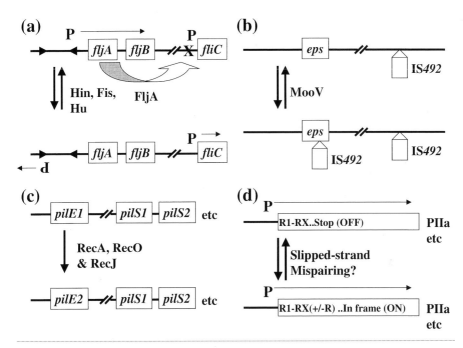

Fig. 1. Examples of genomic rearrangements that control gene expression. (a) Conservative site-specific recombination (inversion) controls flagella type in *Salmonellae*. In the ON orientation, a promoter within the invertible region (P) directs the transcription of *fljA* and *fljB*. FljB is one of two alternative flagellin (FljB or FliC) subunits expressed. FljA represses expression of *fliC* (X). In the OFF orientation, FljA and FljB expression is switched off and FliC expression is activated. Inversion is catalysed by the recombinase Hin, and stimulated by the DNA-binding proteins Fis and HU. (b) Transposition controlling ON↔OFF phase variation in extracellular polysaccharide expression (eps). The insertion and precise excision of the transposable element IS492 in *Pseudoalteromonas atlantica* results in the phase variation in expression of extracellular polysaccharide. Switching in either direction requires the IS element-encoded protein MooV. (c) RecA-dependent recombination controlling pilin antigenic variation in *Neisseria gonorrhoeae*. Non-reciprocal recombination ('gene conversion') results in the transfer of sequences from silent, partial pilin variants (*pilS1*, etc; from *pilS1* in the example shown) to an expression locus (*pilE*) to produce populations that express a diverse collection of pilin types. (d) Insertion or deletion of short sequence repeats as a means of controlling ON↔OFF variation in gene expression. Slipped-strand mispairing is thought to provide a mechanism allowing the insertion/deletion of short nucleotide repeats. In *N. gonorrhoeae*, changes in repeat length within the coding region of the PII family of genes (PIIa, etc.) produce phase variation in expression. Transcription initiating at P produces mRNA that is translated to produce a full-length product in phase ON cells. However, in phase OFF cells, insertion/deletion of a repeat unit (or units) introduces a frameshift mutation that leads to the premature termination of translation.

tion') (Fig. 3). In each case, transposition usually involves the duplication of a short target sequence at the site of insertion to leave the transposable element flanked by direct repeats. Strand cutting and exchange is mediated by specific proteins (transposases), but the DNA synthesis involved appears to rely upon the host cell's normal DNA synthesis and repair pathways.

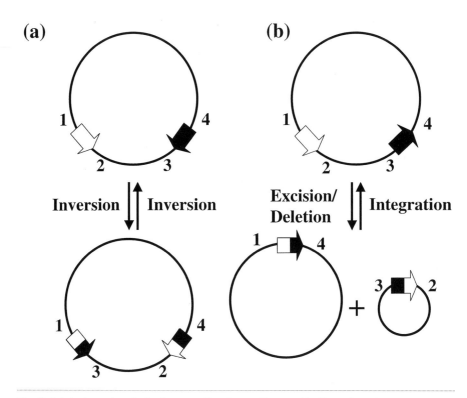

Fig. 2. Substrates and products of conservative site-specific recombination or homologous recombination. (a) Recombination between the inverted repeats (arrowheads) leads to DNA inversion. (b) Intramolecular recombination between the direct repeats (white and black arrows) results in excision or deletion. The products comprise two circles, each containing one of the repeat elements. For some recombination systems, intermolecular recombination allows the reaction to operate in the opposite direction (integration). In both (a) and (b), numbers (1–4) representing reference sequences have been included to allow the changes in spatial organization upon recombination to be followed easily.

Site-specific recombinases and transposases, with one interesting exception, belong to quite separate families of proteins that utilize different reaction mechanisms (see below). The site-specific recombinases characterized to date carry out the strand cleavage and exchange steps via a two-step transesterification pathway involving the formation of covalent tyrosine-DNA (lambda integrase family) or serine-DNA (Hin/resolvase family) linked intermediates. In sharp contrast, characterized transposases release the transposable element from the donor site by hydrolysis, leaving free 3'-OH ends to mount a nucleophilic attack on the target site to produce strand exchange. Additional cleavage of the 5' ends of the transposon leads to 'cut and paste' transposition, with DNA replication producing the characteristic duplication of the target site only. In contrast, in the absence of second strand cleavage, replication duplicates the entire trans-

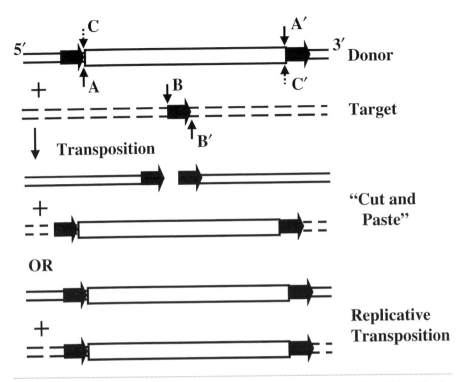

Fig. 3. Replicative and non-replicative ('cut and paste' or 'conservative') transposition. Nicks at the 3′ ends of the transposable element (arrows labelled A and A′), and a staggered break in the target DNA (arrows labelled B and B′), are introduced by transposase. In 'cut and paste' transposition, the 5′ ends of the transposable element are also cut, leading to release of the element from the donor site and the introduction of a double-stranded break into the donor DNA. Ligation of the element into the target site, followed by replication of the gapped DNA, produces direct repeats flanking the element. In replicative transposition, ligation of the free 3′-OH at the ends of the transposable element to the 5′ ends of the target site is followed by replication of the entire element from the remaining free 3′-OH groups by DNA polymerase. Replication forms a cointegrate (fusion of the donor and target DNA with replication of the transposable element; not shown) that is often resolved via a conservative site-specific recombination reaction to yield the products shown.

posable element, as well as the target site, with the formation of a cointegrate ('replicative transposition'). An interesting exception to the rule that recombinases and transposases are quite separate are members of the novel Piv/MooV family. The invertase Piv lacks homology to either the Hin/resolvase family or to the integrase family of site-specific recombinases. Moreover, MooV, encoded by IS492, appears to act as both a transposase *and* as a site-specific recombinase. Thus Piv and other members of this family may employ a different reaction mechanism to achieve recombination than those of better-characterized recombinases (Tobiason *et al.*, 2001). For further details about site-specific recombination and transposition see Hallet & Sherratt (1997).

In contrast to both conservative site-specific recombination and transposition, generalized (or homologous) recombination usually involves long regions of DNA homology (>50 bp). Homologous recombination requires numerous protein factors that include the RecA protein, nucleases, helicases, polymerase and ligase as well as the Ruv proteins required for the branch migration and resolution of Holliday intermediates (for a more detailed overview see Kowalczykowski, 2000). In *E. coli*, the early stages of recombination can follow one of two pathways, termed the RecBCD and RecF pathways. In either pathway, recombination requires that a single-stranded DNA, coated with the RecA protein, invades a homologous duplex to displace one strand and form a D-loop. The RecBCD enzyme, or members of the RecF pathway, are required to prepare the invasive single strand. A key feature of the RecBCD pathway is that the exonuclease activity of the RecBCD enzyme is modified by encounter with a short, asymmetric DNA sequence (5' GCTGGTGG 3') termed Chi, leading to recombination hotspots in the vicinity of such elements (Dixon & Kowalczykowski, 1993).

In addition to recombination, what appears to amount to high frequency, but reversible, mutation has been implicated in the control of gene expression. Thus in some bacteria, changes in the length of short repetitive elements either within coding regions, or in promoters, are often associated with reversible ON↔OFF phase variation in gene expression. Such changes in the repeat elements are generally thought to arise by slipped-strand mispairing (SSM). In this process, local denaturation of the DNA duplex, followed by mispairing between different copies of the repeats, leads to either insertion or deletion of an integral number of repeat units. Such changes might arise either upon replication of the mispaired duplex or upon repair. For systems controlled by slipped-strand mispairing, the frequency of phase switching should typically increase when mismatch repair is defective (Levinson & Gutman, 1987).

In the sections that follow, representative examples of genomic rearrangements that control gene expression in bacteria are described in more detail.

CONSERVATIVE SITE-SPECIFIC RECOMBINATION AS A MEANS OF CONTROLLING GENE EXPRESSION

Site-specific recombinases are widespread in bacteria and their viruses. These proteins play important roles in processes such as chromosome segregation, genome organization and the resolution of cointegrates formed in the course of replicative transposition, as well as in the control of gene expression. Examples of site-specific recombination that control gene expression considered here are listed in Table 1.

The alternate expression of flagellin subunits H1 (FliC) or H2 (FljB) in *Salmonella enterica* serovar Typhimurium, which has been studied extensively, is controlled by the

Table 1. Representative examples of site-specific recombination that regulate gene expression in bacteria and the processes that they control

Organism	Substrate location	Recombination type	Recombinase family	Process controlled	Reference
S. enterica serovar Typhimurium	Chromosomal	Simple inversion	Hin/resolvase	Switching between expression of alternative flagella types	Nanassy & Hughes (2001)
E. coli	Chromosomal	Simple inversion	Integrase	ON→OFF phase variation in type 1 fimbrial expression	Blomfield (2001)
M. lacunata and M. bovis	Chromosomal	Simple inversion	Piv/MooV	Switching between expression of alternative pili types	Tobiason et al. (1999)
Enteric bacteria	Plasmid R64	Complex inversions	Integrase	Alternative C-terminal region of PilV controlling recipient specificity in mating	Komano (1999)
B. subtilis	Chromosomal	Deletion	Hin/resolvase (large serine recombinase subgroup)	Irreversible activation of sigma factor expression in sporulation	Thorpe & Smith (1998)

site-specific inversion of a 995 bp element of DNA (Zieg *et al.*, 1977). In the ON orientation, a promoter within the invertible element directs the transcription of a polycistronic mRNA that encodes both H2 (FljB) and the regulatory protein FljA (Fig. 1a). Since FljB represses expression of the unlinked *fliC* gene, the orientation of the invertible element coordinates the control of *fljA* and *fliC*. The inversion reaction takes place between two 26 bp inverted repeats (*hixL* and *hixR*) that lie at the ends of the invertible element, and the reaction is catalysed by Hin. The recombinase is a member of the Hin/resolvase family, and is one of a series of related proteins that catalyse inversion reactions in bacteria or bacteriophage (Plasterk *et al.*, 1983). In addition to Hin, the inversion is stimulated by the DNA-binding proteins Fis and HU (Johnson & Simon, 1985; Johnson *et al.*, 1986). Fis interacts with a recombinational enhancer element proximal to *hixL* within the invertible region, and HU permits DNA bending between this inverted repeat and the enhancer (Johnson *et al.*, 1986). Recombination is thought to proceed via a series of steps, and to involve the formation of a double-looped nucleoprotein structure (invertasome) with *hixL* and *hixR* in a juxtaposition at synapses (Johnson *et al.*, 1987). The recombination reaction involved has been studied further in considerable biochemical detail, and for a recent publication in this area see Nanassy & Hughes (2001).

Although much attention has been focused on the reaction mechanism controlling the Hin inversion, there has been less interest in its function. However, recent studies by Ikeda and co-workers indicate that expression of H2, as opposed to H1, specifically decreases virulence in systemic mouse infection (Ikeda *et al.*, 2001). Although the reason for this effect is unknown, it seems unlikely that the ability to switch H2 expression OFF is simply a means of avoiding the host's humoral response.

Whereas the activity of Hin in *Salmonella* allows phase switching between two alternative flagellin subunits, other systems that are controlled by DNA inversion produce an ON↔OFF switching in expression of a single structural gene. In *E. coli*, the activity of the FimB and FimE recombinases allows switching between the expression (ON) and non-expression (OFF) of type 1 fimbriae (Dove & Dorman, 1996; Gally *et al.*, 1996). Like the Hin-catalysed system, the *fim* invertible element contains a promoter that directs transcription of the adjacent structural genes in the ON orientation. However, FimB and FimE are members of the integrase family, and the inversion reactions are regulated by IHF (integration host factor) and by Lrp (leucine-responsive regulatory protein), rather than by Fis (Blomfield *et al.*, 1997; Roesch & Blomfield, 1998). FimB and FimE appear to play separate roles in the inversion reactions, and whereas FimB catalyses inversion in either direction, FimE shows a bias for inversion in the ON→OFF direction (Klemm, 1986). Moreover, *fimE* is repressed when the *fim* invertible element is in the OFF orientation, illustrating the complexity of the regulation of this system

(Kulasekara & Blomfield, 1999; Sohanpal *et al.*, 2001). The *fim* switch is controlled by a number of environmental signals, including temperature and levels of the branched-chain amino acids and alanine (Gally *et al.*, 1993). The overall frequency of phase variation in *both* the ON→OFF and in the OFF→ON directions is enhanced by the amino acids, and it seems likely that it is important for the descendants of a single cell to be able to switch phase repeatedly.

In parallel with flagellin phase variation in *Salmonella*, the expression of type IV pilin in *Moraxella lacunata* and *Moraxella bovis* switches between two alternative types. However, in these instances, DNA inversion switches expression between alternative open reading frames that share a common promoter and 5′ translated region, and a variable 3′ (*tfpQ* or *tfpI*) translated region. The 2.1 kb invertible regions are bound by 26 bp inverted repeats (*invL* and *invR*), and the inversion requires the novel Piv recombinase (Marrs *et al.*, 1990). As noted above, Piv does not resemble members of either the lambda integrase or the Hin/resolvase families of site-specific recombinases, but forms part of the Piv/MooV transposase/recombinase family. Piv binds to multiple sequence elements within the invertible region (Tobiason *et al.*, 1999). Since binding to *invL* is unstable, it is possible that, like lambda integrase, efficient recombination requires the interaction of the protein with additional, high-affinity binding sites at a distance from the sites of strand cleavage (Tobiason *et al.*, 1999).

In the examples described above, a recombination reaction takes place between two inverted repeats to bring about a defined switching in gene expression, either between ON or OFF expression states, or to alter expression between two alternative proteins. In principle, the number of protein variants produced by a conservative site-specific recombination system could be increased by combining different invertible elements together. In practice, such 'shufflon' systems do exist, and they have been found to play a key role in generating variant proteins with diverse functional or immunological properties (for a review see Komano, 1999). One of the best characterized shufflons controls variation in the specificity of mating of the incompatibility group I1 plasmid, R64. In this system, seven recombination sites within the 3′ end of the *pilV* gene, encoding the R64 thin pilus, allow fusion of alternate C-terminal coding regions to a single conserved N-terminal region. The inversions are catalysed by the *rci* gene product (an integrase family member), and the recombinase shows a strict specificity for recombination between sites in an inverted, rather than a direct, orientation. Thus the action of Rci can generate up to seven alternative PilV proteins without deletion of DNA.

For the most part, programmed genomic rearrangements that control gene expression are reversible events. Several interesting exceptions to this general rule exist, however, and they represent pathways for terminal differentiation. In *Bacillus subtilis*, sporulation involves the development of a forespore, with an asymmetric division playing a pivotal role in this process (for a review see Kroos *et al.*, 1999). A key step in the pathway involves the activation of an alternative sigma factor, σ^K, within the mother cell. Recombination is controlled by a cascade of events that leads up to the expression of the recombinase (SpoIVCA) in the mother cell, which in turn leads to the irreversible deletion of the 42 kb *skin* element from the coding region of *sigK* to splice the 5′ and 3′ segments of the gene together (Kunkel *et al.*, 1990; Popham & Stragier, 1992). Activation of σ^K within the mother cell leads to production of the coat protein that surrounds the spore. Thus the mother cell directs the final stages of maturation of the spore before undergoing lysis. SpoIVCA belongs to a subgroup within the Hin/resolvase family of recombinases which, although they contain the conserved N-terminal catalytic domain of other members of the family, are unusually large proteins and which share little additional amino acid homology (Thorpe & Smith, 1998).

PHASE VARIATION DETERMINED BY THE INSERTION AND PRECISE EXCISION OF IS ELEMENTS

The insertion of transposable elements into DNA is an important mutagenic process in bacteria. Not only do such insertions disrupt genes, but additional recombination events (such as homologous recombination between elements) provide a route for further genomic rearrangements. It might also be anticipated that the insertion of a transposable element, followed by its precise excision, would provide a common mechanism controlling ON↔OFF phase variation in gene expression. However, most transposable elements show low target site selectivity, limiting the likelihood of insertion into a specific sequence. Moreover, reversion to wild-type by precise excision is usually a very rare event ($<10^{-6}$). Nevertheless, in several instances, insertion and precise deletion of an insertion sequence at a specific target does occur at a detectable frequency.

In situations where precise insertion and deletion of transposable elements occurs within a gene or its regulatory elements, transposition is implicated clearly as a mechanism determining the ON↔OFF phase variation in gene expression. For example, the activity of insertion sequences has been implicated in the phase variation of extracellular polysaccharide in *Pseudoalteromonas atlantica* (IS492), *Neisseria meningitidis* (IS1301) and in *Staphylococcus epidermidis* (IS256) (Bartlett *et al.*, 1988; Hammerschmidt *et al.*, 1996; Ziebuhr *et al.*, 1999).

In the first example listed above, the 1.2 kb element IS492 inserts into, and excises from, the *eps* gene of *P. atlantica* to produce phase variation in synthesis of extracellular polysaccharide (EPS) (Fig. 1b). Transposition of IS492 into *eps* results in the characteristic duplication of the target sequence (5′ CTTGT for this element) as a direct repeat

(Bartlett & Silverman, 1989). Excision of the transposable element from *eps* does not result in the appearance of the element at a new site in the chromosome, but may generate a circle containing a copy of the 5 bp duplicated target separating the ends of the IS element (Bartlett & Silverman, 1989; Perkins-Balding *et al.*, 1999). Thus excision of IS*492* from *eps* appears to be a conservative site-specific recombination, and MooV able to act as both a transposase and a recombinase.

The requirement for IS element-encoded products does not appear to have been investigated in other cases where the precise excision of a transposable element is requisite for the OFF↔ON phase transitions. It would be interesting to determine how the precise excision of IS*1301* and of IS*256* occurs, and if IS*492* (and related elements) plays a more general role in phase variation in *P. atlantica* and other bacteria.

PROGRAMMED REARRANGEMENTS THAT INVOLVE THE RecA PROTEIN

As outlined above, variation in protein expression is achieved in some instances by conservative site-specific recombination. However, protein variation arises in other systems following homologous recombination to generate an impressive array of variants. In the best studied example thus far, the antigenic variation in pilin expression in *Neisseria gonorrhoeae*, exchange involves factors that participate in the generalized recombination pathways yet also clearly requires a specialized reaction mechanism (Fig. 1c). Thus recombination occurs at high frequency ($>10^{-2}$), involves very limited sequence identity (9–30 bp) and requires specialized *cis*-active sequences (and, in all likelihood, specific *trans*-active factors). Moreover, variation in pilin expression is usually determined by the unidirectional transfer of sequences from one of an array of partial, non-expressed pilin copies (*pilS*) to an expression locus (*pilE*) (Meyer *et al.*, 1984; Seifert, 1996).

The transfer reaction requires the RecA protein, which probably participates in heteroduplex formation as a step in the recombination process, as well as RecO- and RecQ-like proteins (Koomey *et al.*, 1987; Mehr & Seifert, 1998). Perhaps surprisingly, the pilin variation is not influenced by mutants that lack the gonococcal homologues of the RecBCD exonuclease (Mehr & Seifert, 1998). Thus pilin antigenic variation appears to require the activity of a *recF*-like pathway, but not a gonococcal equivalent of the *E. coli recBCD* pathway. In addition, recombination is enhanced by the presence of short repeat sequences (Sma/Cla repeats) found at the 3′ ends of both *pilE* and *pilS* (Wainwright *et al.*, 1994). The precise mechanism(s) by which non-reciprocal transfer is achieved remains unclear, but it seems likely that DNA transformation (either with a novel recombinant *pilE* copy, and/or with *pilS*) and unequal crossing over between multiple genome copies are both contributory factors (Seifert, 1996).

In addition to the participation of RecA in the antigenic variation of pilin in the gonococcus, this protein has also been implicated in the control of gene expression in several other bacteria, including *Campylobacter* and *Pseudomonas tolaasii* and others (Dworkin *et al.*, 1997; Sinha *et al.*, 2000).

THE INSERTION AND DELETION OF SHORT NUCLEOTIDE REPEATS AND HOMOPOLYMERIC RUNS CONTROLLING GENE EXPRESSION

In some bacteria, such as *Neisseria*, *Bordetella*, *Haemophilus* and the *Mycoplasma*, changes in the length of short repetitive elements are associated with the reversible ON↔OFF phase variation in gene expression. Surprisingly, this mechanism of phase variable control has yet to be reported in enteric bacteria, despite the fact that the pertinent sequence changes occur when the elements are cloned in *E. coli*. Sequence elements may comprise either a single nucleotide (homopolymeric runs), or contain multiple copies of a short repeat sequence (for a recent review on such elements see van Belkum, 1999). Moreover, repeat elements have been found both in regulatory regions (separating −35 and −10 regions of the promoter), and within open reading frames [leading to frameshifts that produce either full length (OFF→ON switching) or truncated (ON→OFF switching) proteins]. Although it is generally reported that variation in repeat element number occurs via slipped-strand mispairing, the mechanism of control has yet to be demonstrated unambiguously.

One of the first reported examples in which changes in repeat length were shown to control phase variation, that of the Op or PII protein in the gonococcus, remains one of the best characterized (Stern *et al.*, 1986). *N. gonorrhoeae* contains multiple copies of the gene encoding PII, and each of these is capable of switching between expression (ON) and non-expression (OFF) states (Fig. 1d). Phase variation is determined by the addition or loss of 5′ CTCTT repeats within the region of the genes encoding the hydrophobic signal sequence (Stern *et al.*, 1986). Thus a single bacterium is capable of producing a variable number of different PII proteins on the cell surface depending upon which copies of the gene are in or out of frame. Changes in phase of PII involve the gain or loss of an integral number of repeat units, with addition or deletion of a single unit being the most common event. Moreover, the frequency of switching increases in direct proportion to the number of repeats present (Murphy *et al.*, 1989). The phase switching is independent of the RecA protein, and the above characteristics are consistent with the hypothesis that PII phase variation is controlled by slipped-strand mispairing (Murphy *et al.*, 1989). Nevertheless, PII phase variation is not diminished in a mismatch repair deficient mutant (*mutS*), as might be expected for a system controlled by slipped-strand mispairing. However, mismatch repair in *E. coli* requires slipped loops of 4 bp or less to function effectively, and so the latter observation may be misleading (J. Cannon, unpublished results).

In contrast to the situation for PII, the phase variation of the haemoglobin receptors in *Neisseria*, which appears to involve changes in length of homopolymeric runs of guanine residues in both *hmbR* and *hpuA*, is diminished in a *mutS* and *mutL* background (Lewis *et al.*, 1999; Richardson & Stojiljkovic, 2001). Thus, at least in this system, phase variation is likely to be determined by slipped-strand mispairing.

CONCLUSIONS AND FUTURE PERSPECTIVES

The expression of many cell surface factors in pathogenic bacteria is controlled by both phase and antigenic variation. In many instances, such control is determined by programmed genomic rearrangements. The mechanisms of such control are diverse, including site-specific recombination and transposition, as well as homologous recombination and (probably) slipped-strand mispairing. Such systems provide excellent models for studying basic processes, such as protein–DNA interactions and recombination, but it will be of particular interest to learn more about the precise functions of the rearrangements studied. Since gene regulation is adaptive, understanding how various systems are controlled should provide important clues about function, which can in turn be tested in a suitable *in vivo* model. For example, the construction of phase-locked mutants, and their effects on *in vivo* growth and pathogenesis, has provided some exciting observations about flagella phase variation recently (Ikeda *et al.*, 2001). For many types of rearrangement, it is possible to predict that phase or antigenic variation occurs for a particular gene by identifying characteristic nucleotide sequences (such as repetitive elements or the presence of putative recombinase genes) (see Carrick *et al.*, 1998 and Wang *et al.*, 2000, for examples). Many systems identified thus far affect the expression of cell-surface structures (which are most readily detected), but there seems no reason why other cellular factors should not also be affected. Future studies should provide a more complete inventory of genomic rearrangements that control gene expression, as well as their molecular mechanisms and functions in microbial survival and pathogenesis.

ACKNOWLEDGEMENTS

I thank Janne Cannon, Anna Glasgow Karls, Riccardo Manganelli and Marjan van de Woude for helpful comments, and my wife, Barbara, for editorial assistance. This work was supported by grants from the BBSRC (the Biotechnology and Biological Sciences Research Council).

REFERENCES

Bartlett, D. H. & Silverman, M. (1989). Nucleotide sequence of IS*492*, a novel insertion sequence causing variation in extracellular polysaccharide production in the marine bacterium *Pseudomonas atlantica. J Bacteriol* **171**, 1763–1766.

Bartlett, D. H., Wright, M. E. & Silverman, M. (1988). Variable expression of extracellular polysaccharide in the marine bacterium *Pseudomonas atlantica* is controlled by genome rearrangement. *Proc Natl Acad Sci U S A* **85**, 3923–3927.

van Belkum, A. (1999). Short sequence repeats in microbial pathogenesis and evolution. *Cell Mol Life Sci* **56**, 729–734.

Blomfield, I. C. (2001). The regulation of Pap and type 1 fimbriation in *Escherichia coli. Adv Microb Physiol* **45**, 1–49.

Blomfield, I. C., Kulasekara, D. H. & Eisenstein, B. I. (1997). Integration host factor stimulates both FimB- and FimE-mediated site-specific DNA inversion that controls phase variation of type 1 fimbriae expression in *Escherichia coli. Mol Microbiol* **23**, 705–717.

Carrick, C. S., Fyfe, J. A. & Davies, J. K. (1998). *Neisseria gonorrhoeae* contains multiple copies of a gene that may encode a site-specific recombinase and is associated with DNA rearrangements. *Gene* **220**, 21–29.

Dixon, D. A. & Kowalczykowski, S. C. (1993). The recombination hotspot *x* is a regulatory sequence that acts by attenuating the nuclease activity of *E. coli* RecBCD enzyme. *Cell* **73**, 87–96.

Dove, S. L. & Dorman, C. J. (1996). Multicopy *fimB* gene expression in *Escherichia coli*: binding to the inverted repeats *in vivo*, effect of *fimA* transcription and DNA inversion. *Mol Microbiol* **21**, 1161–1173.

Dworkin, J., Shedd, O. L. & Blaser, M. J. (1997). Nested DNA inversion of *Campylobacter fetus* S-layer genes is *recA* dependent. *J Bacteriol* **179**, 7523–7529.

Dybvig, K. (1993). DNA rearrangements and phenotypic switching in prokaryotes. *Mol Microbiol* **10**, 465–471.

Gally, D. L., Bogan, J. A., Eisenstein, B. I. & Blomfield, I. C. (1993). Environmental control of the *fim* switch controlling type 1 fimbrial phase variation in *Escherichia coli* K-12: effects of temperature and media. *J Bacteriol* **175**, 6186–6193.

Gally, D. L., Leathart, J. & Blomfield, I. C. (1996). Interaction of FimB and FimE with the *fim* switch that controls the phase variation of type 1 fimbriation in *Escherichia coli. Mol Microbiol* **21**, 725–738.

Hallet, B. & Sherratt, D. J. (1997). Transposition and site-specific recombination: adapting DNA cut-and-paste mechanisms to a variety of genetic rearrangements. *FEMS Microbiol Rev* **21**, 157–178.

Hammerschmidt, S., Hilse, R., van Putten, J. P. M., Gerardy-Schahn, R., Unkmeir, A. & Frosch, M. (1996). Modulation of cell surface sialic acid expression in *Neisseria meningitidis* via a transposable genetic element. *EMBO J* **15**, 192–198.

Henderson, I. R., Owen, P. & Nataro, J. P. (1999). Molecular switches – the on and off of bacterial phase variation. *Mol Microbiol* **33**, 919–932.

Ikeda, J. S., Schmitt, C. K., Darnell, S. C. & 8 other authors (2001). Flagellar phase variation of *Salmonella enterica* serovar Typhimurium contributes to virulence in the murine typhoid infection model but does not influence *Salmonella*-induced enteropathogenesis. *Infect Immun* **69**, 3021–3030.

Johnson, R. C. & Simon, M. I. (1985). Hin-mediated site-specific recombination requires two 26 bp recombination sites and a 60 bp recombinational enhancer. *Cell* **41**, 781–789.

Johnson, R. C., Bruist, M. F. & Simon, M. I. (1986). Host protein requirements for *in vitro* site-specific DNA inversion. *Cell* **46**, 531–539.

Johnson, R. C., Glasgow, A. C. & Simon, M. I. (1987). Spatial relationship of the Fis binding sites for Hin recombination enhancer activity. *Nature* **329**, 462–465.

Klemm, P. (1986). Two regulatory *fim* genes, *fimB* and *fimE*, control the phase variation of type 1 fimbriae in *Escherichia coli*. *EMBO J* **5**, 1389–1393.

Komano, T. (1999). Shufflons: multiple inversion systems and integrons. *Annu Rev Genet* **33**, 171–191.

Koomey, M., Gotschlich, E. C., Robbins, K., Bergstrom, S. & Swanson, J. (1987). Effects of *recA* mutations on pilus antigenic variation and phase transitions in *Neisseria gonorrhoeae*. *Genetics* **117**, 391–398.

Kowalczykowski, S. C. (2000). Initiation of genetic recombination and recombination-dependent replication. *Trends Biochem Sci* **25**, 156–165.

Kroos, L., Zhang, B., Ichikawa, H. & Yu, Y. T. (1999). Control of sigma factor activity during *Bacillus subtilis* sporulation. *Mol Microbiol* **31**, 1285–1294.

Kulasekara, D. H. & Blomfield, I. C. (1999). The molecular basis for the specificity of *fimE* in the phase variation of type 1 fimbriae of *Escherichia coli* K-12. *Mol Microbiol* **31**, 1171–1181.

Kunkel, B., Losick, R. & Stragier, P. (1990). The *Bacillus subtilis* gene for the development transcription factor sigma K is generated by excision of a dispensable DNA element containing a sporulation recombinase gene. *Genes Dev* **4**, 525–535.

Levinson, G. & Gutman, G. A. (1987). High frequencies of short frameshifts in poly-CA/TG tandem repeats borne by bacteriophage M13 in *Escherichia coli* K-12. *Nucleic Acids Res* **15**, 5323–5338.

Lewis, L. A., Gipson, M., Hartman, K., Ownbey, T., Vaughn, J. & Dyer, D. W. (1999). Phase variation of HpuAB and HmbR, two distinct haemoglobin receptors of *Neisseria meningitidis* DNM2. *Mol Microbiol* **32**, 977–989.

Marrs, C. F., Rozsa, F. W., Hackel, M., Stevens, S. P. & Glasgow, A. C. (1990). Identification, cloning, and sequencing of *piv*, a new gene involved in inverting the pilin genes of *Moraxella lacunata*. *J Bacteriol* **172**, 4370–4377.

Mehr, I. J. & Seifert, H. S. (1998). Differential roles of homologous recombination pathways in *Neisseria gonorrhoeae* pilin antigenic variation, DNA transformation and DNA repair. *Mol Microbiol* **30**, 697–710.

Meyer, T. F., Billyard, E., Haas, R., Storzbach, S. & So, M. (1984). Pilus genes of *Neisseria gonorrhoeae*: chromosomal organization and DNA sequence. *Proc Natl Acad Sci U S A* **81**, 6110–6114.

Murphy, G. L., Connell, T. D., Barritt, D. S., Koomey, M. & Cannon, J. G. (1989). Phase variation of gonococcal protein II: regulation of gene expression by slipped-strand mispairing of a repetitive DNA sequence. *Cell* **56**, 539–547.

Nanassy, O. Z. & Hughes, K. T. (2001). *In vivo* identification of intermediate stages of the DNA inversion reaction catalyzed by the *Salmonella* Hin recombinase. *Genetics* **149**, 1649–1663.

Norris, T. L. & Baumler, A. J. (1999). Phase variation of the *lpf* operon is a mechanism to evade cross-immunity between *Salmonella* serovars. *Proc Natl Acad Sci U S A* **96**, 13393–13398.

Perkins-Balding, D., Duval-Valentin, G. & Glasgow, A. C. (1999). Excision of IS*492* requires flanking target sequences and results in circle formation in *Pseudomonas atlantica*. *J Bacteriol* **181**, 4937–4948.

Plasterk, R. H., Brinkman, A. & van de Putte, P. (1983). Phase variation: evolution of a controlling element. *Science* **209**, 1370–1374.

Popham, D. L. & Stragier, P. (1992). Binding of the *Bacillus subtilis spoIVCA* product to the recombination sites of the element interrupting the σ^K-encoding gene. *Proc Natl Acad Sci U S A* **89**, 5991–5995.

Richardson, A. R. & Stojiljkovic, I. (2001). Mismatch repair and the regulation of phase variation in *Neisseria meningitidis*. *Mol Microbiol* **40**, 645–655.

Roesch, P. L. & Blomfield, I. C. (1998). Leucine alters the interaction of the leucine-responsive regulatory protein (Lrp) with the *fim* switch to stimulate site-specific recombination in *Escherichia coli*. *Mol Microbiol* **27**, 751–761.

Seifert, H. S. (1996). Questions about gonococcal pilus phase- and antigenic variation. *Mol Microbiol* **21**, 433–440.

Sinha, H., Pain, A. & Johnstone, K. (2000). Analysis of the role of *recA* in phenotypic switching of *Pseudomonas tolaasii*. *J Bacteriol* **182**, 6532–6535.

Sohanpal, B., Kulasekara, D. H., Bonnen, A. & Blomfield, I. C. (2001). Orientational control of *fimE* expression in *Escherichia coli*. *Mol Microbiol* **42**, 483–494.

Stern, A., Brown, M., Nickel, P. & Meyer, T. F. (1986). Opacity genes in *Neisseria gonorrhoeae*: control of phase and antigenic variation. *Cell* **47**, 61–71.

Thorpe, H. M. & Smith, M. C. M. (1998). *In vitro* site-specific integration of bacteriophage DNA catalyzed by a recombinase of the resolvase/invertase family. *Proc Natl Acad Sci U S A* **95**, 5505–5510.

Tobiason, D. M., Lenich, A. G. & Glasgow, A. C. (1999). Multiple DNA binding activities of the novel site-specific recombinase, Piv, from *Moraxella lacunata*. *J Biol Chem* **274**, 9698–9706.

Tobiason, D. M., Buchner, J. M., Thei, W. H., Gernert, K. M. & Glasgow Karls, A. C. (2001). Conserved amino acid motifs from the novel Piv/MooV family of transposases and site-specific recombinases are required for catalysis of DNA inversion by Piv. *Mol Microbiol* **39**, 641–651.

Wainwright, L. A., Pritchard, K. H. & Seifert, H. S. (1994). A conserved DNA sequence is required for efficient gonococcal pilin variation. *Mol Microbiol* **13**, 75–87.

Wang, G., Ge, Z., Rasko, D. A. & Taylor, D. E. (2000). Lewis antigens in *Helicobacter pylori*: biosynthesis and phase variation. *Mol Microbiol* **36**, 1187–1196.

van der Woude, M. W., Braaten, B. & Low, D. (1996). Epigenetic phase variation of the *pap* operon in *Escherichia coli*. *Trends Microbiol* **4**, 5–9.

Ziebuhr, W., Krimmer, V., Rachid, S., Lößner, I., Götz, F. & Hacker, J. (1999). A novel mechanism of phase variation of virulence in *Staphylococcus epidermidis*: evidence for control of the polysaccharide intercellular adhesin synthesis by alternating insertion and excision of the insertion sequence element IS*256*. *Mol Microbiol* **32**, 345–356.

Zieg, J., Silverman, M., Hilmen, M. & Simon, M. (1977). Recombinational switch for gene expression. *Science* **196**, 170–172.

Structures of multisubunit DNA-dependent RNA polymerases

Robert D. Finn,[1] Elena V. Orlova,[2] Marin van Heel[3] and Martin Buck[3]

[1] Wellcome Sanger Institute, Wellcome Trust Genome Campus, Hinxton, Cambs CB11 1SA, UK

[2] Department of Crystallography, Birkbeck College, University of London, London WC1E 7HX, UK

[3] Department of Biological Sciences, Imperial College, London SW7 2AZ, UK

INTRODUCTION

Expression of genetic information is as essential to life as DNA replication and transmission. Units of genetic information, **genes**, specify the information necessary to make a particular protein. However, DNA is not the direct template used for protein synthesis. Instead, the cell employs complex machinery to make replicas of genes, called RNA molecules. The process of copying DNA into RNA is known as **transcription** (see Fig. 1).

Transcription generates three major forms of RNA molecules: messenger (m)RNA, transfer (t)RNA and ribosomal (r)RNA. The **mRNA** is involved in carrying the information necessary for protein synthesis (Fig. 1). The tRNA, rRNA and other RNA molecules have either structural or catalytic functions. The enzyme that catalyses the synthesis of RNA from DNA templates is the DNA-dependent RNA polymerase (EC 2.7.7.6). The complexity of the polymerase ranges from the single-subunit phage and plastid enzymes through to the multisubunit enzymes of the archaea, bacteria, plants, fungi and animals.

Transcription is a cyclical process, which has been subdivided into specific steps (Fig. 2). These steps include transcription pre-initiation, initiation, elongation, pausing and termination. Although these events have been characterized biochemically, a structural knowledge of the complex and dynamic interplay of RNA polymerase with DNA, RNA and accessory proteins such as activators of transcription is essential for a detailed understanding of transcription at a molecular level. The *Escherichia coli* RNA

SGM symposium 61: Signals, switches, regulons and cascades: control of bacterial gene expression.
Editors D. A. Hodgson, C. M. Thomas. Cambridge University Press. ISBN 0 521 81388 3 ©SGM 2002.

Fig. 1. Flow of genetic information. The standard flow of genetic information goes from DNA to protein. Transcription occupies a central point between replication (DNA synthesis) and translation (protein synthesis).

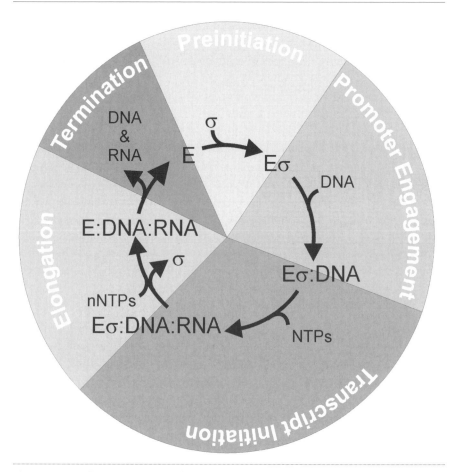

Fig. 2. Functionally distinct segments of the transcription cycle.

polymerase is the prototypical multisubunit DNA-dependent RNA polymerase. In its simplest form, the bacterial RNA polymerase consists of four subunits, two alpha, beta and beta prime ($\alpha_2\beta\beta'$), referred to as **core RNA polymerase** (Burgess *et al.*, 1969). In this form, the core RNA polymerase is both sufficient and necessary for the basic transcription reaction, yet is incapable of recognizing and binding to specific DNA regions. The catalytic function of core RNA polymerase is to carry out RNA chain extension, termed transcription elongation. DNA-sequence-specific initiation of transcription requires an additional protein, called a **sigma (σ) factor** (Fig. 2). The σ factor binds to the core RNA polymerase, to form a modified enzyme known as the **holoenzyme ($\alpha_2\beta\beta'\sigma$)** (Burgess *et al.*, 1969). The holoenzyme is able to locate a discrete control sequence, known as the promoter, near the start of the gene. At the promoter an initial binary DNA–protein complex is formed. Controlled transcription initiation clearly requires a higher complexity of RNA polymerase than that required to carry out RNA synthesis *per se*. Transcription initiation of specific sets of genes is often directed by one of a number of specific sigma subunits (see below). After transcription initiation, the sigma factor is released by RNA polymerase, leaving the core enzyme to carry out transcription elongation in a processive manner (Gelles & Landick, 1998; Landick, 1999). After the gene has been transcribed, signals are conferred to the core RNA polymerase to release the mRNA and DNA, a process called termination (Fig. 2).

Gene expression is a cascade of biochemical events that requires stringent regulation, such that an appropriate response can be adopted to the ever-changing environmental stimuli. The primary control point for gene expression is transcription. Therefore, a detailed understanding of the transcription cycle is necessary to provide insights into regulation of gene expression, and structural studies of the RNA polymerase form an important part of this understanding.

Cryoelectron microscopy (cryo-EM) together with single particle reconstitution has enabled the rapid determination of the structure of *E. coli* RNA polymerase in different functional states to a medium level of resolution (this chapter). A powerful combination of cryo-EM, X-ray crystallography, DNA cross-linking and protein footprinting studies has been used to create model structures of the *E. coli* RNA polymerase at different steps in transcription. Comparisons to the *Saccharomyces cerevisiae* pol II enzyme demonstrate a remarkable conservation in structure between the bacterial and eukaryotic enzymes (Ebright, 2000; Geiduschek & Bartlett, 2000; Minakhin *et al.*, 2001a). Similarities of sequence, as well as structure and function, establish that the principal features of the multisubunit RNA polymerase enzyme have been retained throughout evolution and that common mechanisms of transcription are used among prokaryotes and eukaryotes. The focus of this chapter is on the multisubunit *E. coli* bacterial enzyme.

BACTERIAL RNA POLYMERASE SUBUNITS

The alpha subunit

The α subunit, encoded by the gene *rpoA*, is a 36.5 kDa protein comprising 329 aa. In the core RNA polymerase and holoenzyme, there are two copies of the α subunit. The two α subunits form a dimer that is essential for RNA polymerase assembly. In addition, these α subunits play a role in directing DNA binding to upstream sequences of certain promoters (Estrem *et al.*, 1999). Structurally, the *E. coli* α subunit is the only completely characterized subunit from the *E. coli* core polymerase. The structural analyses arose from the fact that limited proteolysis of the subunit allowed the identification of two structurally distinct regions (Igarashi & Ishihama, 1991; Jeon *et al.*, 1995). These regions are referred to as the N-terminal domain (αNTD, residues 1–239) and the C-terminal domain (αCTD, residues 249–329) (see Fig. 3). These two regions have been shown by NMR (nuclear magnetic resonance) studies to be connected by a 10-amino-acid peptide possessing little or no secondary structure (Fujita *et al.*, 2000; Jeon *et al.*, 1997). Thus the connector has been termed the flexible linker. Biochemical studies demonstrated that the αNTD possessed a dimerization interface and that the αNTD is responsible for interacting with the β and β' subunits (Igarashi *et al.*, 1991). The αCTD is necessary for interacting with upstream promoter DNA and some transcriptional activators (Blatter *et al.*, 1994; Igarashi *et al.*, 1991).

Chronologically, the αCTD structure was the first region to be elucidated, using NMR methods (Jeon *et al.*, 1995). The isolated αCTD was found to function in binding to specific activator regions in the DNA upstream of the promoter itself (termed UP elements) and in interacting with certain activators of transcription. The structure of the αCTD indicated that interaction with the UP element was mediated by a helix–turn–helix (HTH) motif – the best studied protein motif/structure that mediates recognition and binding to DNA. It also showed the prominence of a loop and helix within the structure, close in 3D space, but separated in sequence, which were identified by mutational analysis to interact with activators (Jeon *et al.*, 1995). Recently, a very similar structure to the *E. coli* αCTD was derived, using NMR, from the thermophile *Thermus thermophilus*. Interestingly, the sequence of these two species of αCTD is quite divergent, demonstrating that evolutionary selection pressure has been directed more towards structural conservation than sequence conservation.

The *E. coli* αNTD structure was determined using X-ray crystallography and resolved to 2.5 Å (Zhang & Darst, 1998). The 26 kDa αNTD monomer consists of two structural domains. Two helices from one αNTD domain interact with the corresponding helices from the other monomer to form an extensive hydrophobic core within the

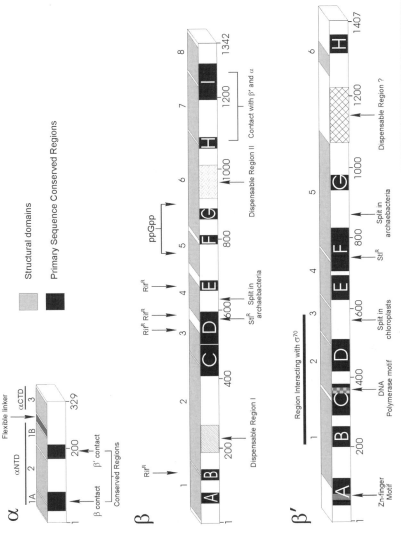

Fig. 3. Schematics of core RNA polymerase subunits. Conserved primary sequence regions are marked by solid black boxes. Grey boxes mark structural domains. For the α subunit, the black bars denote the two regions, αNTD and αCTD. The conserved regions also correspond to the regions interacting with β or β'. For the β subunit, Rif^R and Stl^R are regions to which mutations conferring resistance to rifampicin and streptolydigin, respectively, have been mapped. The shaded regions indicate the two dispensable regions. The conserved regions are labelled A–I. ppGpp marks a region that has been implied to be involved in the stringent response affecting interaction with the effector nucleotide ppGpp. For the β' subunit, Stl^R is defined as above, while the region of β' interacting with σ^{70} is denoted by a solid black bar. Conserved regions are labelled A–H. Other key features are marked and discussed in the text.

αNTD dimer (Zhang & Darst, 1998). The combination of this structure with protein footprinting data (Heyduk *et al.*, 1996) provided insights into how the core RNA polymerase was assembled (Zhang & Darst, 1998). The β subunit first binds to one α subunit, followed by β′ subunit binding predominantly to the other α subunit. More recently, the αNTD dimer structure was solved as part of the complete *Thermus aquaticus* core RNA polymerase (Zhang *et al.*, 1999), clearly demonstrating its role in assembly and illustrating how the αNTDs interact with β and β′. The structure of the *T. aquaticus* αNTD closely resembled that of the *E. coli* αNTD dimer.

The beta subunit

In *E. coli*, the two genes encoding the largest subunits, β and β′, are transcribed from a single operon. The β subunit (1342 aa, 150 kDa) is encoded by *rpoB*, and is the second largest subunit within the *E. coli* enzyme. It has been divided up into nine separate regions (labelled A–I; see Fig. 3) based on sequence alignments between all β subunits and homologous subunits from eukaryotes and archaea (Severinov *et al.*, 1992, 1996). Such sequence alignments also showed the presence of two long insertions (~150 aa) within the *E. coli* β subunit, termed dispensable regions (DR) I and II (Borukhov *et al.*, 1991; Severinov *et al.*, 1992, 1996). DRI is located between the conserved regions βB and βC (Fig. 3), whilst DRII is found between regions βG and βH (Fig. 3). The accommodation of such long insertions is indicative of a modular organization of independently folded discrete domains. The modular organization of the β subunit was further demonstrated by the formation of a functional polymerase, following the introduction of artificial split sites within the *rpoB* gene (Severinov *et al.*, 1996). The two most C-terminal elements, conserved regions βH and βI, are involved in interacting with the β′ and α subunits (Fig. 3). β and β′ combine to form the catalytic core of the enzyme (Heyduk *et al.*, 1996).

Within the *E. coli* polymerase, all of the many mutations conferring resistance to the antibiotic rifampicin were mapped to the β subunit, at four distinct locations (Severinov *et al.*, 1998b). From this observation, it was hypothesized that these regions, although separated in the linear sequence, were likely to be close together in the folded protein structure of the β subunit. This view was proven correct following the determination of the structure of *T. aquaticus* core RNA polymerase co-crystallized with rifampicin (Campbell *et al.*, 2001).

Finally, the β subunit has also been implicated in playing a role in the stringent response. It has been postulated that the stringent response effector molecule guanosine tetraphosphate binds to the β subunit (see Fig. 3) (Ishihama, 2000; Zhou & Jin, 1998).

The beta prime subunit

The *E. coli* β' subunit (1407 aa, 155.2 kDa), encoded by *rpoC*, is the largest subunit. As with β, there are a number of highly conserved regions that exist in the prokaryotic β' subunits and the archaeal and eukaryotic equivalents (A'/A'' and RpbA, respectively) (Archambault & Friesen, 1993). There are eight regions of conserved sequence (A–H) (Severinov *et al.*, 1996) [the β conserved regions (A–I) and β' conserved regions (A–H) do not share any significant sequence similarity to each other], with a mean of ~70 % sequence similarity between the prokaryotic subunits (Fig. 3). Along the linear sequence, a number of nucleic acid binding motifs are located (e.g. region A has a Zn-finger-like motif and region C has similarity to the DNA-binding cleft of DNA polymerase I). Mutations conferring resistance to the antibiotic streptolydigin were all found to localize to region β'F (Severinov *et al.*, 1998a). Similar resistance mutations were also found in βD, leading to the hypothesis that β'F and βD are proximal in space and form a discrete binding pocket for streptolydigin (Severinov *et al.*, 1998a). Mutations affecting pausing and termination were found to be located throughout the β' sequence. Both proximity and mutational studies have implied a strong interaction between β' conserved regions B, C and D (residues 201–477) and σ^{70} (Fig. 3) (Arthur *et al.*, 2000; Arthur & Burgess, 1998; Burgess *et al.*, 1998). Like β, the *E. coli* β' has an inserted region, between regions G and H, that is believed to be a dispensable region (β'DR) (see Fig. 3) (Severinov *et al.*, 1996). Many studies have demonstrated that the C-termini of the β and β' subunits are absolutely necessary for formation of a competent RNA polymerase (Katayama *et al.*, 2000; Mustaev *et al.*, 1997). Once the structure of the core RNA polymerase from *T. aquaticus* was solved, the complexity of the intimacy between β and β' was revealed (Zhang *et al.*, 1999). A key and noteworthy point is that an interaction between β and β' occurs at the active centre, where the Mg^{2+} is chelated (Zaychikov *et al.*, 1996). Here, β regions H and I and β' region D interact to position the -NADFDGD- motif of β'D necessary for chelating the active site Mg^{2+} (Zhang *et al.*, 1999).

Omega

During the early years of *E. coli* RNA polymerase investigation, a critical observation was made. It was found that during the purification of the enzyme, a small protein (110 aa) was found to co-purify. This protein (encoded by the gene *rpoZ*) was termed omega (ω) (Gentry & Burgess, 1986). Protein–protein cross-linking analysis indicated that the ω protein was intimately associated with the C-terminus of the β' subunit (Gentry & Burgess, 1993). Although with a degree of uncertainty, ω was believed for many years to play a role in mediating the stringent response. Two lines of evidence suggested this: *in vitro* transcription analysis (Igarashi *et al.*, 1989) and the fact that ω and SpoT (a key protein in the stringent response) are co-transcribed (Gentry & Burgess, 1989). However, this view was questioned based on the phenotype of *rpoZ* mutants (Gentry *et al.*, 1991). Very recent biochemical and structural evidence has shown that ω wraps around the C-terminus of β' (Minakhin *et al.*, 2001a), playing a role in the fold/stability of the β' subunit (Minakhin *et al.*, 2001a; Mukherjee & Chatterji, 1997; Mukherjee *et al.*, 1999).

The sigma factors

The conversion of core RNA polymerase to the holoenzyme form requires the binding of a σ factor (Burgess *et al.*, 1969). Within *E. coli* there are two classes of σ factor: the σ^{70}-like class and the σ^{N} class (Lonetto *et al.*, 1992). These two classes were defined based upon sequence and functional comparisons. Described below are the archetypal members from each class.

The σ^{70}-like class. The specificity factor responsible for the transcription of the majority of genes used during the exponential growth phase is **sigma-70** (σ^{70}). σ^{70} is encoded by the *rpoD* gene, which produces a single peptide of 613 aa in length and 70.2 kDa in molecular mass. The binding of σ^{70} to the free core RNA polymerase forms the σ^{70} holoenzyme (Burgess *et al.*, 1969). It is only in the holoenzyme form that the RNA polymerase can direct sequence-specific transcription initiation. All σ factors play a direct role in the binding specificity, as it is the σ factors themselves that make the critical DNA-sequence-specific contacts. Members of the σ^{70}-like class of transcription factors all share similarity at the level of sequence and functionality (see Fig. 4). Moreover, a subset of σ factors, the σ^{70} type, are known as the primary σ factors and direct expression of housekeeping genes. All the other σ factors are known as secondary or alternative σ factors. These are subtle divisions based on sequence similarity, but all members share common regions. σ^{70} has been divided into four main regions, with each region further subdivided based on details of their functionality (Gross *et al.*, 1992, 1998).

σ^{70} is unable to bind DNA in the absence of core RNA polymerase. Region 1, localized at the N-terminal end of the σ^{70}, plays a critical role in inhibition of σ^{70} DNA interactions (Callaci *et al.*, 1998; Dombroski *et al.*, 1993) and is therefore termed an autoinhibition region (Fig. 4). Furthermore, subregion 1.2 contains an acidic patch, which is thought to be a key element in inhibition, as this acidic patch would electrostatically repel the DNA. It appears that region 1 undergoes a relative repositioning upon binding the core RNA polymerase (Callaci *et al.*, 1998). Evidence suggests that region 1 is also required for the efficient initiation of transcription, through stabilizing the open complex (Wilson & Dombroski, 1997).

Region 2 of σ^{70} is the most conserved region, indicating its importance (Fig. 4). The subregion 2.1 has a coiled-coil motif that has been demonstrated, via mutagenesis

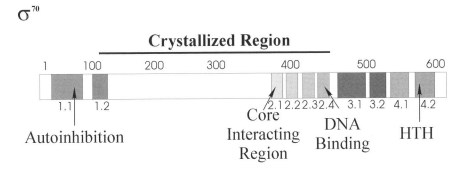

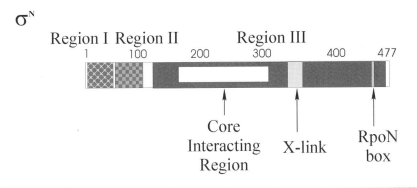

Fig. 4. Schematic of the sigma factors. Top: σ^{70}. The conserved regions are marked. The black bar represents the residues present in the structural fragment of σ^{70}. Bottom: σ^{N} conserved regions and the functions assigned to areas within each region.

experiments, to be pivotal in binding of σ^{70} to the core RNA polymerase (Nagai & Shimamoto, 1997). Region 2.2 also plays a role in core RNA polymerase binding. Regions 2.1 and 2.3 have been extensively implicated in maintaining strand separation during transcription initiation (Waldburger & Susskind, 1994). Region 2.4 (known as the RpoD box, *E. coli* residues 437–440) is necessary for the binding of the −10 promoter hexamer (consensus -TATAAT-) (Fenton *et al.*, 2000). Region 2.4 forms an amphipathic α-helix, with the hydrophobic residues localized on one side of the helix, which faces the core structure of the σ^{70} protein, whilst the hydrophilic residues are all located on the other, solvent-exposed, face of the helix. Cross-linking experiments have shown that this region of σ^{70} binds the non-template strand, during the open complex phase of transcription (transcript initiation in Fig. 2), thereby stabilizing the melted DNA structure. This also renders the template strand available to the catalytic centre of the enzyme.

Information regarding region 3 is not as rich as for the previous two regions. Mutational analysis has implicated region 3 in binding core RNA polymerase and provided evidence that region 3.2 is involved in the stringent response.

Finally, region 4, at the C-terminal end of σ^{70} (Fig. 4), is involved in recognition of the consensus -TTGACA- of the -35 promoter element (Gardella *et al.*, 1989; Keener & Nomura, 1993; Reddy *et al.*, 1997). This protein–DNA interaction is thought to be mediated by a strongly conserved 30 aa HTH-like structure found in subregion 4.2 (Kuldell & Hochschild, 1994; Kim *et al.*, 1995). Regardless of the exact mechanism of interaction, strong evidence from proximity studies using tethered iron chelates (Owens *et al.*, 1998a) and from the fact that mutations within region 4.2 suppress mutations made in the -35 promoter element, support the role of region 4.2 in interacting with the -35 promoter element. Region 4.2 is also involved in binding two types of regulatory proteins: activators (such as CRP at class II promoters) (Kumar *et al.*, 1994) and anti-sigmas (e.g. AsiA and Rsd) (Jishage *et al.*, 2001; Minakhin *et al.*, 2001b).

The availability of a crystal structure of a σ^{70} fragment (residues 114–446) has contributed to unravelling many of the issues regarding the mechanism of σ^{70} usage (Malhotra *et al.*, 1996). The fragment includes region 2.4, and demonstrates how the aromatic residues with the amphipathic helix align on the solvent-exposed side. The side chains of these residues are at an appropriate spacing to allow the intercalation of the side chain between the base pairing of the DNA. Such an interaction is hypothesized to favour disruption of the hydrogen bonds between the bases of the template and non-template strands, thereby leading to stabilization of DNA strand separation and DNA melting (Malhotra *et al.*, 1996).

The σ^{N} class. Historically, σ^{N}, also known as σ^{54} (the product of the *rpoN* gene), was identified as a factor necessary for expression of genes involved in nitrogen metabolism (Merrick, 1992). The *E. coli* σ^{N} is a 54 kDa peptide comprised of 477 aa (reviewed by Buck *et al.*, 2000). σ^{N} is now known to direct the expression of various genes involved in a diverse set of adaptive responses. Although σ^{N} is not usually essential for bacterial life, indeed not every bacterial species has a copy of the *rpoN* gene, σ^{N} is widely distributed amongst bacteria (Studholme & Buck, 2000). Normally, organisms only have one *rpoN* gene, but there are examples where two copies are present in the genome, e.g. *Bradyrhizobium japonicum* (Studholme & Buck, 2000). Based on primary sequence, σ^{N} is distinguished from all other σ factors, and can be divided into three distinct regions (Merrick, 1993). The N-terminal region, termed region I (Fig. 4), is glutamine-rich, and is involved in inhibition of formation of an open complex with σ^{N} specific promoters (Casaz & Buck, 1999; Gallegos & Buck, 2000). Region I has also been shown to be responsible for interacting with activators (Cannon *et al.*, 2000; Gallegos *et al.*, 1999). Region II is poorly conserved (Fig. 4), variable in length, but predominantly acidic. Region III possesses the core RNA polymerase binding determinants and the DNA sequence specific binding determinants (Cannon *et al.*, 1995). At the C-terminal end of region III there is a highly conserved motif, known as the RpoN box (Fig. 4). Parts of region III and region I are postulated to be proximal in space, and it is the presence of region I that seems to mask the activity of the DNA-binding domain (in region III) necessary for DNA strand separation, thereby preventing open complex formation prior to activation. To date, no high-resolution structural information is available for σ^N.

The σ^N-like class of sigma factors is not only distinguishable from the σ^{70} class based on differences in sequence, but also by marked differences in functionality (Buck *et al.*, 2000). σ^N promoters are typified by a GG dinucleotide around -24 and a GC dinucleotide around -12. In addition, σ^N has the ability to bind specifically to DNA, in either the presence or absence of the core RNA polymerase (Buck & Cannon, 1992). Although σ^N can bind to promoter DNA to form stable closed complexes, no transcription occurs without activators (unlike σ^{70}, where there is rapid isomerization to form an open complex after formation of the closed complex). In order to form an open complex, a σ^N-specific activator is required, such as the nitrogen control protein NtrC. Even in the presence of such an activator, isomerization does not occur until the activator hydrolyses ATP. The activators are themselves mechanochemical proteins and are often tightly regulated by a phosphorelay system known as the two-component system (see the chapter by Dixon in this book). Binding sites for these activators are found a long way upstream ($\sim$150 bp) of the promoter site. This is to allow direct interaction between the σ^N holoenzyme and the activator mediated by a DNA-looping event (see Fig. 5).

Transcription by σ^N has many parallels with eukaryotic transcription initiation: the requirement of activators, phosphorylation of the activator, hydrolysis of ATP and DNA looping. Hence, the σ^N holoenzyme is often thought to be a simple model system for the more complex eukaryotic transcription initiation process that uses transcriptional enhancers.

Structural subunit similarity with other multisubunit RNA polymerases

The nuclear multisubunit polymerases of eukaryotes, and the archaeal, bacterial and chloroplast enzymes all follow a common line of evolutionary descent. Five of the yeast pol II subunits are actually shared by pols I and III, and a further five are conserved between the three nuclear enzymes (Archambault & Friesen, 1993). The two largest subunits of bacterial RNA polymerase are homologous to the two largest subunits of

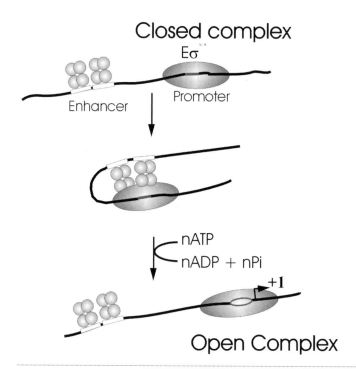

Fig. 5. Schematic of the events that occur during σ^N transcription initiation. Initially the σ^N holoenzyme binds to the promoter DNA to form a closed complex. Unlike σ^{70} holoenzyme, there is no rapid isomerization from the closed complex to the open complex. For isomerization to occur, oligomeric activator protein binds to the enhancer element DNA. The activator then directly interacts with the σ^N holoenzyme via a DNA-looping event mediated by IHF (integration host factor, not shown). The activator then hydrolyses ATP, to enable the formation of an open complex.

pol II (Ebright, 2000) (Fig. 6, homologous subunits). The bacterial alpha subunit N-terminal domain is homologous to Rpb3 and Rpb11, and in both enzymes these sub-units serve as a platform for assembly of the RNA polymerase and as a contact point for transcription activators (Cramer *et al.*, 2000; Zhang *et al.*, 1999; Zhang & Darst, 1998). The omega subunit of the bacterial enzyme is homologous to Rpb6, and is functionally equivalent in both enzymes (Minakhin *et al.*, 2001a). Only the bacterial alpha C-terminal domain has no obvious counterpart in pol II, and in bacteria serves as a point of contact for activators and interacts with a secondary promoter element (see the chapter by Browning and others in this volume).

STRUCTURE OF *E. COLI* RNA POLYMERASE

Cryo-EM (Dubochet *et al.*, 1988) has been used to determine the structures of the *E. coli* core RNA polymerase and the holoenzyme with type members of the two sigma classes bound (see Annex I). Atomic structures of fragments of some subunits of the *E.*

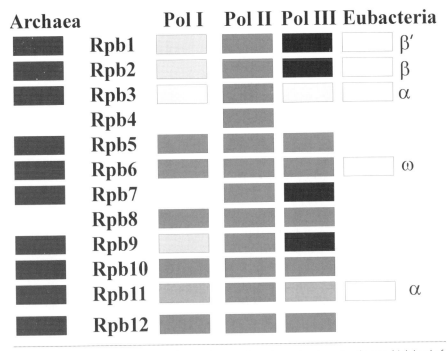

Fig. 6. Multisubunit RNA polymerase from archaea, eukaryotes and prokaryotes shares a high level of functionality and sequence similarity. A solid block represents each subunit. Those subunits shared between the eukaryotic polymerases (pols I, II and III) are shaded the same. Subunits found in the same row share significant sequence similarity. Gaps in a column indicate that no homologous subunit is found in that polymerase.

coli enzyme are known (Table 1), as is the related core enzyme from the thermophile *T. aquaticus*, solved by Darst and co-workers using crystal X-ray diffraction (Zhang *et al.*, 1999). The X-ray crystal structure of *S. cerevisiae* pol II (lacking Rpb4 and 7) was solved by Kornberg and co-workers (Cramer *et al.*, 2001).

The multisubunit RNA polymerase must exist in different functional states as it progresses through the many different events that constitute the steps of a pathway for transcription initiation. Unlike crystal-based structure determination methods, cryo-EM (in combination with single particle analysis) allows structure determination without the need for crystal growth, so is well suited for generating a series of related structures that reflect the various steps in a pathway. Below, the subnanometer structures of three states of the *E. coli* RNA polymerase determined by using cryo-EM and angular reconstitution are described.

Table 1. Known atomic structures of bacterial RNA polymerase subunits

Organism	Structure
Escherichia coli	αCTD
Thermus thermophilus	αCTD
Escherichia coli	αNTD-dimer
Escherichia coli	$\Delta\sigma^{70}$(114–446)
Thermus aquaticus	$\alpha_2\beta\beta'\omega$

Structure of the *E. coli* core enzyme

The *E. coli* core RNA polymerase grossly resembles a crab claw shape (Polyakov *et al.*, 1995), with the β subunit largely forming one pincer and the β' subunit the other (Finn *et al.*, 2000). The α subunit dimer forms the join/pivot between the two pincers. An internal channel, with a diameter of ~32 Å (green line in Fig. 7), runs between the jaws of the claw. This diameter is consistent with the proposed DNA-binding function associated with the channel (Polyakov *et al.*, 1995). The overall dimensions of the *E. coli* RNAp are: 120 Å along the direction of the channel; 150 Å from the back of the complex to the tips of the jaws; and 115 Å crossing between the jaws of the claw and perpendicular to the other two axes (see Fig. 7).

The global positions of each subunit are shown in Fig. 7. The structure in Fig. 7 is resolved to ~12 Å. At this level of resolution, some significant differences are observed between the crystal-derived core RNA polymerase structure of *T. aquaticus* and the cryo-EM-derived *E. coli* core RNA polymerase (Finn *et al.*, 2000). These differences can be largely rationalized based on differences of primary sequence or through subtle domain rearrangements. Therefore, to understand the *E. coli* density map further, a larger dataset was collected and processed (see Annex I). The larger dataset enabled a significant improvement in the resolution of the core RNA polymerase structure. This 8.5 Å core RNA polymerase structure allowed the identification of secondary structure elements, thereby allowing an accurate fit of the *T. aquaticus* structure (Zhang *et al.*, 1999) into the *E. coli* core RNA polymerase density. The fitting methodology allowed modest movements of discrete domains, thereby giving an excellent fit (see Fig. 8). The pseudo-atomic model for the *E. coli* core RNA polymerase permitted the localization of specific residues enabling a more accurate structural-based understanding of the *E. coli* core RNA polymerase (see below). Further, the fitting process has also allowed the positioning of the αCTDs in relation to the αNTDs and the localization of specific structural differences between the *T. aquaticus* and *E. coli* core RNA polymerases (not shown here, R. D. Finn and others, unpublished).

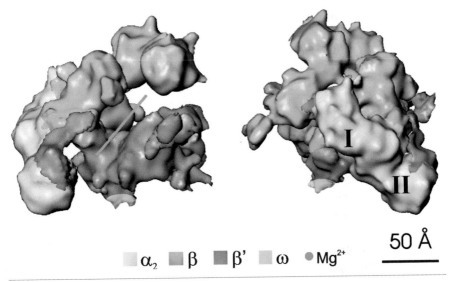

Fig. 7. Structure of core RNA polymerase resolved to 12 Å. Left: a view approximately down the axis of the RNAp channel, illustrating the 'crab claw' shape. Each jaw of the claw consists primarily of β or β' subunit. The green line indicates the orientation of the channel, which is thought to be a site of DNA interaction. The deduced position of the chelated Mg^{2+} ion at the active centre is indicated by the red sphere. Right: the left view is rotated about the vertical axis by 110° to reveal the distinctive shape of the αNTD dimer and βG flap protruding to the left. The locations of the subunits are colour-coded as indicated. The two α subunits, αI and αII, are indicated.

Structure of the *E. coli* σ^{70} holoenzyme

From protein fragmentation and mutation studies, numerous pairwise points of inter-action appear to occur between σ^{70} and the core enzyme (Burgess *et al.*, 1998; Owens *et al.*, 1998b). These suggest σ^{70} region 1.1 contacts the tip of the βG flap (900–909) and that regions 2.1 and 2.2 interact with the β' coiled-coil structure (260–309), the latter to form a complex as stable as the holoenzyme with respect to sigma binding. Furthermore, region 3.1 contacts β (1060–1240). To explain the multiplicity of interac-tions, it is suggested that there are sequential binding interactions that are part of an ordered pathway of binding and rebinding that is important for initiation. After initial binding of σ and core RNA polymerase, it is suggested that the Zn-finger region of β' interacts with σ^{70} regions 3.1–4.2, and regions 4.1 and 4.2 with β.

The structure of the *E. coli* σ^{70} holoenzyme is shown in Fig. 9. This structure closely resembles the *E. coli* core RNA polymerase, but shows one substantial difference. In the σ^{70} holoenzyme, the channel formed by each arm or pincer of the claw has been par-tially filled by extra density which bridges the gap between the tips of the β and β' sub-units. The increased density has been attributed to the bound sigma factor (see Fig. 9).

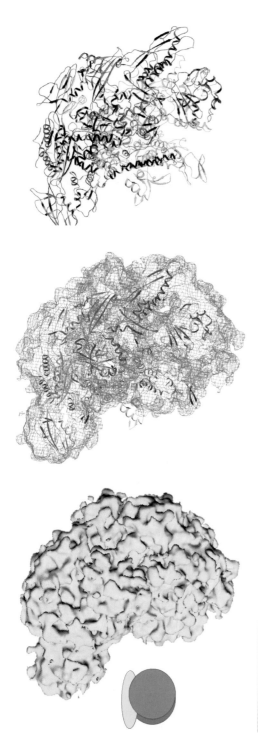

Fig. 8. Schematic for combining high- and low-resolution structures. The ability to interpret the low-resolution (8.5 Å) density maps is pivotal for the development of an understanding of how and which regions of RNA polymerase exhibit structural variability in flexibility, positioning and interactions. To understand the RNA polymerase density maps it has been necessary to accurately combine low- and high-resolution data. The orientation of the polymerase is schematically shown (insert, left) with the subunits coloured as in Fig. 7. Left: the core RNA polymerase is shown as a surface representation (yellow), typical of EM-derived structures. Many structural features are apparent on the surface compared to the structure shown in Fig. 7, but as the amount of detail increases, the ability to interpret the density map from a surface representation decreases. Therefore, to understand the density maps better, fittings of high-resolution structures were carried out to allow the identification of specific subunits, domains and now even helices (see Annex I for details of fittings). To demonstrate the quality of the fittings, the density map (now represented as purple chicken wire) is presented with the atomic co-ordinates of *T. aquaticus* docked into the structure (middle). However, the amount of lattice work within the chicken wire makes it very difficult to see the high-resolution atomic structure. Therefore, the same orientation of the fitted atomic structure is shown on the right. The *T. aquaticus* core RNA polymerase is represented as ribbons. The α and all subunits are coloured yellow and green, respectively, whilst the β' subunit is blue and the β subunit red. By comparing the three RNA polymerases in the panel it is possible to identify specific features on the surface of the density map that correspond to features present in the high-resolution structure. Such high levels of detailed interpretation are necessary to allow the description of the movements of domains between different RNA polymerase functional states, and the mapping of biochemical data onto the density maps allows the proposal of models for other functional states not yet visualized by cryo-EM.

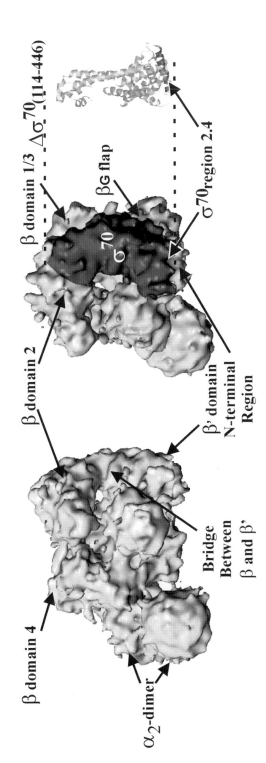

Fig. 9. Structure of σ^{70} holoenzyme resolved to 9 Å. Left: a view of the holoenzyme in a similar orientation to the left view of the core RNA polymerase in Fig. 7. Again, the view is looking down the primary channel formed by the crab claw shaped RNA polymerase. The α_2-dimer is found perpendicular to the plane of the page. The β subunit comprises the upper part of the claw, whilst β' forms the lower part. There is a bridging piece of density between the two arms of the claw; compare with Fig. 7. This bridging piece of density is expected to be σ^{70}. The exact region occupied by σ^{70} is shown in the middle. This view is rotated approximately 110° clockwise about the vertical axis from the left view. σ^{70} is shaded dark-grey compared to the rest of the holoenzyme. The location of σ^{70} was determined, in part, by the docking of a σ^{70} fragment (residues 114–446) into the density. The ribbons representation of the fragment is shown on the right. The helix involved in making the −10 promoter DNA contact is shaded dark-grey.

The presence of σ^{70} increases the dimensions of the polymerase from 120 Å along the channel to ~135 Å. The other two orthogonal dimensions, 150 Å and 115 Å, have remained approximately the same as in the *E. coli* core RNA polymerase.

The σ^{70} binds primarily to conserved regions β'B, β'C and β'D, segments previously implicated in σ^{70} interactions. The docking of the structure of a σ^{70} fragment (residues 144–446) (Malhotra *et al.*, 1996) clearly demonstrates that region 2.1 makes this interaction, with the principal component of the interaction likely to be between the coiled-coil structures in β' and σ^{70}. Region 2.4, the C-terminal helix of the σ^{70} structural fragment, is localized on the outside face of the holoenzyme, close to the β' interaction. This location is supported by geometrical and functional restraints that require the interaction of region 2.1 with the -10 promoter element and the placement of the DNA $+1$ bp within the active centre. The interaction of σ^{70} with conserved regions β'B, β'C and β'D has slightly rotated these regions of β' away from the central channel.

The σ^{70} extends upwards from β' to the bilobed-shaped structure of β, formed by domains 2 and 3. At the interacting region between β domains 2 and 3, σ^{70} makes another substantial contact. The interaction made by σ^{70} and the interaction region between β domains 2 and 3 has modestly repositioned domains 2 and 3, such that they are more separated in the σ^{70} holoenzyme than in the core enzyme. Finally, a separate interaction is formed between σ^{70} and the βG flap. The exact path followed by σ^{70} in the regions beyond those covered by the σ^{70} fragment residues is difficult to trace at the current level of resolution (the σ^{70} fragment covers residues 114–446 out of 613), but the position of the other conserved regions can be approximated by use of the fitting. Fitting the *E. coli* core RNA polymerase pseudo-atomic model followed by mapping onto the model the functional data derived from the protein fragmentation and mutation studies has indicated where the remaining regions of σ^{70} might be (details are complex, hence not shown).

Structure of the *E. coli* σ^{N} holoenzyme

As already mentioned, σ^{70} and σ^{N} are unrelated by sequence. However, two important properties are shared: they bind the same core RNA polymerase and they both bind promoter DNA. The cryo-EM structure of the σ^{N} holoenzyme has revealed that the σ^{N} holoenzyme and σ^{70} holoenzyme are structurally very similar, particularly in terms of the shape of the two different sigma factors (compare Figs 9 and 10). The interacting surfaces between the core RNA polymerase and the two sigma subunits appear to be conserved. Currently, no high-resolution structural model for σ^{N} is available, nor are the functional data regarding protein–protein interactions as detailed for σ^{N} as they are for σ^{70}, hence it is difficult to accurately localize individual regions of σ^{N}. However, the locality and shape of σ^{N} deduced from the EM density maps (Fig. 10) are supported by

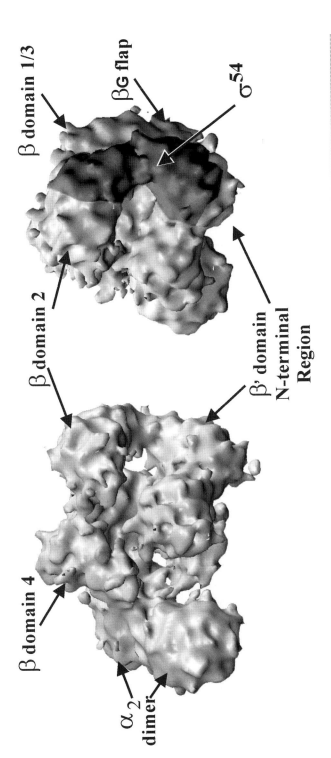

β domain 1/3

βG flap

β domain 2

σ⁵⁴

β' domain
N-terminal
Region

β domain 4

α₂
dimer

Fig. 10. Structure of σ^N holoenzyme resolved to 9 Å. Left: a view of the holoenzyme in a similar orientation to the left view of the core RNA polymerase in Fig. 7. Again, the view is looking down the primary channel formed by the crab claw shaped RNA polymerase. The α_2-dimer is found perpendicular to the plane of the page. The β subunit comprises the upper part of the claw, whilst β' forms the lower part. Note, as with the σ^{70} holoenzyme, there is a bridging piece of density between the two arms of the claw, which is not present in the core RNA polymerase. This bridging piece of density is expected to be σ^N. The exact region occupied by σ^N is shaded dark-grey compared to the rest of the holoenzyme. σ^N is shown on the right. This view is approximately 110° from that of the left view.

both peptide fragmentation studies (Wigneshweraraj *et al.*, 2000) and an envelope shape of σ^N derived using small angle X-ray scattering (Svergun *et al.*, 2000).

The structure of the σ^N holoenzyme is very close to that of the σ^{70} holoenzyme, suggesting that the activator requirement of the σ^N holoenzyme to form open complexes is probably due to an inhibitory effect within σ^N, rather than σ^N causing some inhibitory conformational change within the core enzyme.

MODEL STRUCTURES OF DNA–ENZYME COMPLEXES

Decades of structural and functional studies have indicated that cellular RNA polymerases are very complex molecular machines. Many regulatory signals are sent and received by RNA polymerase and their intricate integration is vital in the appropriate regulation of gene expression. Structures illustrating the effect on the basic catalytic subunit of binding a specificity factor have already been outlined above. The discussion now turns to the dynamic interaction that RNA polymerase makes with DNA. Results from detailed DNA interaction studies using the *E. coli* core RNA polymerase and σ^{70} holoenzyme have revealed the likely organization of the elongating transcription complex and the open promoter complex.

Transcription initiation – closed promoter complex

The information used in modelling the downstream DNA onto the *E. coli* enzyme was derived from a study using photochemical cross-linking between the RNA polymerase and promoter of an archaeon, *Pyrococcus furiosus* (Bartlett *et al.*, 2000). However, given the similarity of sequence in the largest subunits from archaea, prokaryotes and eukaryotes, the derived data can usefully be applied to the *E. coli* system. This model can also be extended to include interactions upstream of the transcription start site, since the promoter DNA must make the sequence-specific interactions with the σ factor (Fig. 11). Thus modelling suggests that the upstream DNA passes in over σ^{70}, where sequence-specific interactions are made between region 4.2 and the -35 promoter element, and region 2.4 and the -10 promoter element. The promoter bp at position $+1$ is located proximal to the active site Mg^{2+}. The downstream DNA passes through the active site, along the main channel, and out over the cleft or groove formed by the C-terminal portion of the β' and capped by β domain 2.

Transcription initiation – open complex

In the open promoter complex (Fig. 11), DNA is orientated with its downstream end interacting with β' (821–1407) and upstream end with β (643–989) and β' (1–580) (Naryshkin *et al.*, 2000). The downstream end of the open complex DNA bubble is within the active centre cleft, with the rest of the bubble arising from the floor of the cleft in an arrangement whereby the trajectory of the DNA has turned through 90°.

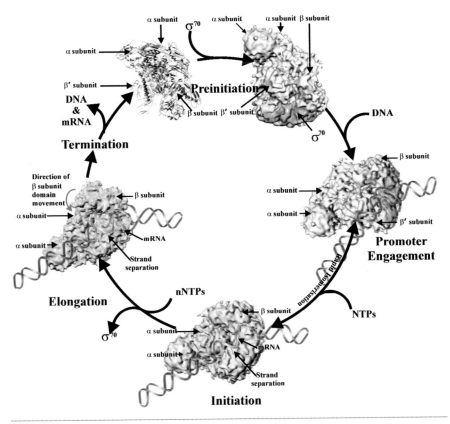

Fig. 11. A three-dimensional transcription cycle. The same transcription cycle as shown in Fig. 2 is now shown with the cryo-EM-derived structures (core RNA polymerase and σ^{70} holoenzyme) and subsequent models derived from the fitting of high-resolution data, followed by docking of promoter DNA. The free core RNA polymerase structure is represented as ribbons. The σ^{70} holoenzyme is divided into two regions: the core RNA polymerase (green) and the σ^{70}-specific region (red). Double-stranded DNA is represented by a dark-blue ribbon, single-stranded DNA by a light-blue ribbon, and mRNA as a red ribbon. In all cases, the subunit locations are marked. More specific details are found under the relevant sections within this chapter.

The +1 position is at the floor of the cleft, and about 20 Å from the active site Mg^{2+}. The 3' end of the RNA is adjacent to the Mg^{2+}, with RNA bases 2–6 nt from the 3' end interacting with β (235–643) (Naryshkin *et al.*, 2000), which includes the site of rifampicin binding (Campbell *et al.*, 2001). The RNA 5' end is directed to a channel of β (643–989) and β' (951–1342) and β' (1–580) (Naryshkin *et al.*, 2000). Throughout its length, the non-template DNA strand is found in a deep narrow groove involving β (235–643) as a wall structure and β (1–235) as the groove floor (Naryshkin *et al.*, 2000). Contacts between σ^{70} and DNA in the open complex are mapped to the expected regions 4, for -35 contact, and 2, for -10 interactions.

Although disordered in the *T. aquaticus* crystal structure, the position of the αCTD maps to consecutive minor grooves of upstream DNA, in the open complex, and is restricted (as a consequence of protein–protein contacts) to one groove when the activator protein CRP is present. The DNA interactions seen with and without CRP are otherwise unchanged, consistent with the view that CRP functions to recruit (stabilizes DNA binding) holoenzyme to promoters and does not otherwise change interactions. To account for the distance range of αCTD to DNA contacts, a bend in the DNA behind the RNA polymerase is proposed.

Transcription elongation

Compared to the open complex, a closely related organization of the elongation complex is evident, but here the σ is absent (Fig. 11). The model of the ternary elongating complex (TEC) presented here is largely based on the previously described heterogeneous model of the *T. aquaticus* core RNA polymerase structure and the *E. coli* cross-linking data (Korzheva *et al.*, 2000). Again, the downstream end of the DNA is within a deep cleft delineated by β′ and capped by β. Effectively this structure is a sliding clamp over the DNA, and is evident too in the initiating complex and with the yeast pol II-elongating complex (Bushnell *et al.*, 2001). Interestingly, in the hybrid model previously described, a region of β (domain 2) is believed to rotate down towards the downstream DNA (Korzheva *et al.*, 2000), thereby clamping the DNA even more securely into the cleft. From this cleft, the DNA enters the primary channel, formed by β and β′, where the template and non-template DNA strands have been separated, or melted. The template strand (outer downstream edge) is then placed within the catalytic centre (Korzheva *et al.*, 2000; Nudler *et al.*, 1998). The RNA–DNA hybrid is formed here, with the Mg^{2+} of the active site at the downstream end of the hybrid. Binding studies suggest that there is an RNA-binding site for the last two RNA bases (at −7 and −8) of the 8–9 bp hybrid of the 12–13 bp transcription bubble. The front edge of the hybrid-binding site appears to be formed by the β′D region, the catalytic centre and the βD loop sequence. Close to the Mg^{2+} and front (downstream) edge, there is the formation of a secondary channel by β′, which seems to function for RNA exit during backtracking and/or pausing. The RNA–DNA hybrid extends, back through the enzyme towards the region of the β subunit where the rifampicin-resistant mutations are localized (Campbell *et al.*, 2001) and then to the β′C rudder structure. The hybrid is terminated at the β′C rudder. The protein–DNA interactions at the rear part of the hybrid are associated with RNA binding and its displacement from the hybrid. DNA strand separation seems to be mediated by the rudder structure. The RNA exit path maps to underneath a flap formed by the βG domain loop structure. Meanwhile, the non-template strand is held apart from the template strand by the interface between the characteristic bilobed feature of β, formed by β domains 2 and 3. Three further melted base pairs of DNA occur before the transcription bubble closes on the upstream outer edge. It is in this region on non-hybrid, melted DNA where the DNA undergoes a bend of approximately 90°. This angle was evident from the elongating pol II complex and deduced from atomic force microscopy studies on active TEC. Thus this bend in the DNA is a conserved architectural feature of the TEC. This TEC model is elegantly supported by the complementary nature of the DNA/protein electrostatic charge distributions. In the TEC, the position of upstream DNA is not well defined, and may be only weakly associated with the polymerase. The DNA is believed to pass out of the front of the polymerase over the βG flap, a structure implicated in pausing and termination.

Termination

At the end of the gene or operon, the transcription complex of core RNA polymerase, DNA and RNA needs to be dissociated to allow termination of chain growth. Termination has not been as widely investigated as the other stages of the transcription cycle and little structural information is available, hence termination will only be briefly discussed here. The biochemical characterization of termination has been reviewed (see Mooney *et al.*, 1998). Termination can occur through intrinsic signals within the transcription complex; typically a specific stem–loop structure is formed by a specific sequence in the RNA (known as factor-independent termination). Another method of termination, factor-dependent termination) is via the action of termination factors (e.g. Rho). In the case of rho-dependent termination, rho binds to the RNA and exhibits ATPase and helicase activities. It is thought that the action of the ATP-dependent helicase causes the disruption of the transcription complexes. After termination, whichever method of termination is used, the RNA and DNA are released by the core RNA polymerase. The core RNA polymerase is then recycled and binds a sigma factor again, thereby completing the cycle (Fig. 11).

SUMMARY

The complementary combination of high-resolution structural data with lower resolution cryo-EM-derived data and protein–nucleic acid cross-linking data has enlarged our understanding of the *E. coli* core RNA polymerase and two distinct holoenzymes. Although functionally distinct with respect to activation mechanisms, the σ^N and σ^{70} holoenzymes show a high degree of structural similarity that most likely reflects the common contribution of core RNA polymerase as a binding surface for sigma and a requirement for a specific restricted geometry when complexes with promoter DNA are formed (see Figs 9 and 10). Further, the near complete pseudo-atomic structures of the core RNA polymerase and σ^{70} holoenzyme have allowed the detailed mapping of functional data, thereby producing models of the complexes found at various stages in transcription (Fig. 11). Although these models show the likely relationship between the RNA polymerase, DNA and RNA, there is the limitation that there must be further conformational changes that occur between the sequential steps of the transcription cycle, and these changes have not been modelled. By extending the number of complexes studied so far by cryo-EM to include binary complexes of DNA–holoenzyme and potentially activator–holoenzyme complexes, there will be an even greater understanding of the interplay of the RNA polymerase and DNA and its regulatory factors.

ANNEX I – METHOD OF STRUCTURE DETERMINATION USING CRYO-EM

Cryo-EM of individual non-crystalline molecules (termed single particles) together with angular reconstitution is becoming a routine method for determining the three-dimensional (3D) structure of large (>300 kDa) biological macromolecules. A solution of highly purified sample, in this case RNA polymerase, is taken and embedded in amorphous ice by rapid freezing (Dubochet *et al.*, 1988). This method of sample preparation maintains the biological specimen in its native hydrated state. Images of the vitrified sample are then taken, each at a certain defocus, at near liquid nitrogen temperatures ($\sim -180\,^{\circ}C$) in an electron microscope using a low-dose protocol ($\sim 10\ e^-\ Å^{-1}$) (see Fig. 12). Good micrographs are selected by optical diffraction and subsequently digitized with a high-resolution densitometer.

These digitized micrographs are then processed within the IMAGIC-5 software system (van Heel *et al.*, 1996). The angular reconstitution methodology used within the IMAGIC-5 software system exploits the random orientations of particles within a vitreous ice matrix, allowing the direct extraction of 3D information from the images. Images of single particles are first picked from the digitized micrograph. Contrast transfer function (CTF) correction is then applied to each individual image (Finn *et al.*, 2000). The CTF corrected images are subsequently centred during an alignment procedure. Multivariate statistical analysis and classification (van Heel, 1984) are used to sort the dataset into groups of alike images. These groups are then averaged to obtain relatively noise-free characteristic views of the RNA polymerase. The primary principle underlying the angular reconstitution technique is that any two projections (2D images) of the same 3D object have a common 1D line between the projections. Using this principle within the IMAGIC-5 angular reconstitution algorithm, it is possible to assign Euler angles to selected characteristic views. Having calculated the relative orientations of each of selected characteristic views, 3D reconstructions are calculated with the exact filter back-projection algorithm (Finn *et al.*, 2000 and references therein). The procedures of alignment, multivariate statistical analysis, classification and angular reconstitution are applied iteratively to refine the 3D structure until no further improvement can be made (see Fig. 12). The resolution of the final 3D reconstruction is determined by the 3σ Fourier Shell criterion (Harauz & van Heel, 1986).

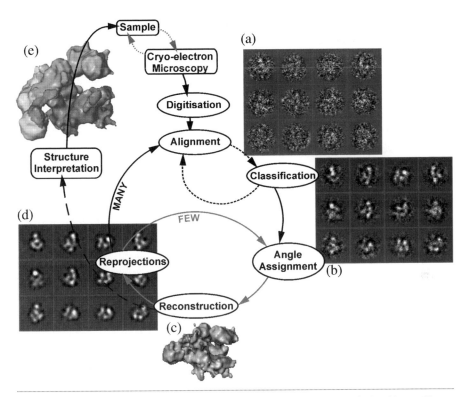

Fig. 12. Overview of image processing. Highly purified protein samples are vitrified and imaged in an electron microscope. Images of the sample are digitized, followed by data collection (boxing out) of single particles (a). The images of single particles are very noisy and therefore have to be averaged. To average alike images together, the images are aligned, and then classified into groups of ~17 alike images. These are then averaged together to produce classes (b). The relative orientation of each class with respect to each other is then determined (angle assignment). Classes are then used to reconstruct a three-dimensional volume (c). The structure is then reprojected (d) either to refine the alignment or for angular assignment. The whole procedure is applied in an iterative manner, until a stable structure is reached. This structure is then interpreted using a variety of different tools (e) to deduce functional information.

Having obtained a 3D reconstruction, often referred to as a density map, there is the need to interpret this density map. This is primarily achieved by docking atomic resolution structures into the density map and by comparing two EM-derived density maps. The docking of atomic structures into the density maps was achieved using three methods: manual fitting, real space refinement and reciprocal space refinements. The initial fittings of atomic structures (see Table 1) were carried out using manual methods followed by real space refinements. The real space refinements were carried out using ESSENS (part of RAVE – http://rose.bmc.uu.se). ESSENS performs a real-space convolution between a density map and a structural fragment. This gave an approximate fit, allowing the structure to be interpreted at a global level. However, given the resolution of the present structures, a more precise level of fitting was sought. This was achieved by using a reciprocal space refinement. This methodology uses structure factors directly, which is both fast and precise. To calculate the structure factors the electron density map was Fourier-transformed into a P1 cell (reflections hkl). This allowed the use of classical crystallographic refinement programs and packages (e.g. XPLOR, CCP4) to carry out rigid body refinement.

Using such fitting methods for the core RNA polymerase, it became evident that modest domain movements between the atomic structure of the core RNA polymerase and density map had occurred. Thus to optimize the fit, discretely folded domains were defined as independent rigid bodies, connected by linkers. Dihedral angles and bond distances were defined as nuclear Overhauser enhancement (NOE) type restraints across the linker residues, thereby allowing flexibility of the linkers, yet maintaining peptide connectivity and geometry. The final step was to apply homology modelling, using MODELLER 6a, to the precisely fitted atomic model from *T. aquaticus*, thereby producing a quasi-atomic model for the *E. coli* core RNA polymerase. This has allowed comparison of an atomic model to an EM structure, derived using single particle analysis, to an unprecedented level. DNA was docked into the *E. coli* core RNA polymerase and σ^{70} holoenzyme structures using a similar NOE-distance restraint derived from DNA–protein cross-linking data.

ACKNOWLEDGEMENTS

This work was supported by a BBSRC project grant to M. B. and M. v. H. The authors would also like to thank Bruno Klaholz for informative discussions on the fitting of atomic structures to EM density and Patricia Burrows for help with preparation of the manuscript.

REFERENCES

Archambault, J. & Friesen, J. D. (1993). Genetics of eukaryotic RNA polymerases I, II and III. *Microbiol Rev* **57**, 703–724.

Arthur, T. M. & Burgess, R. R. (1998). Localization of a sigma70 binding site on the N terminus of the *Escherichia coli* RNA polymerase beta' subunit. *J Biol Chem* **273**, 31381–31387.

Arthur, T. M., Anthony, L. C. & Burgess, R. R. (2000). Mutational analysis of beta '260-309, a sigma 70 binding site located on *Escherichia coli* core RNA polymerase. *J Biol Chem* **275**, 23113–23119.

Bartlett, M. S., Thomm, M. & Geiduschek, E. P. (2000). The orientation of DNA in an archaeal transcription initiation complex. *Nat Struct Biol* **7**, 782–785.

Blatter, E. E., Ross, W., Tang, H., Gourse, R. L. & Ebright, R. H. (1994). Domain organization of RNA polymerase alpha subunit: C-terminal 85 amino acids constitute a domain of dimerization and DNA binding. *Cell* **78**, 889–896.

Borukhov, S., Severinov, K., Kashlev, M., Lebedev, A., Bass, I., Rowland, G. C., Lim, P. P., Glass, R. E., Nikiforov, V. & Goldfarb, A. (1991). Mapping of trypsin cleavage and antibody-binding sites and delineation of a dispensable domain in the beta subunit of *Escherichia coli* RNA polymerase. *J Biol Chem* **266**, 23921–23926.

Buck, M. & Cannon, W. (1992). Specific binding of the transcription factor sigma-54 to promoter DNA. *Nature* **358**, 422–424.

Buck, M., Gallegos, M. T., Studholme, D. J., Guo, Y. & Gralla, J. D. (2000). The bacterial enhancer-dependent sigma(54) (sigma(N)) transcription factor. *J Bacteriol* **182**, 4129–4136.

Burgess, R. R., Travers, A. A., Dunn, J. J. & Bautz, E. K. F. (1969). Factor stimulating transcription by RNA polymerase. *Nature* **221**, 43–44.

Burgess, R. R., Arthur, T. M. & Pietz, B. C. (1998). Interaction of *Escherichia coli* sigma 70 with core RNA polymerase. *Cold Spring Harbor Symp Quant Biol* **63**, 277–287.

Bushnell, D. A., Cramer, P. & Kornberg, R. D. (2001). Selenomethionine incorporation in *Saccharomyces cerevisiae* RNA polymerase II. *Structure* **9**, R11–R14.

Callaci, S., Heyduk, E. & Heyduk, T. (1998). Conformational changes of *Escherichia coli* RNA polymerase sigma70 factor induced by binding to the core enzyme. *J Biol Chem* **273**, 32995–33001.

Campbell, E. A., Korzheva, N., Mustaev, A., Murakami, K., Nair, S., Goldfarb, A. & Darst, S. A. (2001). Structural mechanism for rifampicin inhibition of bacterial rna polymerase. *Cell* **104**, 901–912.

Cannon, W., Missailidis, S., Smith, C., Cottier, A., Austin, S., Moore, M. & Buck, M. (1995). Core RNA polymerase and promoter DNA interactions of purified domains of sigma N: bipartite functions. *J Mol Biol* **248**, 781–803.

Cannon, W. V., Gallegos, M. T. & Buck, M. (2000). Isomerization of a binary sigma-promoter DNA complex by transcription activators. *Nat Struct Biol* **7**, 594–601.

Casaz, P. & Buck, M. (1999). Region I modifies DNA-binding domain conformation of sigma-54 within the holoenzyme. *J Mol Biol* **285**, 507–514.

Cramer, P., Bushnell, D. A., Fu, J., Gnatt, A. L., Maier-Davis, B., Thompson, N. E., Burgess, R. R., Edwards, A. M., David, P. R. & Kornberg, R. D. (2000). Architecture of RNA polymerase II and implications for the transcription mechanism. *Science* **288**, 640–649.

Cramer, P., Bushnell, D. A. & Kornberg, R. D. (2001). Structural basis of transcription: RNA polymerase II at 2.8 angstrom resolution. *Science* **292**, 1863–1876.

Dombroski, A. J., Walter, W. A. & Gross, C. A. (1993). Amino-terminal amino acids modulate sigma-factor DNA-binding activity. *Genes Dev* **7**(12A), 2446–2455.

Dubochet, J., Adrian, M., Chang, J. J., Homo, J. C., Lepault, J., McDowall, A. W. & Schultz, P. (1988). Cryo-electron microscopy of vitrified specimens. *Q Rev Biophys* **21**, 129–228.

Ebright, R. H. (2000). RNA polymerase: structural similarities between bacterial RNA polymerase and eukaryotic RNA polymerase II. *J Mol Biol* **304**, 687–698.

Estrem, S. T., Ross, W., Gaal, T., Chen, Z. W., Niu, W., Ebright, R. H. & Gourse, R. L. (1999). Bacterial promoter architecture: subsite structure of UP elements and interactions with the carboxy-terminal domain of the RNA polymerase alpha subunit. *Genes Dev* **13**, 2134–2147.

Fenton, M. S., Lee, S. J. & Gralla, J. D. (2000). *Escherichia coli* promoter opening and −10 recognition: mutational analysis of sigma70. *EMBO J* **19**, 1130–1137.

Finn, R. D., Orlova, E. V., Gowen, B., Buck, M. & van Heel, M. (2000). *Escherichia coli* RNA polymerase core and holoenzyme structures. *EMBO J* **19**, 6833–6844.

Fujita, N., Endo, S. & Ishihama, A. (2000). Structural requirements for the interdomain linker of alpha subunit of *Escherichia coli* RNA polymerase. *Biochemistry* **39**, 6243–6249.

Gallegos, M. T. & Buck, M. (2000). Sequences in sigma(54) region I required for binding to early melted DNA and their involvement in sigma-DNA isomerisation. *J Mol Biol* **297**, 849–859.

Gallegos, M. T., Cannon, W. V. & Buck, M. (1999). Functions of the sigma(54) region I in trans and implications for transcription activation. *J Biol Chem* **274**, 25285–25290.

Gardella, T., Moyle, H. & Susskind, M. M. (1989). A mutant *Escherichia coli* sigma 70 subunit of RNA polymerase with altered promoter specificity. *J Mol Biol* **206**, 579–590.

Geiduschek, E. P. & Bartlett, M. S. (2000). Engines of gene expression. *Nat Struct Biol* **7**, 437–439.

Gelles, J. & Landick, R. (1998). RNA polymerase as a molecular motor. *Cell* **93**, 13–16.

Gentry, D. R. & Burgess, R. R. (1986). The cloning and sequence of the gene encoding the omega subunit of *Escherichia coli* RNA polymerase. *Gene* **48**, 33–40.

Gentry, D. R. & Burgess, R. R. (1989). *rpoZ*, encoding the omega subunit of *Escherichia coli* RNA polymerase, is in the same operon as *spoT*. *J Bacteriol* **171**, 1271–1277.

Gentry, D. R. & Burgess, R. R. (1993). Cross-linking of *Escherichia coli* RNA polymerase subunits: identification of beta' as the binding site of omega. *Biochemistry* **32**, 11224–11227.

Gentry, D., Xiao, H., Burgess, R. & Cashel, M. (1991). The omega subunit of *Escherichia coli* K-12 RNA polymerase is not required for stringent RNA control in vivo. *J Bacteriol* **173**, 3901–3903.

Gross, C. A., Lonetto, M. & Losick, R. (1992). Bacterial sigma factors. In *Transcription Regulation*, pp. 129–176. Edited by K. R. Yamamoto & S. L. McKnight. Cold Spring Harbor, NY: Cold Spring Harbor Laboratory.

Gross, C. A., Chan, C., Dombroski, A., Gruber, T., Sharp, M., Tupy, J. & Young, B. (1998). The functional and regulatory roles of sigma factors in transcription. *Cold Spring Harbor Symp Quant Biol* **63**, 141–155.

Harauz, G. & van Heel, M. (1986). Resolution criteria for three dimensional reconstruction. *Optik* **73**, 146–156.

van Heel, M. (1984). Multivariate statistical classification of noisy images (randomly oriented biological macromolecules). *Ultramicroscopy* **13**, 165–184.

van Heel, M., Harauz, G., Orlova, E. V., Schmidt, R. & Schatz, M. (1996). A new generation of the IMAGIC image processing system. *J Struct Biol* **116**, 17–24.

Heyduk, T., Heyduk, E., Severinov, K., Tang, H. & Ebright, R. H. (1996). Determinants of RNA polymerase alpha subunit for interaction with beta, beta', and sigma subunits: hydroxyl-radical protein footprinting. *Proc Natl Acad Sci U S A* **93**, 10162–10166.

Igarashi, K. & Ishihama, A. (1991). Bipartite functional map of the *E. coli* RNA polymerase alpha subunit: involvement of the C-terminal region in transcription activation by cAMP-CRP. *Cell* **65**, 1015–1022.

Igarashi, K., Fujita, N. & Ishihama, A. (1989). Promoter selectivity of *Escherichia coli* RNA polymerase: omega factor is responsible for the ppGpp sensitivity. *Nucleic Acids Res* **17**, 8755–8765.

Igarashi, K., Fujita, N. & Ishihama, A. (1991). Identification of a subunit assembly domain in the alpha subunit of *Escherichia coli* RNA polymerase. *J Mol Biol* **218**, 1–6.

Ishihama, A. (2000). Functional modulation of *Escherichia coli* RNA polymerase. *Annu Rev Microbiol* **54**, 499–518.

Jeon, Y. H., Negishi, T., Shirakawa, M., Yamazaki, T., Fujita, N., Ishihama, A. & Kyogoku, Y. (1995). Solution structure of the activator contact domain of the RNA polymerase alpha subunit. *Science* **270**, 1495–1497.

Jeon, Y. H., Yamazaki, T., Otomo, T., Ishihama, A. & Kyogoku, Y. (1997). Flexible linker in the RNA polymerase alpha subunit facilitates the independent motion of the C-terminal activator contact domain. *J Mol Biol* **267**, 953–962.

Jishage, M., Dasgupta, D. & Ishihama, A. (2001). Mapping of the Rsd contact site on the sigma 70 subunit of *Escherichia coli* RNA polymerase. *J Bacteriol* **183**, 2952–2956.

Katayama, A., Fujita, N. & Ishihama, A. (2000). Mapping of subunit-subunit contact surfaces on the beta' subunit of *Escherichia coli* RNA polymerase. *J Biol Chem* **275**, 3583–3592.

Keener, J. & Nomura, M. (1993). Dominant lethal phenotype of a mutation in the −35 recognition region of *Escherichia coli* sigma 70. *Proc Natl Acad Sci U S A* **90**, 1751–1755.

Kim, S. K., Makino, K., Amemura, M., Nakata, A. & Shinagawa, H. (1995). Mutational analysis of the role of the first helix of region 4.2 of the sigma 70 subunit of *Escherichia coli* RNA polymerase in transcriptional activation by activator protein PhoB. *Mol Gen Genet* **248**, 1–8.

Korzheva, N., Mustaev, A., Kozlov, M., Malhotra, A., Nikiforov, V., Goldfarb, A. & Darst, S. A. (2000). A structural model of transcription elongation. *Science* **289**, 619–625.

Kuldell, N. & Hochschild, A. (1994). Amino acid substitutions in the −35 recognition motif of sigma 70 that result in defects in phage lambda repressor-stimulated transcription. *J Bacteriol* **176**, 2991–2998.

Kumar, A., Grimes, B., Fujita, N., Makino, K., Malloch, R. A., Hayward, R. S. & Ishihama, A. (1994). Role of the sigma 70 subunit of *Escherichia coli* RNA polymerase in transcription activation. *J Mol Biol* **235**, 405–413.

Landick, R. (1999). Shifting RNA polymerase into overdrive. *Science* **284**, 598–599.

Lonetto, M., Gribskov, M. & Gross, C. A. (1992). The sigma 70 family: sequence conservation and evolutionary relationships. *J Bacteriol* **174**, 3843–3849.

Malhotra, A., Severinova, E. & Darst, S. A. (1996). Crystal structure of a sigma 70 subunit fragment from *E. coli* RNA polymerase. *Cell* **87**, 127–136.

Merrick, M. J. (1992). Regulation of nitrogen fixation genes in free-living and symbiotic bacteria. In *Biological Nitrogen Fixation*, pp. 835–876. Edited by G. Stacey, R. H. Burris & H. J. Evans. New York & London: Chapman & Hall.

Merrick, M. J. (1993). In a class of its own – the RNA polymerase sigma factor sigma 54 (sigma N). *Mol Microbiol* **10**, 903–909.

Minakhin, L., Bhagat, S., Brunning, A., Campbell, E. A., Darst, S. A., Ebright, R. H. & Severinov, K. (2001a). Bacterial RNA polymerase subunit omega and eukaryotic RNA polymerase subunit RPB6 are sequence, structural, and functional homologs and promote RNA polymerase assembly. *Proc Natl Acad Sci U S A* **98**, 892–897.

Minakhin, L., Camarero, J. A., Holford, M., Parker, C., Muir, T. W. & Severinov, K. (2001b). Mapping the molecular interface between the sigma(70) subunit of *E. coli* RNA polymerase and T4 AsiA. *J Mol Biol* **306**, 631–642.

Mooney, R. A., Artsimovitch, I. & Landick, R. (1998). Information processing by RNA polymerase: recognition of regulatory signals during RNA chain elongation. *J Bacteriol* **180**, 3265–3275.

Mukherjee, K. & Chatterji, D. (1997). Studies on the omega subunit of *Escherichia coli* RNA polymerase – its role in the recovery of denatured enzyme activity. *Eur J Biochem* **247**, 884–889.

Mukherjee, K., Nagai, H., Shimamoto, N. & Chatterji, D. (1999). GroEL is involved in activation of *Escherichia coli* RNA polymerase devoid of the omega subunit in vivo. *Eur J Biochem* **266**, 228–235.

Mustaev, A., Kozlov, M., Markovtsov, V., Zaychikov, E., Denissova, L. & Goldfarb, A. (1997). Modular organization of the catalytic center of RNA polymerase. *Proc Natl Acad Sci U S A* **94**, 6641–6645.

Nagai, H. & Shimamoto, N. (1997). Regions of the *Escherichia coli* primary sigma factor sigma70 that are involved in interaction with RNA polymerase core enzyme. *Genes Cells* **2**, 725–734.

Naryshkin, N., Revyakin, A., Kim, Y., Mekler, V. & Ebright, R. H. (2000). Structural organization of the RNA polymerase-promoter open complex. *Cell* **101**, 601–611.

Nudler, E., Gusarov, I., Avetissova, E., Kozlov, M. & Goldfarb, A. (1998). Spatial organization of transcription elongation complex in *Escherichia coli*. *Science* **281**, 424–428.

Owens, J. T., Chmura, A. J., Murakami, K., Fujita, N., Ishihama, A. & Meares, C. F. (1998a). Mapping the promoter DNA sites proximal to conserved regions of sigma 70 in an *Escherichia coli* RNA polymerase-lacUV5 open promoter complex. *Biochemistry* **37**, 7670–7675.

Owens, J. T., Miyake, R., Murakami, K., Chmura, A. J., Fujita, N., Ishihama, A. & Meares, C. F. (1998b). Mapping the sigma70 subunit contact sites on *Escherichia coli* RNA polymerase with a sigma70-conjugated chemical protease. *Proc Natl Acad Sci U S A* **95**, 6021–6026.

Polyakov, A., Severinova, E. & Darst, S. A. (1995). Three-dimensional structure of *E. coli* core RNA polymerase: promoter binding and elongation conformations of the enzyme. *Cell* **83**, 365–373.

Reddy, B. V., Gopal, V. & Chatterji, D. (1997). Recognition of promoter DNA by subdomain 4.2 of *Escherichia coli* sigma 70: a knowledge based model of -35 hexamer interaction with 4.2 helix-turn-helix motif. *J Biomol Struct Dyn* **14**, 407–419.

Severinov, K., Mustaev, A., Kashlev, M., Borukhov, S., Nikiforov, V. & Goldfarb, A. (1992). Dissection of the beta subunit in the *Escherichia coli* RNA polymerase into domains by proteolytic cleavage. *J Biol Chem* **267**, 12813–12819.

Severinov, K., Mustaev, A., Kukarin, A., Muzzin, O., Bass, I., Darst, S. A. & Goldfarb, A. (1996). Structural modules of the large subunits of RNA polymerase. *J Biol Chem* **271**, 27969–27974.

Severinov, K., Markov, D., Severinova, E., Nikiforov, V., Landick, R., Darst, S. A. & Goldfarb, A. (1998a). Streptolydigin-resistant mutants in an evolutionary conserved region of the beta′ subunit of *Escherichia coli* RNA polymerase. *J Biol Chem* **270**, 23926–23929.

Severinov, K., Mustaev, A., Severinova, E., Kozlov, M., Darst, S. A. & Goldfarb, A. (1998b). The beta subunit Rif-cluster I is only angstroms away from the active center of *Escherichia coli* RNA polymerase. *J Biol Chem* **270**, 29428–29432.

Studholme, D. J. & Buck, M. (2000). The biology of enhancer-dependent transcriptional regulation in bacteria: insights from genome sequences. *FEMS Microbiol Lett* **186**, 1–9.

Svergun, D. I., Malfois, M., Koch, M. H., Wigneshweraraj, S. R. & Buck, M. (2000). Low resolution structure of the sigma54 transcription factor revealed by X-ray solution scattering. *J Biol Chem* **275**, 4210–4214.

Waldburger, C. & Susskind, M. M. (1994). Probing the informational content of *Escherichia coli* sigma 70 region 2.3 by combinatorial cassette mutagenesis. *J Mol Biol* **235**, 1489–1500.

Wigneshweraraj, S. R., Fujita, N., Ishihama, A. & Buck, M. (2000). Conservation of sigma-core RNA polymerase proximity relationships between the enhancer-independent and enhancer-dependent sigma classes. *EMBO J* **19**, 3038–3048.

Wilson, C. & Dombroski, A. J. (1997). Region 1 of sigma70 is required for efficient isomerization and initiation of transcription by *Escherichia coli* RNA polymerase. *J Mol Biol* **267**, 60–74.

Zaychikov, E., Martin, E., Denissova, L., Kozolv, M., Markovtsov, V., Kashlev, M., Heumann, H., Nikiforov, V., Goldfarb, A. & Mustaev, A. (1996). Mapping of catalytic residues in the RNA polymerase active center. *Science* **273**, 107–109.

Zhang, G. & Darst, S. A. (1998). Structure of the *Escherichia coli* RNA polymerase alpha subunit amino-terminal domain. *Science* **281**, 262–266.

Zhang, G., Campbell, E. A., Minakhin, L., Richter, C., Severinov, K. & Darst, S. A. (1999). Crystal structure of *Thermus aquaticus* core RNA polymerase at 3.3 A resolution. *Cell* **98**, 811–824.

Zhou, Y. N. & Jin, D. J. (1998). The *rpoB* mutants destabilizing initiation complexes at stringently controlled promoters behave like "stringent" RNA polymerases in *Escherichia coli. Proc Natl Acad Sci U S A* **95**, 2908–2913.

The ECF sigma factors of *Streptomyces coelicolor* A3(2)

Mark S. B. Paget,[1] Hee-Jeon Hong,[2] Maureen J. Bibb[2] and Mark J. Buttner[2]

[1]School of Biological Sciences, University of Sussex, Brighton BN1 9QG, UK

[2]Department of Molecular Microbiology, John Innes Centre, Colney, Norwich NR4 7UH, UK

INTRODUCTION

In bacteria, gene expression is controlled primarily at the level of transcription initiation. Control can be achieved through the use of DNA-binding proteins (repressors and activators) that affect the efficiency of initiation, but also through the use of alternative forms of RNA polymerase with different promoter recognition characteristics. The promoter specificity of the RNA polymerase holoenzyme depends on the nature of the σ subunit that associates with the core enzyme. This key role of σ in promoter recognition suggests a mechanism for the coordinate control of gene expression using alternative forms of σ and different subsets of promoters, an idea that was first proposed as soon as the role of σ was established (Burgess *et al.*, 1969). It is now clear that most, if not all, bacteria use alternative σ subunits to control gene expression, and that these σ factors fall into two distinct families: the σ^N (or σ^{54}) family, which is discussed in the preceding chapter, and the σ^{70} family. The σ^{70} family includes those σ factors that, broadly speaking, are related in sequence and domain organization to the primary *Escherichia coli* σ factor, σ^{70}. Although the overall architecture of members of the σ^{70} family appears to be conserved, the σ^{70} family can be divided into several phylogenetically distinct subfamilies (Lonetto *et al.*, 1992). Members of each subfamily are often involved in the control of related functions, such as the heat-shock response, flagella biosynthesis, or sporulation.

The ECF subfamily of σ factors

In the late 1980s, biochemical analysis of RNA polymerase from *Streptomyces coelicolor* and *E. coli* led to the identification of two σ factors that were particularly small in

SGM symposium 61: Signals, switches, regulons and cascades: control of bacterial gene expression.
Editors D. A. Hodgson, C. M. Thomas. Cambridge University Press. ISBN 0 521 81388 3 ©SGM 2002.

size. In *E. coli*, σ^E (21.7 kDa) was shown to account for transcription of the gene encoding the heat-shock σ factor, σ^{32}, at high temperatures (Erickson & Gross, 1989). In *S. coelicolor*, another small σ factor, also named σ^E (20.4 kDa), was shown to direct *in vitro* transcription from one of four promoters (*dagAp2*) of the agarase-encoding gene *dagA* (Buttner *et al.*, 1988). The cloning of the gene encoding *S. coelicolor* σ^E several years later using a reverse genetics approach revealed that it belonged, together with *E. coli* σ^E, to a new subfamily of the σ^{70} family (Lonetto *et al.*, 1994). Members of this new subfamily were sufficiently different from the previously known σ factors that in many cases they were not identified as σ factors by standard similarity searching methods. As a consequence, several members of the subfamily were present in the protein databases, but their biochemical role was unrecognized. Each had been identified by genetic means, each had a known positive regulatory role, but with no biochemical understanding of mechanism. These included AlgU from *Pseudomonas aeruginosa*, CarQ from *Myxococcus xanthus* and FecI from *E. coli*. The available information about the roles of these σ factors at the time suggested that they functioned as effector molecules responding to extracytoplasmic stimuli, and that they often controlled extracytoplasmic functions, and for this reason, the new subfamily was named the ECF subfamily (Lonetto *et al.*, 1994). For example, *E. coli* σ^E is involved in sensing and responding to protein misfolding in the extracytoplasmic space (Ades *et al.*, 1999), *M. xanthus* σ^{CarQ} activates the synthesis of membrane-localized carotenoids in response to light (Gorham *et al.*, 1996), and *E. coli* FecI activates the citrate-dependent iron(III) transport system in response to citrate and iron in the periplasmic space (Härle *et al.*, 1995). The characteristically small size of ECF σ factors ($\sim$20–30 kDa) is accounted for by the absence of most or all of both regions 1 and 3 (Lonetto *et al.*, 1994). For a detailed review of σ domain structure and function see Lonetto *et al.* (1992, 1994).

Since the initial discovery of the ECF subfamily, hundreds of new members have been discovered in a wide variety of Gram-negative and Gram-positive bacteria, mostly through genome sequencing projects. Indeed, for several bacteria, including *Bacillus subtilis*, *Mycobacterium tuberculosis* and *S. coelicolor*, ECF σ factors represent the major class of σ factors. It is striking that relatively few ECF σ factors were discovered by traditional genetic approaches. For example, in *B. subtilis* there are seven ECF σ factor genes, none of which was discovered genetically. This seems to imply that either they are functionally redundant or they control the expression of genes not pertinent to normal laboratory culture conditions (or both).

The genome sequence of *S. coelicolor* has revealed an astonishing 51 ECF σ factors from a total of 65 σ factors, implying that these proteins play a major role in transcriptional regulation in *Streptomyces*. In order to understand the physiological roles of these ECF σ factors, it will be necessary to elucidate the signals to which they respond, to characterize the regulatory mechanisms involved in their activation, and to identify their regulons (the genes under their control). The aim of this review is to summarize current understanding of the biological roles and regulation of the three ECF σ factors (σ^E, σ^R and σ^{BldN}) that have been studied in detail in *S. coelicolor*. For each of these three σ factors, the mechanism controlling σ factor activity is different, variously involving *de novo* synthesis, pro-σ processing, and anti-σ factor-directed control. These examples serve to illustrate the fascinating variety of regulatory systems that exist in bacteria to ensure that σ factors are recruited to core RNA polymerase only when appropriate.

THE σ^E PATHWAY FOR SENSING AND RESPONDING TO CELL ENVELOPE STRESS

Since the initial cloning of the *sigE* gene (Lonetto *et al.*, 1994), extensive analysis suggests that σ^E is part of a signal transduction pathway that allows *S. coelicolor* to sense and respond to changes in the integrity of its cell envelope (Paget *et al.*, 1999a, b). A model for the pathway is shown in Fig. 1. The signal transduction system is composed of four proteins, encoded in an operon: σ^E itself; CseA, a negative regulator of undefined biochemical function; CseB, a response regulator; and CseC, a sensor histidine protein kinase with two predicted transmembrane helices; (Cse = control of sigma E). Expression of σ^E activity is governed at the level of *sigE* transcription by the CseB/CseC two-component signal transduction system. In response to signals that originate in the cell envelope when it is under stress, the sensor kinase, CseC, becomes autophosphorylated at His-271, and, in accordance with the known mechanism for other two-component regulatory systems, this phosphate is then transferred to Asp-55 in the response regulator, CseB. Phospho-CseB activates the promoter of the *sigE* operon, and σ^E is recruited by core RNA polymerase to transcribe genes with cell-envelope-related functions, including a putative operon of 12 genes likely to specify cell wall glycan synthesis.

Evidence for the model

sigE null mutants were extremely sensitive to cell wall hydrolytic enzymes, and had an altered cell wall muropeptide profile, suggesting that *sigE* is required for normal cell wall integrity. Importantly, the *sigE* mutant was sensitive to both muramidases (for example, lysozyme) and amidases, which cut the peptidoglycan backbone and the peptide side chain, respectively, suggesting that the defect in the *sigE* mutant envelope allowed hydrolytic enzymes increased access, rather then specifically altering their target sites (Paget *et al.*, 1999a). Mg^{2+} ions are known to have stabilizing effects on cell envelopes, and *sigE* null mutants required millimolar levels of Mg^{2+} for normal growth and sporulation, forming crenellated colonies, sporulating poorly, and overproducing the blue antibiotic actinorhodin in its absence (Paget *et al.*, 1999a).

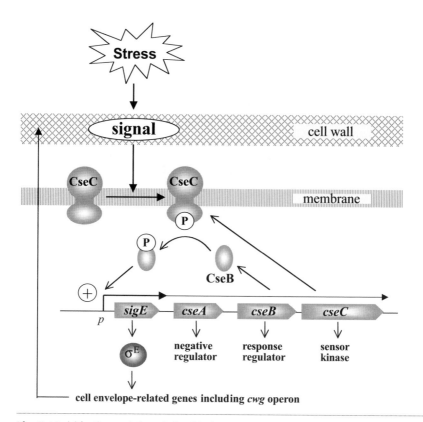

Fig. 1. Model for the regulation of σ^E activity in response to signals from the cell envelope. Expression of σ^E activity is governed at the level of *sigE* transcription by the CseB/CseC two-component signal transduction system. In response to signals that originate in the cell envelope when it is under stress, the sensor kinase, CseC, becomes autophosphorylated at His-271, and, in accordance with the known mechanism for other two-component regulatory systems, this phosphate is then transferred to Asp-55 in the response regulator, CseB. Phospho-CseB activates the promoter of the *sigE* operon, and σ^E is recruited by core RNA polymerase to transcribe genes with cell-envelope-related functions, including a putative operon of 12 genes likely to specify cell wall glycan synthesis.

Most transcripts from the *sigE* promoter terminate immediately downstream of *sigE*, but about 10% read through into the downstream genes (Paget *et al.*, 1999b). Analysis of the activity of the *sigE* promoter in different mutant backgrounds was highly informative. The *sigE* promoter was found to be inactive in a constructed *cseB* null mutant, such that *cseB* mutants lack σ^E. This observation explained why *cseB* and *sigE* mutants had the same phenotype (Paget *et al.*, 1999b). In contrast, the *sigE* promoter was substantially up-regulated in a *sigE* null mutant, suggesting that the cell envelope defects in *sigE* mutants are sensed by CseC, which responds by increasing the level of phospho-CseB in the cytoplasm in a futile attempt to increase expression of *sigE* and hence expression of the cell-envelope-related genes under σ^E control (Paget *et al.*, 1999b).

What signal is sensed by CseC?

The exact nature of the signal recognized by the sensor kinase is known for relatively few two-component systems. In order to better understand the nature of the signal sensed by CseC, a screening system was developed to test for compounds that induced the *sigE* promoter (H.-J. Hong, M. S. B. Paget, E. Leibovitz & M. J. Buttner, unpublished). The *sigE* promoter was placed upstream of a plasmid-borne kanamycin-resistance gene to yield a construct that conferred a basal level of kanamycin resistance on the host. A wide selection of antibiotics was then tested to see which increased kanamycin resistance above the basal level in a plate assay. In agreement with the proposed role for σ^E in controlling cell envelope integrity, antibiotics that target the cell envelope induce *sigE* expression. These included certain β-lactam antibiotics and, most effectively, glycopeptide antibiotics such as vancomycin and teicoplanin. 'Negative control' antibiotics that target the ribosome (e.g. thiostrepton, streptomycin) or DNA gyrase (novobiocin) did not induce *sigE* expression. In addition to antibiotics, lysozyme was also found to induce *sigE* expression, making it highly unlikely that CseC senses these inducers directly.

It is important to note that the *sigE* gene is transcribed under all growth conditions tested, implying that the CseB/CseC signal transduction system may be responding to changes in cell envelope metabolism that occur during 'normal' growth, which are amplified by the effects of antibiotics and enzymes that target the cell envelope. Accordingly, CseC could be activated by the accumulation of an intermediate in peptidoglycan degradation or biosynthesis, analogous to the control of β-lactam-inducible β-lactamase gene expression in many Gram-negative bacteria (Jacobs *et al.*, 1997). Alternatively, it is conceivable that CseC might be responding to some physical characteristic of the cell envelope (e.g. turgor). The KdpD/KdpE sensor kinase/response regulator pair of *E. coli* (Walderhaug *et al.*, 1992; Sugiura *et al.*, 1994) has been proposed to sense and respond to physical changes in the cell envelope.

CseA has a negative role in *sigE* expression

The gene immediately downstream from *sigE*, *cseA*, appears to play a negative role in *sigE* expression. The basal level of transcription from the *sigE* promoter was substantially higher in a constructed, in-frame *cseA* deletion mutant, and the maximal level of transcription from *sigEp* following induction with vancomycin was also several fold higher than in the wild-type (H.-J. Hong, M. S. B. Paget, E. Leibovitz & M. J. Buttner, unpublished). Although CseA has no similarity with any other proteins in the databases, its first 21 N-terminal amino acids (MAVFVALGVSLAGCGTGGTGA) are predicted to form a single transmembrane domain. Since CseA cannot function as a σ^E-specific anti-σ factor (σ^E does not direct transcription from the *sigE* promoter), perhaps it modulates the CseB/CseC signal transduction pathway, for example as an inhibitor of the kinase activity of CseC, or as a CseB-specific phosphatase.

σ^E directs transcription of a putative operon of 12 genes likely to specify cell wall glycan synthesis

Although σ^E was discovered by virtue of its ability to direct transcription of *dagAp2 in vitro*, when genetic analysis of *sigE* began, the activity of this promoter was found to be unaffected in a constructed *sigE* null mutant (Paget *et al.*, 1999a). Presumably this reflects relaxed promoter specificity *in vitro*, and the existence of a closely related ECF σ that recognizes *dagAp2 in vivo*. The first bona fide σ^E-dependent promoter identified was *hrdDp1* (Paget *et al.*, 1999a; Kang *et al.*, 1997), one of two promoters of the *hrdD* gene, which itself encodes a σ factor. However, this discovery was relatively uninformative because the physiological function of σ^{HrdD} is unknown (*hrdD* null mutants have no apparent phenotype; Buttner *et al.*, 1990). To identify further σ^E-dependent promoters, computer-searching methods were used to identify sequences in the emerging *S. coelicolor* genome sequence that closely resemble the *hrdDp1* promoter (GCAAC – 17 bp – CGTCT). An initial search identified a perfect match lying upstream of 12 genes that are likely to form an operon (H.-J. Hong, M. S. B. Paget & M. J. Buttner, unpublished). The predicted functions of the enzymes encoded by this operon strongly suggest that the operon specifies the synthesis of a species of cell wall glycan (hence the operon has been named *cwg*). High-resolution S1 nuclease mapping showed that the putative -10 and -35 sequences identified by computer searching do indeed correspond to a bona fide promoter, and that the *cwg* promoter is induced by vancomycin in a *sigE*-dependent manner (H.-J. Hong, M. S. B. Paget & M. J. Buttner, unpublished). Thus a set of genes under σ^E control has been identified that has a clear cell-envelope-related function, and transcription of these genes has been shown to be induced by vancomycin and, presumably therefore, other cell-wall-targeted antibiotics and enzymes. A constructed mutant in which the *cwg* operon was deleted did not show any of the phenotypes associated with *sigE* mutants, showing that other, as yet unknown, σ^E target genes play critical roles in maintaining cell envelope integrity.

THE σ^R PATHWAY FOR SENSING AND RESPONDING TO OXIDATIVE STRESS

σ^R was the second ECF σ factor to be discovered in *S. coelicolor*. Like σ^E, it was first identified in purified RNA polymerase holoenzyme preparations isolated from liquid-grown cultures (Kang *et al.*, 1997; Paget *et al.*, 1998). The role of σ^R as a key regulator of the oxidative stress response was discovered after phenotypic analysis of a constructed *sigR* null mutant. This mutant was sensitive to oxidizing agents such as the superoxide-generating, redox cycling compounds menadione and plumagin, and was particularly sensitive to a thiol-specific oxidant called diamide. The cytoplasm of all organisms is a reducing environment where thiol groups are maintained in their reduced state. The diamide-sensitive phenotype suggested that *sigR* mutants may be unable to respond to adverse changes in the thiol–disulphide redox balance, a condition termed disulphide stress (Åslund & Beckwith, 1999). This hypothesis was borne out by the demonstration of lowered levels of cytoplasmic disulphide reductase activity in *sigR* mutants (Paget *et al.*, 1998). The major system for controlling the thiol–disulphide balance in *Streptomyces* spp. is the thioredoxin system, which consists of the disulphide reductase thioredoxin and its reactivating enzyme thioredoxin reductase (Aharonowitz *et al.*, 1993; Cohen *et al.*, 1993). These enzymes use the reducing power of NADPH to remove unwanted disulphide bonds in oxidized cellular proteins, and to reduce enzymes, such as ribonucleotide reductase, that form disulphide bonds at their active site as part of their catalytic cycle. Reconstituted RNA polymerase holoenzyme containing purified σ^R initiated transcription from *trxBp1*, one of the two promoters that transcribe *trxBA*, the operon that encodes thioredoxin reductase and thioredoxin. Most importantly, *trxBp1* activity was rapidly and massively induced by the addition of the thiol-specific oxidizing agent diamide to wild-type mycelium, but remained uninduced in the *sigR* null mutant (Paget *et al.*, 1998).

Regulation of σ^R activity

The second σ^R target promoter to be identified, *sigRp2*, lay upstream of its own structural gene, *sigR*, thereby establishing a positive feedback loop for its own synthesis (Paget *et al.*, 1998). It thus became clear that, in order to prevent an upward spiral of σ^R synthesis, there must be a negative regulator in place to ensure that σ^R is only switched on when necessary and to ensure that its activity is effectively switched off when the disulphide stress has been dealt with. This key negative regulator was identified as RsrA (regulator of sigR), a σ^R-specific anti-σ factor that is encoded by the gene lying immediately downstream of *sigR*. Anti-σ factors are proteins that inhibit σ factor activity either by binding to it and preventing its interaction with core RNA polymerase, or by binding to the σ factor when it is part of the holoenzyme form, thereby preventing promoter binding (Hughes & Mathee, 1998; Helmann, 1999). Purified RsrA can bind tightly to σ^R and inhibit σ^R-directed transcription *in vitro*. However, RsrA can only perform this function when the *in vitro* conditions are sufficiently reducing. In the absence of strong thiol-reducing agents such as dithiothreitol (DTT), RsrA can neither bind to σ^R nor inhibit σ^R-directed transcription (Kang *et al.*, 1999). Moreover, if *rsrA* is deleted from the *S. coelicolor* chromosome, σ^R target promoters are constitutively expressed at the fully induced level (Paget *et al.*, 2001a). In other words, the regulation of σ^R activity by disulphide stress appears to be mediated solely by RsrA, with RsrA itself acting as the direct sensor of the thiol–disulphide redox status of the cell. Indeed, unlike σ^R, which contains no cysteines, RsrA, a protein of only 105 residues, contains seven cysteines and rapidly forms intramolecular disulphide bonds in the absence of thiol-reducing compounds (Kang *et al.*, 1999). A model for how RsrA regulates σ^R activity is presented in Fig. 2. σ^R protein is present in the hyphae all the time, but σ^R activity is not, because, in the absence of oxidative stress, RsrA sequesters σ^R in an

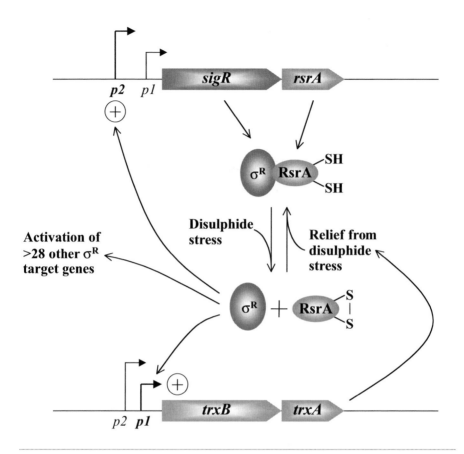

Fig. 2. Model for the regulation of σ^R activity in response to disulphide stress. The thiol–disulphide status of *S. coelicolor* is controlled by a novel regulatory system consisting of a σ factor, σ^R, and RsrA, a redox-sensitive, σ^R-specific, anti-σ factor. Under reducing conditions, RsrA binds to σ^R and prevents it from activating transcription. Exposure to disulphide stress induces the formation of one or more intramolecular disulphide bonds in RsrA, which causes it to lose its affinity for σ^R, releasing σ^R to activate transcription of >30 genes and operons, including *trxBA*. Increased *trxBA* expression in turn leads to the thioredoxin-dependent reduction of oxidized RsrA back to its σ^R-binding conformation, thereby shutting off σ^R-dependent transcription. In addition, σ^R positively autoregulates expression of the *sigR–rsrA* operon. As a consequence, disulphide stress not only activates σ^R post-translationally, but also induces its *de novo* synthesis.

RsrA : σ^R complex. σ^R is released during oxidative stress as a direct consequence of the inactivation of RsrA through intramolecular disulphide bond formation. σ^R is then free to associate with core RNA polymerase and activate transcription of its target genes, including *trxBA* and other thiol–disulphide oxidoreductase genes (see 'The σ^R regulon' below). At least *in vitro*, oxidized RsrA is a direct biochemical substrate for purified thioredoxin, the product of the *trxA* gene (Kang *et al.*, 1999). If the thioredoxin system

```
RsrA  (SC7E4.14)    18  LYEFLDKEMPDSDCVKFEHHFEECSPCLEKY   48
RstA  (SCH24.13c)   10  IADLAEGLLPTTRTTEVRQHLESCELCADVY   40
RsuA  (SCE59.12c)   13  VGAYALGILDDAEATAFEAHLATCEWCAQQL   43
SCM10.32            10  VGAYALGVLDEAEAFRFEDHLMECPRCAAQV   40
SCF56.17            10  TGAYALHALPDDEREAFERHLAGCATCEQEA   40
SCI11.11c            9  AAPYALDALEGAERVRFERHLEGCARCAAEV   39
SC6F11.06c          71  LGAWALAACSAPEAAAVEEHLGECDSCADEA  101
SCD84.14            19  LRAYARGELAAPALWSTDAHLTACATCRGVL   49
SCE46.08            29  VEAYADGQLTGAHRMQVAAHIACCWACSGSL   59
SCJ12.08             9  LVELALGHASGEADVGALRHAASCPRCREEL   39
SCP8.11             21  LSALVDGELGHDARERVLAHVATCPKCKAEV   51
SCE46.06c           24  LQSYLDGETDEVTARRVAAHLEDCRRCGLEA   54
```

Fig. 3. Alignment of the HisXXXCysXXCys motif in RsrA and 11 other putative ZAS anti-σ factors from *S. coelicolor*. All of the genes encoding these proteins are located near (typically downstream and immediately adjacent to) genes encoding ECF σ factors.

also reduces (reactivates) RsrA *in vivo*, this would allow it to rebind σ^R and shut down the response, thereby creating a simple homeostatic feedback loop in which the σ^R regulon is regulated in response to changes in the thiol–disulphide redox status of the hyphae.

This model raises several important questions, including the exact nature of the redox event that inactivates RsrA. Attempts to identify which of the seven cysteines in RsrA form the disulphide bond switch have not been straightforward. In principle, the loss of a cysteine residue that is involved in inactivating RsrA might be expected to lock RsrA in a constitutively active conformation, causing it to bind σ^R irrespective of the redox conditions. However, the individual substitution of each of seven RsrA cysteines did not reveal such mutants. Four of the cysteines in RsrA could be substituted, individually or collectively, still leaving a protein that could both inhibit σ^R activity and release it during disulphide stress. The remaining three individual cysteine mutants (C11, C41 and C44) had no σ^R-binding activity, preventing analysis of their ability to sense redox (Paget *et al.*, 2001a). There is now good evidence to suggest that, in their reduced state, these three cysteines play an important role in the σ^R-binding activity of RsrA by coordinating a zinc cofactor (see below).

The ZAS family of anti-σ factors

Since the discovery of *rsrA*, many related genes have been uncovered by genome sequencing in both Gram-positive and Gram-negative bacteria. Although the sequence similarity between the products of these genes is often very low, certain residues are highly conserved, especially an invariant HisXXXCysXXCys motif (see, for example, Fig. 3). Furthermore, each *rsrA*-related gene is located near (typically downstream and immediately adjacent to) an ECF σ-factor gene, strongly suggesting that the corresponding pair of proteins interact. Metal content analysis of RsrA (Paget *et al.*, 2001a) and ChrH (an RsrA-related anti-σ factor from *Rhodobacter sphaeroides*; see below) (Newman *et al.*, 2001) revealed that they are zinc metalloproteins. This, together with the absolute conservation of HisXXXCysXXCys, a potential zinc-binding motif, strongly suggests that all RsrA-related proteins are likely to bind zinc. This new family of proteins was therefore named the ZAS (zinc-binding anti-σ factor) family of anti-σ factors (Paget *et al.*, 2001a).

The redox regulation of RsrA is not a paradigm for all ZAS anti-σ factors

Importantly, although all RsrA-related anti-σ factors probably bind zinc, it is already clear that their activities are likely to be regulated in diverse ways, so the regulation of RsrA activity by a reversible thiol–disulphide redox switch is not a paradigm for the whole family. Thus a gene encoding a ZAS anti-σ factor lies immediately downstream of the *sigW* gene in *B. subtilis*, but σ^W-dependent gene expression is not induced by diamide and the known σ^W target genes have no obvious connection to thiol–disulphide metabolism (Huang *et al.*, 1999; Cao *et al.*, 2001; Wiegert *et al.*, 2001; J. Helmann, pers. comm.). Similarly, the ZAS anti-σ factor ChrR controls the activity of σ^E in *Rhodobacter sphaeroides*, but σ^E directs expression of the cytochrome c_2 structural gene (Newman *et al.*, 1999, 2001). Further, deletion of *chrR* or the σ^E-encoding *rpoE* does not affect the resistance of *R. sphaeroides* to diamide, and diamide does not induce σ^E-dependent gene expression (Newman *et al.*, 2001; T. Donohue & J. Newman, pers. comm.). Eleven of the 51 ECF σ factors in *S. coelicolor* are encoded by genes located near (typically upstream and immediately adjacent to) *zas* genes, and are therefore likely to be regulated by a ZAS anti-σ factor (Fig. 3). Several of these proteins differ from RsrA in having predicted transmembrane helices C-terminal to the HisXXXCysXXCys motif, suggesting that these ZAS proteins may regulate their cognate σ factor in response to extracytoplasmic signals.

The σ^R regulon

Searches for further σ^R target genes were made possible by the generation of a consensus target promoter sequence (GGAAT – 18 bp – GTT) using for comparison *trxBp1* and *sigRp2*, together with the sequence of *hrdDp2*, another promoter recognized by σ^R *in vitro*. Computer searches showed that this sequence occurred more than 60 times in the *S. coelicolor* genome, although only 34 of these were appropriately positioned just upstream from a gene. Each of these 34 sequences was examined experimentally for promoter activity; including *sigRp2*, *trxBp1* and *hrdDp2*, 30 were bona fide promoters that were induced by diamide in a σ^R-dependent manner (Paget *et al.*, 2001b). More than half of the σ^R target genes associated with these promoters have no known biological function.

Unsurprisingly, several σ^R target genes are likely to play important roles in thiol metabolism, including a second thioredoxin, *trxC*, and a glutaredoxin-like gene. Together with the *trxBA* operon, the induction of these genes by σ^R presumably helps to restore the thiol–disulphide balance following disulphide stress. Apart from cysteine thiols in proteins, low-molecular-mass thiols are also likely to become oxidized during disulphide stress, and the induction of the σ^R targets *cysM* and *moeB* is likely to act to restore levels of reduced cysteine and the dithiol-containing cofactor molybdopterin, respectively (Paget *et al.*, 2001b). Unlike Gram-negative bacteria and eukaryotes that use the cysteine-containing tripeptide glutathione as their major thiol buffer, *Streptomyces* and mycobacteria use a structurally unrelated, sugar-containing monothiol compound called mycothiol (Newton *et al.*, 1996). Although no target genes were found that were predicted to play a role in mycothiol biosynthesis, the *sigR* mutant was found to have significantly lowered levels of mycothiol (Paget *et al.*, 2001b). The root cause of diamide sensitivity in *sigR* mutants could therefore be due to any one of these σ^R-dependent mechanisms for coping with disulphide stress, or a combination of all of them.

At least three σ^R targets encode ribosome-associated products, including *relA*, *ssrA* and the ribosomal protein gene *rpmE*, suggesting that ribosome composition and function are modified in response to disulphide stress (Paget *et al.*, 2001b). RelA catalyses the production of ppGpp when ribosomes stall due to an uncharged tRNA entering the ribosome A-site. This intracellular signalling molecule then elicits the stringent response by selectively inhibiting transcription of rRNA genes, thereby acting to slow growth (Cashel *et al.*, 1996; Chatterji & Ojha, 2001). In *Streptomyces* spp., ppGpp also elicits antibiotic production in response to nutritional stress, and plays a role in differentiation (Chakraburtty & Bibb, 1997). *ssrA* encodes an unusual small stable tRNA–mRNA hybrid called tmRNA, which also acts when ribosomes stall, either at a rare codon or when ribosomes reach the end of a 3′ truncated mRNA that lacks a stop codon. tmRNA rescues the ribosome by acting as a surrogate mRNA to tag the nascent peptide with a hydrophobic tag that targets the protein for degradation (Keiler *et al.*, 1996; Roche & Sauer, 1999; Karzai *et al.*, 2000). It is tempting to speculate that disulphide stress inhibits some aspect of the translation process causing ribosomes to stall. A possible ribosomal target for disulphide stress is the product of the σ^R target gene *rpmE*, ribosomal protein L31, which contains a CysXXCys motif. The induction of *relA* and *ssrA* may then provide pathways to rescue stalled ribosomes and to slow ribosome production and growth, respectively, thereby focusing available resources on stress survival.

Another interesting σ^R target, *rbpA*, encodes a newly discovered RNA polymerase-binding protein, which may well be a novel low-molecular-mass RNA polymerase subunit (Paget *et al.*, 2001b). RbpA appears to exist only in the actinomycetes, including the mycobacteria. Although the role of RbpA is not known, the induction of *rbpA* transcription by σ^R suggests that the composition and function of RNA polymerase may also be modified in response to disulphide stress. Like the ribosome subunit L31, RbpA contains a CysXXCys motif, suggesting that it too may undergo thiol–disulphide redox reactions and may be a target of disulphide stress.

It should be noted that the method used to identify σ^R target promoters means that there may be many other, unidentified targets having promoter sequences that differ slightly from the consensus sequence used in the computer searches. The total σ^R regulon may therefore be considerably larger than the current total. Nonetheless, the identification of 30 genes and operons under σ^R control is a very significant step towards understanding the cellular response to disulphide stress in *S. coelicolor*.

Is σ^R a checkpoint in development?

A completely unexpected consequence of *rsrA* inactivation was a block in sporulation, and there is some circumstantial evidence to suggest that *S. coelicolor* may use σ^R as a checkpoint to inhibit development under conditions of oxidative stress, which may make sporulation undesirable. *S. coelicolor* differentiates on solid agar plates by forming aerial hyphae that grow out of the aqueous environment of the substrate mycelium into the air. These multigenomic aerial hyphae eventually undergo synchronous septation to produce chains of unigenomic exospores. Developmental mutants that are unable to raise an aerial mycelium have a shiny appearance on agar plates and are termed 'bald' (*bld*) mutants. Mutants that raise an aerial mycelium in the normal way but are unable to complete the developmental process by sporulating are termed white (*whi*) mutants, because the colonies fail to develop the characteristic grey pigment associated with mature spores.

A constructed *rsrA* mutant had a classical 'early' white phenotype, forming aerial mycelium, but failing to initiatiate sporulation septation. In contrast, a constructed *sigR rsrA* double mutant sporulated normally, showing that the inability of the *rsrA* single mutant to sporulate was a consequence of uncontrolled σ^R activity. One possible explanation for these observations is that the high level of free σ^R out-competes a sporulation-specific σ factor, such as σ^{WhiG} (Chater *et al.*, 1989), for core RNA polymerase (Paget *et al.*, 2001a). However, recent analogous experiments with σ^U and RsuA, another ECF σ factor : ZAS anti-σ factor pair in *S. coelicolor*, provided circumstantial evidence against this model (Gehring *et al.*, 2001). Disruption of *rsuA* caused a bald phenotype, but a *sigU rsuA* double mutant developed normally, again showing that the block in differentiation was a consequence of uncontrolled σ activity. As pointed out by Gehring *et al.* (2001), it seems unlikely that σ^R and σ^U could differentially compete with different σ factors, one required for aerial mycelium formation and one required for spore formation.

An alternative hypothesis is that the developmental phenotype of the *rsrA* null mutant is physiologically significant, that σ^R directs transcription of a 'sporulation inhibitor gene(s) ', and that *S. coelicolor* uses this mechanism as a checkpoint to arrest development under conditions of disulphide stress, which make sporulation undesirable (Gehring *et al.*, 2001; Paget *et al.*, 2001a). If this latter hypothesis is valid, it should be possible to identify mutations in the proposed 'sporulation inhibitor gene' that suppress the white phenotype of *rsrA* mutants, provided that there is only one σ^R target gene that mediates the arrest of development, and that this gene is non-essential. However, the four *rsrA* suppressor mutations characterized to date all map to *sigR* (Paget *et al.*, 2001a).

The σ^R–RsrA system also exists in pathogenic actinomycetes

The σ^R–RsrA system appears to exist in other actinomycetes. It is certainly present in mycobacteria, where it is named σ^H–RshA (Fernandes *et al.*, 1999; Paget *et al.*, 1998; I. Smith, pers. comm.), and analysis of the near-complete genome sequence of *Corynebacterium diphtheriae* (http://www.sanger.ac.uk/Projects/C_diphtheriae/) suggests that it also exists in this important actinomycete pathogen (M. S. B. Paget, unpublished). Of the 30 *S. coelicolor* σ^R target genes and operons so far identified, 13 of the homologues in *M. tuberculosis* have sequences upstream that resemble the consensus for σ^R-dependent promoters and may therefore be regulated by σ^H in *M. tuberculosis* (Paget *et al.*, 2001b). These include homologues of the *S. coelicolor* genes *sigR*, *trxBA*, *ssrA*, *rpmE* and *rbpA*. These observations make it likely that the σ^H–RshA system contributes to the well known resistance of *M. tuberculosis* to oxidative killing by white blood cells during human infection.

THE σ^{BldN} PATHWAY TO AERIAL MYCELIUM FORMATION

Unlike σ^E and σ^R, which were discovered biochemically, σ^{BldN} was identified genetically in a screen for new genes involved in morphological differentiation (Ryding *et al.*, 1999; Bibb *et al.*, 2000). Two NTG-induced point mutants were isolated in the gene encoding σ^{BldN}, the two mutants having strikingly different phenotypes. One, R650, had a white colony phenotype, and microscopic examination showed that the colony produced aberrant spores that were longer than those of the wild-type. The second, R112, had a more severe phenotype, producing substantially less aerial mycelium than the parental strain and only very rare spore chains, sometimes showing highly irregular sporulation septum placement (Ryding *et al.*, 1999). Shotgun complementation of

R650 and R112, followed by subcloning and sequencing, showed that this new developmental gene encoded an ECF σ factor (Bibb *et al.*, 2000). That both these mutants retained partial σ^{BldN} activity became clear when a constructed null mutant was found to have a bald phenotype, devoid of aerial hyphae. Therefore, the gene was named *bldN*. Sequence analysis of the two NTG-induced *bldN* mutant alleles revealed that the more 'severe' mutant, R112, carries a mutation in the ribosome-binding site and presumably produces reduced amounts of wild-type σ^{BldN}, while in the 'weak' mutant, R650, the σ^{BldN} produced carries a glycine to aspartate substitution in region 2.1 (Bibb *et al.*, 2000). In other σ factors, region 2.1 has been implicated in the interaction of σ with core RNA polymerase (Burgess & Anthony, 2001), and it is therefore likely that the mutant σ^{BldN} produced by R650 interacts less efficiently with core RNA polymerase than the wild-type protein.

Control of *bldN* transcription

The *bldN* promoter is temporally regulated, showing little or no activity during vegetative growth, but increasing dramatically during aerial mycelium formation and remaining highly active during sporulation (Bibb *et al.*, 2000). Clues as to the mechanism that controls this temporal regulation in *S. coelicolor* have come from the analysis of *bldN* transcription in other *bld* mutants. No *bldN* transcripts were detectable in *bldG* and *bldH* mutant backgrounds, indicating that *bldN* expression depends on these two genes, either directly or indirectly (Fig. 4; Bibb *et al.*, 2000). *bldH* has not been characterized, but *bldG* encodes a homologue of the SpoIIAA anti-anti-σ factor from *B. subtilis*, implying that the role of *bldG* is indirect. Anti-anti-σ factors are proteins that inhibit the activity of anti-σ factors, thereby stimulating the activity of its cognate σ factor. One possibility, therefore, is that *bldG* mutants have reduced activity of the σ factor that is required for transcription of the *bldN* promoter, caused by the uncontrolled activity of the respective anti-σ factor.

In contrast to the wild-type, *bldN* transcripts were readily detectable during vegetative growth in a *bldD* mutant, indicating that *bldD* acts to repress *bldN* transcription during vegetative growth (Fig. 4; Elliot *et al.*, 2001). *In vitro* biochemical experiments showed that this effect is direct; BldD is a repressor of the *bldN* promoter, binding to two operator sites, one either side of the transcription start site (Elliot *et al.*, 2001). Interestingly, BldD also represses transcription of another key developmental gene, *whiG*, during vegetative growth (Elliot *et al.*, 2001), and of the development-specific promoter (*p2*) of the *sigH* gene in vegetative hyphae (Kelemen *et al.*, 2001), suggesting that one of BldD's roles is to prevent premature expression of developmental genes.

Investigations by Yamazaki *et al.* (2000), working on the orthologue of σ^{BldN} in *Streptomyces griseus*, have raised some intriguing possibilities for another mechanism

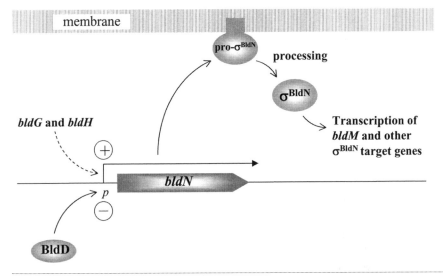

Fig. 4. Model for the regulation of σ^{BldN} activity during development. σ^{BldN} activity is regulated at the level of transcription of the *bldN* gene and by post-translational processing of the primary translation product, pro-σ^{BldN}. The *bldN* promoter shows little or no activity during vegetative growth, but is dramatically up-regulated during differentiation. This developmental control is mediated in part by BldD, which binds the *bldN* promoter and represses *bldN* transcription during vegetative growth. In contrast, the products of the *bldG* and *bldH* genes are required, directly or indirectly, for the activation of *bldN* transcription during development. The primary translation product of the *bldN* gene is a pro-σ factor, which is processed to a smaller, mature form through the proteolytic removal of an unusual N-terminal extension. The amino acid sequence of this N-terminal extension suggests that it might cause pro-σ^{BldN} to associate with the membrane. Release of mature σ^{BldN} allows the activation of its target genes, which include *bldM*.

by which *bldN* transcription might be regulated. In *S. griseus*, the γ-butyrolactone signalling molecule A-factor (2-isocapryloyl-3R-hydroxymethyl-γ-butyrolactone) triggers a regulatory cascade required for both aerial mycelium formation and production of the antibiotic streptomycin (Horinouchi & Beppu, 1994). A-factor causes expression of a transcriptional activator called AdpA, which induces streptomycin biosynthesis by activating transcription of *strR*, the gene encoding the pathway-specific activator of the streptomycin cluster (Ohnishi *et al.*, 1999). Until recently, no targets for AdpA have been identified to explain the morphological defects of an *adpA* mutant. However, Yamazaki *et al.* (2000) isolated new AdpA-binding sites from *S. griseus* chromosomal DNA, one of which was the promoter of an ECF σ factor gene they named *adsA* (<u>A</u>dpA-<u>d</u>ependent σ factor), the *S. griseus* orthologue of *bldN*. As is true for *S. coelicolor bldN*, transcription of *S. griseus adsA* begins approximately at the time of aerial mycelium formation, and disruption of *adsA* also results in loss of aerial mycelium formation. Neither *S. coelicolor bldN* nor *S. griseus adsA* is required for antibiotic production.

S. coelicolor does not produce A-factor, but it does produce several closely related γ-butyrolactone molecules (Efremenkova *et al.*, 1985; Kawabuchi *et al.*, 1997; Takano *et al.*, 2000). These molecules are involved in a signalling pathway for antibiotic production, and there is evidence to suggest that some of them may also be involved in morphological development in *S. coelicolor*. The predicted AdpA-binding site is not clearly conserved in the promoter region of *S. coelicolor bldN*, but there is a very close relative of *adpA* in the *S. coelicolor* genome sequence, and it will be interesting to see whether it has a role in the control of *bldN* transcription.

Post-translational processing of σ^{BldN}

Most ECF σ factors either completely lack conserved region 1 or have only a few residues upstream of region 2.1 (Lonetto *et al.*, 1994). σ^{BldN} is unusual in having an N-terminal extension of approximately 86 amino acids that is not present in other σ factors (Bibb *et al.*, 2000). Using a combination of immunoblotting and mutational analysis of the N-terminal extension, we have obtained substantial evidence that the primary translation product of the *bldN* gene is a pro-σ factor, which is processed to a smaller, mature form through the proteolytic removal of most of the N-terminal extension (Fig. 4; M. J. Bibb & M. J. Buttner, unpublished). During *B. subtilis* development, the mother-cell-specific σ factors σ^E and σ^K are synthesized as inactive pro-σ factors that are subsequently activated by proteolysis of the N-terminal 29 and 20 amino acids, respectively, by membrane-localized proteases (Errington, 1996; Stragier & Losick, 1996). In both cases, the activation of this processing event is triggered by signals derived from the forespore, and this 'crosstalk' serves to coordinate the divergent programs of gene expression between the two cellular compartments within the sporangium (Errington, 1996; Stragier & Losick, 1996). The pro sequences of both pro-σ^E and pro-σ^K promote membrane association, whereas the mature forms of these proteins are found in the cytoplasm associated with core RNA polymerase (Hofmeister, 1998; Ju & Haldenwang, 1999; Ju *et al.*, 1997; Zhang *et al.*, 1998). The putative pro sequence of σ^{BldN} contains a stretch of 20 hydrophobic amino acids (YAVPALAAAAV-PAGPCYALA). It will be interesting in the future to determine if pro-σ^{BldN} is membrane-associated, to identify the pro-σ^{BldN} protease, and to define the signals responsible for triggering the processing event.

The σ^{BldN} regulon

To date, only one σ^{BldN} target gene has been identified (Bibb *et al.*, 2000). Given the involvement of σ^{BldN} in the control of aerial mycelium formation, it seemed likely that other *bld* genes might be regulated by σ^{BldN} and would therefore have promoter sequences related to the consensus sequences of other ECF σ factors. Analysis of the promoter regions of known *bld* genes revealed a possible ECF consensus-like promoter upstream of *bldM*. *bldM* encodes an apparently typical member of the FixJ subfamily of response regulators, although, surprisingly, aspartate-54, the putative site of phosphorylation, is not required for BldM function (Molle & Buttner, 2000). Transcript mapping experiments identified two promoters, one of which, *bldMp1*, corresponded to the putative ECF σ factor consensus-like sequence. Like the *bldN* promoter, *bldMp1* was developmentally regulated, being inactive during vegetative growth, but strongly up-regulated during aerial mycelium formation and sporulation. Furthermore, *bldMp1* was inactive in a *bldN* null mutant and was recognized by reconstituted σ^{BldN}-containing holoenzyme *in vitro*, showing that *bldM* is a direct biochemical target for σ^{BldN} holoenzyme (Bibb *et al.*, 2000).

Overlapping promoter specificity between ECF σ factors

Prior to the discovery of the ECF subfamily, sequence similarity had already been noted between the *E. coli* σ^E target *rpoHp3* and the *S. coelicolor* σ^E target *dagAp2* (Erickson & Gross, 1989). Following the characterization of many more promoters under the control of different ECF σ factors, it became clear that there was a significant degree of sequence conservation between them. This fact, together with the existence of multiple ECF σ factors in many bacteria, suggested that some promoters might be recognized by more than one ECF σ *in vivo*, and it is now clear that this is indeed the case. For example, of the 30 σ^R target promoters known, at least 13 retained some activity in a *sigR* null mutant. Furthermore, this σ^R-independent transcription was constitutive for some promoters but stimulated by diamide (but with delayed kinetics) for others, implying that it represented more than one additional ECF σ factor (Paget *et al.*, 2001b). What are the key DNA sequence features that allow some σ^R target promoters to be recognized by additional holoenzymes forms while other promoters are recognized uniquely by σ^R? Analysis of the 30 known σ^R target promoters indicates that most promoters that are recognized by additional σ factors contain the −10 sequence CGTT, whereas those recognized only by σ^R have the −10 sequence TGTT or GGTT. Although the importance of the −10 region of σ^R target promoters in σ selectivity has not been proven, Helmann and colleagues have demonstrated that this region plays a critical role in σ selectivity between two ECF σ factors in *B. subtilis*. Single or double nucleotide changes in the −10 region of σ^X or σ^W target promoters switched their recognition characteristics such that promoters that were usually recognized by σ^W were recognized by σ^X, and vice versa (Qiu & Helmann, 2001). Recognition of a single promoter by multiple holoenzyme forms provides a very attractive mechanism for integrating different signal transduction pathways at single promoter elements. Overlapping specificity may be particularly useful in stress responses because different physical insults can often lead to the same physiological stress. For example, both oxidative stress and heat shock can induce protein misfolding. Nevertheless, target promoter sequence constraints must presumably ensure that, within the total subfamily of 51 ECF σ factors in *S. coelicolor*, each individual ECF σ factor has a distinct regulon and a distinct biological role. The future identification of the complete regulons for each of these ECF σ factors using DNA microarrays will begin to address these intriguing issues.

CONCLUSIONS

The ECF subfamily of σ factors has emerged as a major class of regulatory proteins in *Streptomyces* spp. Detailed analysis of just three of these proteins – σ^E, σ^R and σ^{BldN} – has already revealed their involvement in a fascinating range of biological processes and shown that control of their activity can be exerted at several different levels, variously involving *de novo* synthesis, pro-σ processing, and anti-σ factor-directed regulation. Understanding the role and regulation of each of the remaining 48 ECF σ factors promises to be an absorbing task.

REFERENCES

Ades, S. E., Connolly, L. E., Alba, B. M. & Gross, C. A. (1999). The *Escherichia coli* σ^E-dependent extracytoplasmic stress response is controlled by the regulated proteolysis of an anti-σ factor. *Genes Dev* **13**, 2449–2461.

Aharonowitz, Y., Av-Gay, Y., Schreiber, R. & Cohen, G. (1993). Characterization of a broad-range disulphide reductase from *Streptomyces clavuligerus* and its possible role in β-lactam antibiotic biosynthesis. *J Bacteriol* **175**, 623–629.

Åslund, F. & Beckwith, J. (1999). Bridge over troubled waters: sensing stress by disulfide bond formation. *Cell* **96**, 751–753.

Bibb, M. J., Molle, V. & Buttner, M. J. (2000). σ^{BldN}, an extracytoplasmic function RNA polymerase sigma factor required for aerial mycelium formation in *Streptomyces coelicolor* A3(2). *J Bacteriol* **182**, 4606–4616.

Burgess, R. R. & Anthony, L. (2001). How sigma docks to RNA polymerase and what sigma does. *Curr Opin Microbiol* **4**, 126–131.

Burgess, R. R., Travers, A. A., Dunn, J. J. & Bautz, E. K. F. (1969). Factor stimulating transcription by RNA polymerase. *Nature* **221**, 43–46.

Buttner, M. J., Smith, A. M. & Bibb, M. J. (1988). At least three different RNA polymerase holoenzymes direct transcription of the agarase gene (*dagA*) of *Streptomyces coelicolor* A3(2). *Cell* **52**, 599–607.

Buttner, M. J., Chater, K. F. & Bibb, M. J. (1990). Cloning, disruption and transcriptional analysis of three RNA polymerase sigma factor genes of *Streptomyces coelicolor* A3(2). *J Bacteriol* **172**, 3367–3378.

Cao, M., Bernat, B. A., Wang, Z., Armstrong, R. N. & Helmann, J. D. (2001). FosB, a cysteine-dependent fosfomycin resistance protein under the control of σ^W, an extracytoplasmic-function σ factor in *Bacillus subtilis*. *J Bacteriol* **183**, 2380–2383.

Cashel, M., Gentry, D. R., Hernandez, V. J. & Vinella, D. (1996). The stringent response. In *Escherichia coli and Salmonella: Cellular and Molecular Biology*, pp. 1458–1496. Edited by F. C. Neidhardt, R. Curtiss, III, J. L. Ingraham, E. C. C. Lin, K. B. Low, B. Magasanik, W. S. Reznikoff, M. Riley, M. Schaechter & H. E. Umbarger. Washington, DC: American Society for Microbiology.

Chakraburtty, R. & Bibb, M. J. (1997). The ppGpp synthetase (*relA*) of *Streptomyces coeli-*

color A3(2) plays a conditional role in antibiotic production and morphological differentiation. *J Bacteriol* **179**, 5854–5861.

Chater, K. F., Bruton, C. J., Plaskitt, K. A., Buttner, M. J., Mendez, C. & Helmann, J. D. **(1989).** The developmental fate of *Streptomyces coelicolor* hyphae depends on a gene product homologous with the motility sigma factor of *Bacillus subtilis. Cell* **59**, 133–143.

Chatterji, D. & Ojha, A. K. **(2001).** Revisiting the stringent response, ppGpp and starvation signalling. *Curr Opin Microbiol* **4**, 160–165.

Cohen, G., Yanko, M., Mislovati, M., Argaman, A., Schreiber, R., Av-Gay, Y. & Aharonowitz, Y. **(1993).** Thioredoxin-thioredoxin reductase system of *Streptomyces clavuligerus*: sequences, expression and organization of the genes. *J Bacteriol* **175**, 5159–5167.

Efremenkova, O. V., Anisova, L. N. & Bartoshevich, Y. E. **(1985).** Regulators of differentiation in actinomycetes. *Antibiot Med Biotekhnol* **9**, 687–707.

Elliot, M. A., Bibb, M. J., Buttner, M. J. & Leskiw, B. K. **(2001).** BldD is a direct regulator of key developmental genes in *Streptomyces coelicolor* A3(2). *Mol Microbiol* **40**, 257–269.

Erickson, J. W. & Gross, C. A. **(1989).** Identification of the σ^E subunit of *Escherichia coli* RNA polymerase: a second alternative σ factor involved in high-temperature gene expression. *Genes Dev* **3**, 1462–1471.

Errington, J. **(1996).** Determination of cell fate in *Bacillus subtilis. Trends Genet* **12**, 31–34.

Fernandes, N. D., Wu, Q. L., Kong, D., Puyang, X., Garg, S. & Husson, R. N. **(1999).** A mycobacterial extracytoplasmic sigma factor involved in survival following heat shock and oxidative stress. *J Bacteriol* **181**, 4266–4274.

Gehring, A. M., Yoo, N. J. & Losick, R. **(2001).** An RNA polymerase sigma factor that blocks morphological differentiation by *Streptomyces coelicolor* A3(2). *J Bacteriol* **183**, 5991–5996.

Gorham, H. C., McGowan, S. J., Robson, P. R. H. & Hodgson, D. A. **(1996).** Light-induced carotogenesis in *Myxococcus xanthus*: light-dependent membrane sequestration of ECF sigma factor CarQ by anti-sigma factor CarR. *Mol Microbiol* **19**, 171–186.

Härle, C., Kim, I., Angerer, A. & Braun, V. **(1995).** Signal transfer through three compartments: transcription initiation of the *Escherichia coli* ferric citrate transport system from the cell surface. *EMBO J* **14**, 1430–1438.

Helmann, J. D. **(1999).** Anti-sigma factors. *Curr Opin Microbiol* **2**, 135–141.

Hofmeister, A. **(1998).** Activation of the proprotein transcription factor pro-σ^E is associated with its progression through three patterns of subcellular localization during sporulation in *Bacillus subtilis. J Bacteriol* **180**, 2426–2433.

Horinouchi, S. & Beppu, T. **(1994).** A-factor as a microbial hormone that controls cellular differentiation and secondary metabolism in *Streptomyces griseus. Mol Microbiol* **12**, 859–864.

Huang, X., Gaballa, A., Cao, M. & Helmann, J. D. **(1999).** Identification of target promoters for the *Bacillus subtilis* extracytoplasmic function σ factor, σ^W. *Mol Microbiol* **31**, 361–371.

Hughes, K. T. & Mathee, K. **(1998).** The anti-sigma factors. *Annu Rev Microbiol* **52**, 231–286.

Jacobs, C., Frére, J.-M. & Normark, S. **(1997).** Cytosolic intermediates for cell wall biosynthesis and degradation control inducible β-lactam resistance in Gram-negative bacteria. *Cell* **88**, 823–832.

Ju, J. & Haldenwang, W. G. **(1999).** The "pro" sequence of the sporulation-specific transcription factor σ^E directs it to the mother cell side of the sporulation septum. *J Bacteriol* **181**, 6171–6175.

Ju, J., Luo, T. & Haldenwang, W. G. (1997). *Bacillus subtilis* pro-σ^E fusion protein localises to the forespore septum and fails to be processed when synthesised in the forespore. *J Bacteriol* **179**, 4888–4893.

Kang, J.-G., Hahn, M.-Y., Ishihama, A. & Roe, J.-H. (1997). Identification of sigma factors for growth phase-related promoter selectivity of RNA polymerases from *Streptomyces coelicolor* A3(2). *Nucleic Acids Res* **25**, 2566–2573.

Kang, J.-G., Paget, M. S. B., Seok, Y.-J., Hahn, M.-Y., Bae, J.-B., Kleanthous, C., Buttner, M. J. & Roe, J.-H. (1999). RsrA, an anti-sigma factor regulated by redox change. *EMBO J* **18**, 4292–4298.

Karzai, A. W., Roche, E. D. & Sauer, R. T. (2000). The SsrA-SmpB system for protein tagging, directed degradation and ribosome rescue. *Nat Struct Biol* **7**, 449–455.

Kawabuchi, M., Hara, Y., Nihira, T. & Yamada, Y. (1997). Production of butyrolactone autoregulators by *Streptomyces coelicolor* A3(2). *FEMS Microbiol Lett* **157**, 81–85.

Keiler, K. C., Waller, P. R. & Sauer, R. T. (1996). Role of a peptide tagging system in degradation of proteins synthesized from damaged messenger RNA. *Science* **271**, 990–993.

Kelemen, G. H., Viollier, P. H., Tenor, J., Marri, L., Buttner, M. J. & Thompson, C. J. (2001). A connection between stress and development in the multicellular prokaryote *Streptomyces coelicolor* A3(2). *Mol Microbiol* **40**, 804–814.

Lonetto, M., Gribskov, M. & Gross, C. A. (1992). The sigma 70 family: sequence conservation and evolutionary relationships. *J Bacteriol* **174**, 3843–3849.

Lonetto, M. A., Brown, K. L., Rudd, K. E. & Buttner, M. J. (1994). Analysis of the *Streptomyces coelicolor sigE* gene reveals the existence of a subfamily of eubacterial RNA polymerase σ factors involved in the regulation of extracytoplasmic functions. *Proc Natl Acad Sci U S A* **91**, 7573–7577.

Molle, V. & Buttner, M. J. (2000). Different alleles of the response regulator gene *bldM* arrest *Streptomyces coelicolor* development at distinct stages. *Mol Microbiol* **36**, 1265–1278.

Newman, J. D., Falkowski, M. J., Schilke, B. A., Anthony, L. C. & Donohue, T. J. (1999). The *Rhodobacter sphaeroides* ECF sigma factor, σ^E, and the target promoters *cycAP3* and *rpoEP1*. *J Mol Biol* **294**, 307–320.

Newman, J. D., Anthony, J. R. & Donohue, T. J. (2001). The importance of zinc-binding to the function of *Rhodobacter sphaeroides* ChrR as an anti-sigma factor. *J Mol Biol* **313**, 485–499.

Newton, G. L., Arnold, K., Price, M. S., Sherill, C., Delcardayre, S. B., Aharonowitz, Y., Cohen, G., Davies, J., Fahey, R. C. & Davis, C. (1996). Distribution of thiols in microorganisms: mycothiol is a major thiol in most actinomycetes. *J Bacteriol* **178**, 1990–1995.

Ohnishi, Y., Kameyama, S., Onaka, H. & Horinouchi, S. (1999). The A-factor regulatory cascade leading to streptomycin biosynthesis in *Streptomyces griseus*: identification of a target gene for the A-factor receptor. *Mol Microbiol* **34**, 102–111.

Paget, M. S. B., Kang, J.-G., Roe, J.-H. & Buttner, M. J. (1998). σ^R, an RNA polymerase sigma factor that modulates expression of the thioredoxin system in response to oxidative stress in *Streptomyces coelicolor* A3(2). *EMBO J* **17**, 5776–5782.

Paget, M. S. B., Chamberlin, L., Atrih, A., Foster, S. J. & Buttner, M. J. (1999a). Evidence that the extracytoplasmic function sigma factor, σ^E, is required for normal cell wall structure in *Streptomyces coelicolor* A3(2). *J Bacteriol* **181**, 204–211.

Paget, M. S. B., Leibovitz, E. & Buttner, M. J. (1999b). A putative two-component signal transduction system regulates σ^E, a sigma factor required for normal cell wall integrity in *Streptomyces coelicolor* A3(2). *Mol Microbiol* **33**, 97–107.

Paget, M. S. B., Bae, J.-B., Hahn, M.-Y., Li, W., Kleanthous, C., Roe, J.-H. & Buttner, M. J. (2001a). Mutational analysis of RsrA, a zinc-binding anti-sigma factor with a thiol-disulphide redox switch. *Mol Microbiol* **39**, 1036–1047.

Paget, M. S. B., Molle, V., Cohen, G., Aharonowitz, Y. & Buttner, M. J. (2001b). Defining the disulphide stress response in *Streptomyces coelicolor* A3(2): identification of the σ^R regulon. *Mol Microbiol* **42**, 1007–1020.

Qiu, J. & Helmann, J. D. (2001). The −10 region is a key promoter specificity determinant for the *Bacillus subtilis* extracytoplasmic-function sigma factors σ^X and σ^W. *J Bacteriol* **183**, 1921–1927.

Roche, E. D. & Sauer, R. T. (1999). SsrA-mediated peptide tagging caused by rare codons and tRNA scarcity. *EMBO J* **18**, 4579–4589.

Ryding, N. J., Bibb, M. J., Molle, V., Findlay, K. C., Chater, K. F. & Buttner, M. J. (1999). New sporulation loci in *Streptomyces coelicolor* A3(2). *J Bacteriol* **181**, 5419–5425.

Stragier, P. & Losick, R. (1996). Molecular genetic analysis of sporulation in *Bacillus subtilis*. *Annu Rev Genet* **30**, 297–341.

Sugiura, A., Hirokawa, K., Nakashima, K. & Mizuno, T. (1994). Signal-sensing mechanisms of the putative osmosensor KdpD in *Escherichia coli*. *Mol Microbiol* **14**, 929–938.

Takano, E., Nihira, T., Hara, Y., Jones, J. J., Gershater, C. J., Yamada, Y. & Bibb, M. (2000). Purification and structural determination of SCB1, a gamma-butyrolactone that elicits antibiotic production in *Streptomyces coelicolor* A3(2). *J Biol Chem* **275**, 11010–11016.

Walderhaug, M. O., Polarek, J. W., Voelkner, P., Daniel, J. M., Hesse, J. E., Altendorf, K. & Epstein, W. (1992). KdpD and KdpE, proteins that control expression of the *kdpABC* operon, are members of the two-component sensor-effector class of regulators. *J Bacteriol* **174**, 2152–2159.

Wiegert, T., Homuth, G., Versteeg, S. & Schumann, W. (2001). Alkaline shock induces the *Bacillus subtilis* σ^W regulon. *Mol Microbiol* **41**, 59–71.

Yamazaki, H., Ohnishi, Y. & Horinouchi, S. (2000). An A-factor dependent extracytoplasmic function sigma factor (σ^{AdsA}) that is essential for morphological development in *Streptomyces griseus*. *J Bacteriol* **182**, 4596–\4605.

Zhang, B., Hofmeister, A. & Kroos, L. (1998). The prosequence of pro-σ^K promotes membrane association and inhibits RNA polymerase core binding. *J Bacteriol* **180**, 2434–2441.

Secrets of bacterial transcription initiation taught by the *Escherichia coli* FNR protein

Douglas Browning,[1] David Lee,[1] Jeffrey Green[2] and
Stephen Busby[1]

[1]School of Biosciences, The University of Birmingham, Birmingham B15 2TT, UK

[2]Department of Molecular Biology and Biotechnology, The University of Sheffield, Sheffield S10 2TN, UK

TRANSCRIPTION ACTIVATION BY RECRUITMENT

In many bacteria, the number of genes is in excess of the number of functional RNA polymerase holoenzyme (RNAP) molecules and thus the distribution of RNAP between the different promoters must be strictly controlled and must adapt to changing growth conditions. One of the principal mechanisms of this distribution is the use of transcription regulatory proteins to activate or repress transcription initiation at particular promoters in response to specific signals. Over the past decade, much has been learned about the mechanisms of action of these proteins, and it has become apparent that, in many cases, regulation is due to direct interactions with the transcription apparatus that recruit RNAP to target promoters. This begs the question of the nature of these interactions and how the various contacts are organized at target promoters. Amongst the most intensively studied transcription activators are the *Escherichia coli* FNR protein and the related cAMP receptor protein (CRP). The aim of this short article is to outline recent progress in understanding how FNR and CRP activate transcription, to highlight the differences between them, and to describe some interesting complications in the FNR story. These complications may well provide pointers to understanding the functions of other players from the microbial repertoire, and should be useful in interpreting whole genome information from different micro-organisms.

FNR AND ANAEROBIC REGULATION OF GENE EXPRESSION

FNR was discovered as the product of the *fnr* gene, a locus where mutations caused reductions in the expression of fumarate reductase and nitrate reductase. It was quickly realized that FNR was a transcription activator and that it was responsible for inducing

SGM symposium 61: Signals, switches, regulons and cascades: control of bacterial gene expression.
Editors D. A. Hodgson, C. M. Thomas. Cambridge University Press. ISBN 0 521 81388 3 ©SGM 2002.

the expression of many genes in response to anaerobiosis. Much of our knowledge concerning FNR and its role has come from studies by John Guest and colleagues in Sheffield and is summarized by Spiro & Guest (1990) and Guest *et al.* (1996).

Two crucial observations underpin our understanding of the structure and function of FNR. First, FNR is related to the *E. coli* cAMP receptor protein (CRP) and is a member of a large family of transcription factors distributed across many species (reviewed by Spiro, 1994). Apart from the N-terminal 29-amino-acid extension that contains three crucial cysteines (C20, C23 and C29), the primary structure of FNR is highly related to the primary structure of CRP. It is known that CRP is functional as a homodimer and the high-resolution structure of the cAMP–CRP complex bound to its target has been determined (Schultz *et al.*, 1991). Thus although high-resolution structural information has not been obtained for FNR, it has been possible to model, with confidence, the structure of FNR and its interactions with DNA targets.

Second, FNR contains an iron–sulphur cluster and this plays a key role in sensing anaerobiosis (reviewed by Kiley & Beinert, 1998). For CRP, it was known that the binding of the ligand, cAMP (which is produced in response to certain stresses), triggers a change to a conformation that is able to bind to specific base sequences at target promoters (reviewed by Kolb *et al.*, 1993). For FNR, in anaerobic conditions, one $[4Fe–4S]^{2+}$ cluster is incorporated into each FNR subunit. This [4Fe–FNR] form appears to be the active complex, and its formation involves several of the cysteine side chains in the N-terminal region of FNR. Inactivation of FNR by oxygen during aerobic growth is due to the disassembly of the $[4Fe–4S]^{2+}$ clusters of FNR, initially to $[2Fe–2S]^{2+}$ clusters and ultimately to apo-FNR (Jordan *et al.*, 1997; Khoroshilova *et al.*, 1997). Thus for both FNR and CRP, activity is modulated by ligand-induced conformational changes. Interestingly, CRP is a stable dimer in both its active and inactive state. In contrast, inactive FNR is mainly monomeric and it dimerizes upon activation via the incorporation of the $[4Fe–4S]^{2+}$ cluster. Consistent with this, some mutants that stabilize the dimerization interface result in aerobic activity of FNR (Kiley & Reznikoff, 1991). These mutants are equivalent to the well-characterized *crp**** mutants that permit the activity of CRP in the absence of cAMP (reviewed by Kolb *et al.*, 1993).

FNR-DEPENDENT PROMOTERS

The consensus binding site for the CRP dimer is a 22 bp sequence consisting of two 11 bp sequences, 5′-AAA<u>TGTGA</u>TCT-3′, organized as an inverted repeat. Each subunit of the CRP dimer binds to one of the 11 bp elements, the most important contacts being with the central TGTGA motif (underlined). Analysis of several FNR-dependent promoters identified a similar binding target for FNR, but with a central TTTGA motif (Spiro & Guest, 1987a; Jayaraman *et al.*, 1988, 1989). It was subsequently shown that the activity of a promoter could be switched from dependence on CRP to dependence on FNR simply by changing TGTGA to TTTGA (Bell *et al.*, 1989; Zhang & Ebright, 1990). Structural studies on CRP:DNA complexes, together with genetic investigations, had led to models of how the recognition helix of CRP recognizes the TGTGA motif. Based on mutational analysis and the similarities between FNR and CRP, plausible models for FNR:DNA interactions have also been proposed (Spiro *et al.*, 1990) and FNR can be engineered to recognize DNA sites for CRP, for example at the *E. coli lac* promoter (Spiro & Guest, 1987b).

Sequence analysis has shown that most FNR-dependent promoters in *E. coli* carry the DNA site for FNR centred near position −41 (i.e. 41 bp upstream from the transcript start point) and thus bound FNR must overlap the target promoter −35 hexamer. Interestingly, however, at a small number of FNR-dependent and FNR-repressed promoters, DNA sites for FNR are found at locations further upstream. To investigate this, Wing *et al.* (1995) constructed a series of promoters carrying the same DNA site for FNR located at different positions upstream of the same core promoter element. The resulting promoters could be activated by FNR when the DNA site for FNR was located near positions −41, −61, −71, −82 and −92. This was very similar to the situation previously found for CRP with promoters carrying a DNA site for CRP located at different upstream positions (Gaston *et al.*, 1990; Ushida & Aiba, 1990). Thus FNR and CRP share the ability to activate transcription from different locations at promoters.

FNR FUNCTIONS AS AN 'AMBIDEXTROUS' ACTIVATOR AT CLASS II PROMOTERS

At many activator-dependent promoters, the cognate activator binds to a target that overlaps the promoter −35 region. At such promoters (often referred to as Class II promoters), in the ternary activator:RNAP:promoter complex, the 'Class II' activator binds adjacent to Region 4 of the RNAP σ subunit, which is responsible for recognition of the −35 promoter element. Examples of such activators are the bacteriophage λ cI protein and the *E. coli* PhoB protein, both of which activate transcription by making a direct contact with Region 4 of the RNAP σ subunit (Busby & Ebright, 1994; Rhodius & Busby, 1998). In the case of CRP, bound near position −41 at Class II promoters, it has been found that the situation is more complex. Footprinting studies had shown that the C-terminal domains of the two RNAP α subunits (αCTD) bind to DNA upstream of the bound CRP dimer (Belyaeva *et al.*, 1996). Genetic analysis showed that activation by CRP is driven by two independent contacts with RNAP involving two discrete activating regions on the surface of CRP (Figs 1a, 2a; reviewed by Busby & Ebright, 1997). First, Activating Region 1, a surface-exposed β turn in the upstream subunit of the bound CRP dimer, interacts with one of the upstream-bound αCTDs. Second,

a) CRP

b) FNR

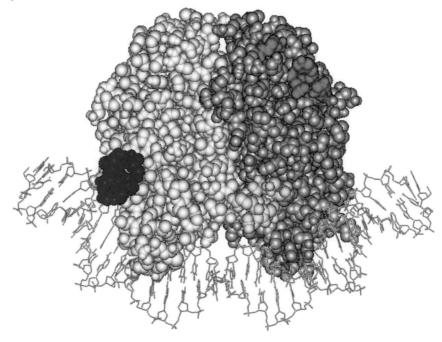

Activating Region 2, a surface-exposed positively charged region in the downstream subunit of the bound CRP dimer, interacts with the N-terminal domain (NTD) of one of the RNAP α subunits. Surprisingly, interactions between CRP and Region 4 of the RNAP σ subunit appear to play little or no role at Class II CRP-dependent promoters. However, CRP mutants have been found with enhanced ability to activate transcription specifically at Class II promoters and these mutants contain single amino acid substitutions that create a surface-exposed activating region (Activating Region 3; Fig. 1a) that interacts directly with Region 4 of the RNAP σ subunit (Lonetto et al., 1998; Rhodius & Busby, 2000a, b). Activating Region 3 consists of a negatively charged surface-exposed β turn and is functional only in the downstream subunit of the bound CRP dimer (Williams et al., 1996).

Mutational analysis has been used to identify the activating regions of FNR that interact with RNA polymerase during transcription activation at promoters where the DNA site for FNR overlaps the promoter −35 element (Williams et al., 1991; Bell & Busby, 1994; Lamberg & Kiley, 2000). These studies show that activation by FNR at Class II promoters is driven by two independent contacts with RNAP, involving two discrete activating regions on the surface of FNR (Figs 1b, 2a). First, a surface in the upstream subunit of the bound FNR dimer that is equivalent to Activating Region 1 of CRP interacts with αCTD (Wing et al., 1995, 2000; Williams et al., 1997). Second, a negatively charged surface-exposed β turn in the downstream subunit of the FNR dimer, which is equivalent to Activating Region 3 of CRP, interacts directly with Region 4 of the RNAP σ subunit (Wing et al., 1995; Lonetto et al., 1998; Lamberg et al., 2002). Thus at Class II promoters, CRP uses Activating Region 1 and Activating Region 2 to contact αCTD and αNTD, respectively, whilst, in contrast, FNR uses Activating Region 1 to contact αCTD and Activating Region 3 to contact Region 4 of the σ subunit (Fig. 2a). The full significance of the difference is, as yet, unclear. Interestingly, just as a functional Activating Region 3 can be created in CRP, a functional equivalent of Activating

Fig. 1. Activating Surfaces of CRP and FNR. (a) Space filling model of the CRP dimer bound to a target site on the DNA. In the left-hand CRP subunit, the amino acid side chains of Activating Region 1 (residues 156–164) are shaded dark-blue. In the right-hand CRP subunit, the amino acid side chains of Activating Region 2 (residues His19, His21, Glu96 and Lys101) are shaded dark-red. The amino acid side chains of the cryptic Activating Region 3 (residues 52–58) are shaded light-yellow in the right-hand CRP subunit. (b) Space filling model of the FNR dimer bound to a target site on the DNA. In the left-hand FNR subunit, the amino acid side chains of Activating Region 1 (Arg72, Ser73, Thr118, Met120 and Ser187) are shaded dark-blue. In the right-hand FNR subunit, the amino acid side chains of Activating Region 3 (residues Lys60, Thr82, Glu83, Gly85, Asp86, Glu87 and Gln88) are shaded dark-yellow. The key amino acid side chain of the cryptic Activating Region 2 (residues Glu47) is shaded light-red in the right-hand FNR subunit. The figures are derived from the structures presented by Schultz et al. (1991).

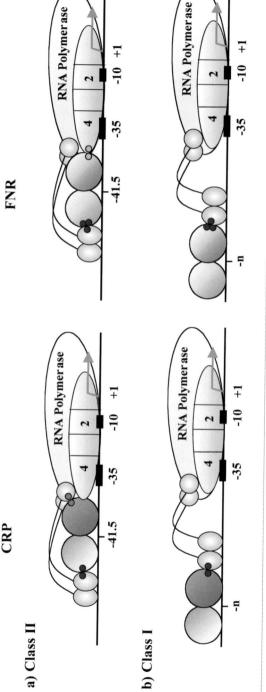

Fig. 2. Transcription activation at Class I and Class II promoters. The figure shows the interactions and the organization of different activator and RNAP subunits at Class II (a) and Class I (b) CRP-dependent and FNR-dependent promoters. CRP and FNR are illustrated, respectively, as grey and light-blue dimers. The interactions of Region 2 and Region 4 of the RNAP σ subunit, respectively, with promoter -10 and -35 elements are shown. The RNAP αCTD and αNTD domains are illustrated as spheres joined by a line that indicates the flexible inter-domain linker. The colour scheme for the different activating regions and their interactions is exactly as in Fig. 1. (a) Class II promoters activated by CRP (left) and FNR (right). For CRP, the interaction between Activating Region 1 and αCTD is indicated by blue dots, whilst the interaction between Activating Region 2 and αNTD is indicated by red dots. For FNR, the interaction between Activating Region 1 and αCTD is indicated by blue dots, whilst the interaction between Activating Region 3 and Region 4 of σ is indicated by yellow dots. (b) Class I promoters activated by CRP (left) and FNR (right). The interactions between Activating Region 1 of CRP/FNR and αCTD are indicated by blue dots.

Region 2 can be created in FNR (Fig. 1b). This was done by starting with FNR derivatives that were defective in transcription activation and then selecting for mutants better able to activate transcription at Class II FNR-dependent promoters (Li *et al.*, 1998; Ralph *et al.*, 2001).

FNR CAN FUNCTION AS A CLASS I ACTIVATOR

At many activator-dependent promoters, the cognate activator binds to a target that is located immediately upstream of the promoter −35 region. At such promoters (often referred to as Class I promoters), in the ternary activator : RNAP : promoter complex, the 'Class I' activator contacts αCTD, which is bound to the segment of DNA upstream of the −35 promoter element (Busby & Ebright, 1994). Spiro & Guest (1987b) showed that FNR could be mutated so that it could activate transcription at the *E. coli lac* promoter. Zhang & Ebright (1990) then found that the *lac* promoter could be activated by FNR provided that the DNA site for CRP was changed to a DNA site for FNR. Since the activator-binding site at the *lac* promoter is centred at position −61.5, the implication of these results was that FNR, like CRP, can function as a Class I-type transcription activator by interacting with αCTD. This point was confirmed by Wing *et al.* (1995), who demonstrated that transcription could be activated by FNR binding to a single DNA site for FNR located near positions −61, −71, −82 or −92. At Class I CRP-dependent promoters, it is known that activation is due to an interaction between Activating Region 1 of the downstream subunit of the CRP dimer and a target in αCTD, the 287 determinant, that was defined by genetic analysis (Figs 2b, 3a; reviewed by Busby & Ebright, 1999). In parallel studies, it was found that Activating Region 1 of FNR, in the downstream subunit of the FNR dimer, is also required for Class I FNR-dependent transcription activation (Fig. 2b). It was shown that the crucial amino acids of Activating Region 1 are T118 and S187 and thus Activating Region 1 of FNR is larger than Activating Region 1 of CRP (Wing *et al.*, 1995; Williams *et al.*, 1997). Further genetic analysis established that the target for Activating Region 1 of FNR in αCTD is adjacent to, but distinct from, the 287 determinant (Fig. 3b; Williams *et al.*, 1997; Lee *et al.*, 2000). Recently, FNR mutants carrying arginine side chains at positions 118 or 187 were made (Lee *et al.*, 2000). As expected, these mutants were defective in Class I FNR-dependent transcription activation. Mutants in αCTD that restored Class I FNR-dependent activation were then found: the locations of the substitutions in these mutants identify the surface of αCTD that can interact with either R118 or R187. Thus in ternary FNR : RNAP : promoter complexes, residues K304 and D305 in αCTD are located close to T118 in FNR, and residues R317 and L318 in αCTD are located close to S187 in FNR.

a) Proposed targets of

 AR1 of CRP

b) Proposed targets of

 AR1 of FNR

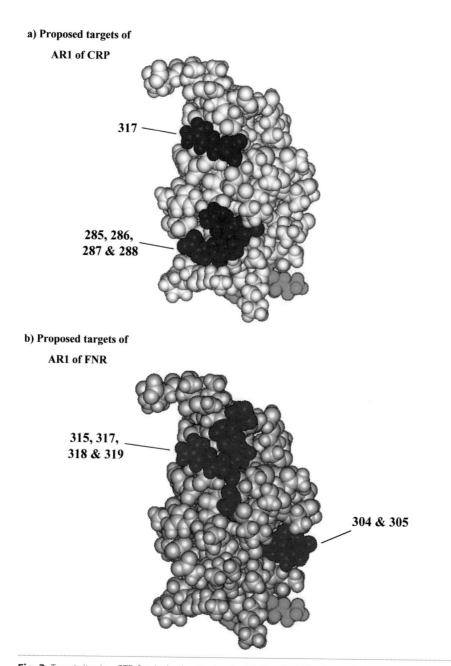

Fig. 3. Target sites in αCTD for Activating Region 1 of CRP or FNR. The figure shows space filling models of αCTD derived from the work of Jeon *et al.* (1995). Side chains of residues that have been implicated in contacts with CRP (a: residues 285, 286, 287, 288 and 317) or FNR (b: residues 304, 305, 315, 317, 318 and 319) are shaded in dark-blue. Residues of the '265 determinant' that is involved in contact with DNA are shaded in green.

a)

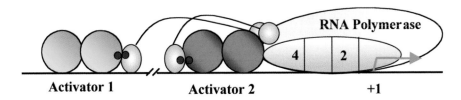

b)

Fig. 4. Activation and repression at complex promoters. The figure illustrates the interactions and the organization of different activator and RNAP subunits at promoters regulated by two transcription factors. The same conventions are used as in Fig. 2. (a) Promoter that is co-dependent on activation by a Class I activator (yellow) and a Class II activator (red). The different interactions with αCTD are shown by blue dots. (b) The *E. coli yfiD* promoter that is regulated by two molecules of FNR (light-blue). The promoter is activated by the downstream-bound FNR and inhibited by the upstream-bound FNR that interacts directly with the downstream-bound FNR (interactions shown by dark dots).

FNR CAN FUNCTION SYNERGISTICALLY WITH OTHER ACTIVATORS

Recent studies with CRP and other Class II activators have suggested a simple mechanism to explain how the activity of a promoter can be coupled to two different transcription activators (Savery *et al.*, 1996). The idea is that one of the activators, a Class II activator, functions by contacting Region 4 of the RNAP σ subunit or αNTD, and that this causes αCTD to be displaced upstream. A second activator, a Class I activator, then binds upstream and makes a contact with the displaced αCTD. The model (illustrated in Fig. 4a) envisages that, at some promoters, transcription activation will be contingent on separate interactions involving both the Class I and Class II activators. Several examples of such co-dependent activation have now been described (reviewed by Busby & Ebright, 1999), and, in some cases, one of the activators is FNR (which can function as either the Class I or the Class II activator). For example, the *E. coli ansB* promoter is co-regulated by FNR binding at position −41.5 as a Class II activator and CRP binding further upstream at position −91.5 as a Class I activator (Scott *et al.*, 1995). Other promoters have been described where FNR is the Class I activator and CRP is the Class II activator and thus the roles of FNR and CRP are reversed (Savery *et al.*, 1996).

REPRESSION BY FNR: TANDEM-BOUND FNR CAN DOWN-REGULATE FNR ACTIVITY

Many bacterial transcription activators can also function as repressors if they are positioned at certain locations at target promoters, for example, overlapping the −10 or −35 hexamer elements. Thus it was found that the binding of FNR to a single appropriately placed site is sufficient to down-regulate the expression of an otherwise constitutive promoter (Williams *et al.*, 1998). However, many naturally occurring promoters that are repressed by FNR do not appear to use this simple mechanism. For example, FNR-dependent repression of the *E. coli ndh* promoter requires binding of FNR to tandem sites located at positions −50.5 and −90.5, and the upstream site is essential for full repression (Meng *et al.*, 1997). Another interesting example is the *E. coli yfiD* promoter that is activated by FNR binding to a target site at position −40.5 but repressed when FNR binds to a second upstream site at position −93.5 (Green *et al.*, 1998). The lesson from these examples is that tandem-bound FNR dimers are able to interact in some way to form a repression complex (illustrated in Fig. 4b). A surface-exposed determinant of FNR that is essential for this interaction was identified and it was found that this determinant overlaps with Activating Region 1 of FNR (Green & Marshall, 1999). Thus, for example, at the *yfiD* promoter, interactions between these determinants on the tandem-bound FNR dimers suppresses the Class II activation due to FNR bound at position −40.5. Since FNR has a lesser affinity for the upstream site than for the downstream site, this provides a simple mechanism for microaerobic induction of the *yfiD* promoter (Marshall *et al.*, 2001). As the concentration of oxygen drops, the *yfiD* promoter is induced as FNR binds first to the activatory site at position −41.5. At lower concentrations of oxygen, FNR then occupies the upstream site at position −93.5 and the promoter is repressed.

The ability of tandem-bound FNR dimers to repress transcription initiation appears not to be shared by CRP. Although a single CRP dimer can function as a repressor when appropriately positioned, tandem-bound CRP dimers activate rather than repress transcription. The reason for this is that CRP can function as both a Class I and a Class II activator and thus, provided they are appropriately placed, two CRP dimers can each make productive interactions with αCTD of RNAP via Activating Region 1 (Belyaeva *et al.*, 1998; Langdon & Hochschild, 1999). In contrast, FNR has acquired an additional functional determinant on the surface that carries Activating Region 1. Thus at promoters carrying tandem-bound FNR dimers, the surface of FNR containing Activating Region 1 is used to drive the formation of a repression structure, rather than participating in interactions with αCTD that result in transcription activation. For reasons that are not understood, FNR appears to be hard-wired to prevent tandem-bound FNR molecules from co-activating transcription.

MODULATION OF FNR-DEPENDENT ACTIVATION AT PROMOTERS IN THE Nar MODULON

Many bacterial promoters that are induced by oxygen starvation are also regulated by other physiological signals. Of these, the most studied is the presence of nitrate ions, which can be used by many bacteria as an alternative electron acceptor in the absence of oxygen. Thus many FNR-dependent promoters are also induced by nitrate (and nitrite). In *E. coli*, this regulation is ensured by NarL and NarP, homologous response-regulator proteins whose activity is triggered by the presence of nitrate or nitrite ions in the environment (reviewed by Darwin & Stewart, 1996). Although there is great diversity in the organization of promoters that are activated by FNR, NarL and NarP, in most cases, the DNA site for FNR is located near position −41, whilst the site(s) for NarL and NarP are found further upstream (note that NarL and NarP recognize the same target heptamer sequences). The mechanisms by which the activity of target promoters is coupled to both FNR and NarL/NarP are not understood. One possibility is that FNR and NarL/NarP make independent contacts with RNAP similar to those described above for promoters co-dependent on a Class I and a Class II activator. However, recent studies with the *E. coli nir* and *nrf* promoters have indicated that this is not the case and have suggested a novel mechanism for co-regulation. The *nir* and *nrf* promoters control expression of NADH-dependent and formate-dependent nitrite reductases, respectively (Harborne *et al.*, 1992; Hussain *et al.*, 1994). Both promoters are regulated by FNR binding to a single target located at position −41.5, and are further regulated by NarL and NarP binding further upstream (Bell *et al.*, 1990; Darwin *et al.*, 1993). In the case of the *nir* promoter, nitrite and nitrate-dependent induction is due to NarP or NarL binding to a symmetrically arranged pair of heptamer sequence elements centred at position −69.5 (Tyson *et al.*, 1993). In the case of the *nrf* promoter, the activatory site for NarP and NarL binding is centred at position −74.5 (Tyson *et al.*, 1994). The key observation, made by Wu *et al.* (1998), is that many deletions, insertions and point mutations at the *nir* promoter, upstream of the DNA site for FNR, relieve the requirement of NarL and NarP for activation, resulting in a fully active FNR-dependent promoter. To explain this, Wu *et al.* (1998) suggested that other factors must bind to the promoter segment upstream of the DNA site for FNR, that these factors must suppress FNR-dependent activation, and that the role of NarL and NarP is to counteract the suppression. Parallel experiments with the *nrf* promoter indicated that a similar mechanism was operational, although in this case, the situation is

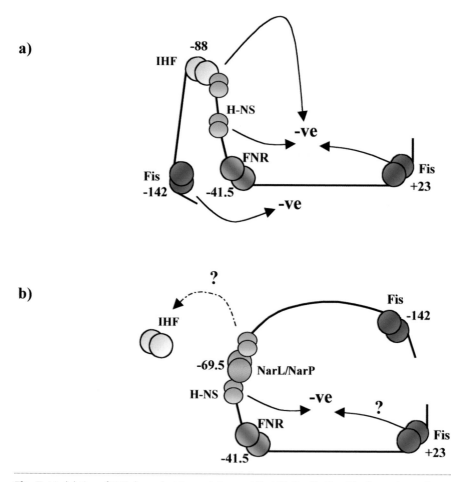

Fig. 5. Modulation of FNR-dependent transcription and NarL/NarP activation. The figure shows the proposed nucleo-protein complex formed at the *E. coli nir* promoter in anaerobic conditions in either the absence (a) or presence (b) of nitrite or nitrate. (a) The binding of Fis, IHF and H-NS to their respective sites inhibits FNR-dependent transcription at the *nir* promoter (indicated as −ve). (b) The binding of activated NarL/NarP displaces IHF and alters the architecture of the *nir* promoter. This enables RNA polymerase to make activatory contacts with upstream sequences. Note that repression by H-NS is not counteracted by NarL/NarP and thus H-NS is shown as a permanent component of *nir* promoter architecture.

complicated by the divergent *acs* promoter (Kumari *et al.*, 2000). In a recent study, Browning *et al.* (2000) identified Fis and IHF as the factors that suppress FNR-dependent activation at the *nir* promoter and also showed that *nir* promoter activity is down-regulated by H-NS. Parallel studies have identified a similar role for IHF at the *nrf* promoter (D. Browning, unpublished). Thus although we are still ignorant of many details, we can conclude that FNR-dependent activation, at least at two promoters, is suppressed by proteins of the bacterial folded chromosome. It seems most likely that this repression is due to the formation of some kind of local structure in the bacterial chromosome, and the role of NarL and NarP appears to be to reconfigure this structure (illustrated in Fig. 5 for the *E. coli nir* promoter).

PROSPECTS

It would be a mistake to suppose that, because so much is known about the *E. coli* CRP protein, there is little or no point in studying FNR. Perhaps the most interesting aspect of FNR is that although it shares many similarities with CRP, its 'big brother', it has many unique features that, presumably, have evolved because of the particular niche fulfilled by FNR in cellular metabolism. Although the significance of many of these features is not really understood, it is clear that many of the lessons learned from studying FNR are going to be important in the dissection of other systems. To date, our understanding of FNR has been retarded by the difficulty of setting up systems for reconstituting transcription activation *in vitro*. However, slow progress has been made and, at last, FNR is now beginning to yield its secrets concerning how it is regulated, how it interacts with the transcription apparatus, and how it functions co-ordinately with other regulatory systems. It may be some time before structural data on the *E. coli* protein are available. However, in the meantime, because of the pivotal role of FNR and its homologues in many adaptive responses, we can be sure that interest will be maintained for the foreseeable future.

ACKNOWLEDGEMENTS

We thank the many colleagues and coworkers who gave us access to their results prior to publication, and we apologize to those whose important results we have not mentioned due to lack of space. Most of the work described here from Birmingham and Sheffield has been supported by very generous funding from the UK BBSRC. We are especially grateful to John Guest and Jeff Cole for introducing us to this research topic and for their enthusiastic support.

REFERENCES

Bell, A. & Busby, S. (1994). Location and orientation of the Activating Region of FNR. *Mol Microbiol* **11**, 383–390.

Bell, A., Gaston, K., Cole, J. & Busby, S. (1989). Cloning of binding sequences for the *Escherichia coli* transcription activators, FNR and CRP: location of bases involved in discrimination between FNR and CRP. *Nucleic Acids Res* **17**, 3865–3874.

Bell, A., Cole, J. & Busby, S. (1990). Molecular genetic analysis of an FNR-dependent anaerobically-inducible *Escherichia coli* promoter. *Mol Microbiol* **4**, 1753–1763.

Belyaeva, T., Bown, J., Fujita, N., Ishihama, A. & Busby, S. (1996). Location of the C-terminal domain of the RNA polymerase alpha subunit in different complexes at the *Escherichia coli* galactose operon regulatory region. *Nucleic Acids Res* **24**, 2243–2251.

Belyaeva, T., Rhodius, V., Webster, C. & Busby, S. (1998). Transcription activation at promoters carrying tandem DNA sites for the *Escherichia coli* cyclic AMP receptor protein: organisation of the RNA polymerase α subunits. *J Mol Biol* **277**, 789–804.

Browning, D., Cole, J. & Busby, S. (2000). Suppression of FNR-dependent transcription activation at the *Escherichia coli nir* promoter by Fis, IHF and H-NS: modulation of transcription initiation by a complex nucleo-protein assembly. *Mol Microbiol* **37**, 1258–1269.

Busby, S. & Ebright, R. (1994). Promoter structure, promoter recognition and transcription activation in prokaryotes. *Cell* **79**, 743–746.

Busby, S. & Ebright, R. (1997). Transcription activation at Class II CAP-dependent promoters. *Mol Microbiol* **23**, 853–859.

Busby, S. & Ebright, R. (1999). Transcription activation by catabolite activator protein (CAP). *J Mol Biol* **293**, 199–213.

Darwin, A. & Stewart, V. (1996). The NAR modulon systems: nitrate and nitrite regulation of anaerobic gene expression. In *Regulation of Gene Expression in Escherichia coli*, pp. 343–359. Edited by E. C. C. Lin & A. Simon Lynch. Austin, TX: R. G. Landes.

Darwin, A., Hussain, H., Griffiths, L., Grove, J., Sambongi, Y., Busby, S. & Cole, J. (1993). Regulation and sequence of the structural gene for cytochrome c_{552} from *E. coli*, not a hexahaem but a 50kDa tetrahaem nitrite reductase. *Mol Microbiol* **9**, 1255–1265.

Gaston, K., Bell, A., Kolb, A., Buc, H. & Busby, S. (1990). Stringent spacing requirements for transcription activation by CRP. *Cell* **62**, 733–743.

Green, J. & Marshall, F. (1999). Identification of a surface of FNR overlapping Activating Region 1 that is required for repression of gene expression. *J Biol Chem* **274**, 10244–10248.

Green, J., Baldwin, M. & Richardson, J. (1998). Down regulation of *Escherichia coli yfiD* expression by FNR occupying a site at −93.5 involves the AR1-containing face of FNR. *Mol Microbiol* **29**, 1113–1123.

Guest, J. R., Green, J., Irvine, A. S. & Spiro, S. (1996). The FNR modulon and FNR-regulated gene expression. In *Regulation of Gene Expression in Escherichia coli*, pp. 317–342. Edited by E. C. C. Lin & A. Simon Lynch. Austin, TX: R. G. Landes.

Harborne, N., Griffiths, L., Busby, S. & Cole, J. (1992). Transcriptional control, translation and function of the products of the five open reading frames of the *E. coli nir* operon. *Mol Microbiol* **6**, 2805–2813.

Hussain, H., Grove, J., Griffiths, L., Busby, S. & Cole, J. (1994). A seven gene operon essential for formate-dependent nitrite reduction to ammonia by enteric bacteria. *Mol Microbiol* **12**, 153–163.

Jayaraman, P.-S., Gaston, K., Cole, J. & Busby, S. (1988). The *nirB* promoter of *E. coli*: location of nucleotide sequences essential for regulation by oxygen, the FNR protein and nitrite. *Mol Microbiol* **2**, 527–530.

Jayaraman, P.-S., Cole, J. & Busby, S. (1989). Mutational analysis of the nucleotide sequence at the FNR-dependent *nirB* promoter of *Escherichia coli*. *Nucleic Acids Res* **17**, 135–145.

Jeon, Y., Negishi, T., Shirakawa, M., Yamazaki, T., Fujita, N., Ishihama, A. & Kyogoku, T. (1995). Solution structure of the activator contact domain of the RNA polymerase α subunit. *Science* **270**, 1495–1497.

Jordan, P., Thomson, A. J., Ralph, E. T., Guest, J. R. & Green, J. (1997). FNR is a direct oxygen sensor having a biphasic response curve. *FEBS Lett* **416**, 349–352.

Khoroshilova, N., Popescu, C., Munck, E., Beinert, H. & Kiley, P. (1997). Iron-sulfur disassembly in the FNR protein of *Escherichia coli* by O_2: [4Fe 4S] to [2Fe 2S] conversion with loss of biological activity. *Proc Natl Acad Sci U S A* **94**, 6087–6092.

Kiley, P. & Beinert, H. (1998). Oxygen sensing by the global regulator, FNR: the role of the iron-sulfur cluster. *FEMS Microbiol Rev* **22**, 341–352.

Kiley, P. & Reznikoff, W. (1991). Fnr mutants that activate gene expression in the presence of oxygen. *J Bacteriol* **173**, 16–22.

Kolb, A., Busby, S., Buc, H., Garges, S. & Adhya, S. H. (1993). Transcriptional regulation by cAMP and its receptor protein. *Annu Rev Biochem* **62**, 749–795.

Kumari, S., Beatty, C., Browning, D., Busby, S., Simel, E., Hovel-Miner, G. & Wolfe, A. (2000). Regulation of acetyl-CoA synthetase in *Escherichia coli*. *J Bacteriol* **182**, 4173–4179.

Lamberg, K. & Kiley, P. (2000). FNR-dependent activation of the class II *dmsA* and *narG* promoters of *Escherichia coli* requires FNR activating regions 1 and 3. *Mol Microbiol* **38**, 817–827.

Lamberg, K. E., Luther, C., Weber, K. D. & Kiley, P. J. (2002). Characterization of Activating Region 3 from *Escherichia coli* FNR. *J Mol Biol* **315**, 275–283.

Langdon, R. & Hochschild, A. (1999). A genetic method for dissecting the mechanism of transcriptional activator synergy by identical activators. *Proc Natl Acad Sci U S A* **96**, 12673–12678.

Lee, D., Wing, H., Savery, N. & Busby, S. (2000). Analysis of interactions between Activating Region 1 of *Escherichia coli* FNR protein and the C-terminal domain of the RNA polymerase α subunit: use of alanine scanning and suppression genetics. *Mol Microbiol* **37**, 1032–1040.

Li, B., Wing, H., Lee, D., Wu, H.-C. & Busby, S. (1998). Transcription activation by *Escherichia coli* FNR protein: similarities to, and differences from, the CRP paradigm. *Nucleic Acids Res* **26**, 2075–2081.

Lonetto, M., Rhodius, V., Lamberg, K., Kiley, P., Busby, S. & Gross, C. (1998). Identification of a contact site for different transcription activators in Region 4 of the *Escherichia coli* RNA polymerase σ^{70} subunit. *J Mol Biol* **284**, 1353–1365.

Marshall, F., Messenger, S., Wyborn, N., Guest, J., Wing, H., Busby, S. & Green, J. (2001). A novel promoter architecture for microaerobic activation by the anaerobic transcription factor FNR. *Mol Microbiol* **39**, 747–753.

Meng, W., Green, J. & Guest, J. (1997). FNR-dependent repression of *ndh* gene expression requires two upstream FNR-binding sites. *Microbiology* **143**, 1521–1532.

Ralph, E. T., Scott, C., Jordan, P. J., Thomson, A. J., Guest, J. R. & Green, J. (2001). Anaerobic acquisition of [4Fe 4S] clusters by the inactive FNR(C20S) variant and restoration of activity by second-site amino acid substitutions. *Mol Microbiol* **39**, 1199–1211.

Rhodius, V. & Busby, S. (1998). Positive activation of gene expression. *Curr Opin Microbiol* **1**, 152–159.

Rhodius, V. & Busby, S. (2000a). Transcription activation by the *Escherichia coli* cyclic AMP receptor protein: determinants within Activating Region 3. *J Mol Biol* **299**, 295–310.

Rhodius, V. & Busby, S. (2000b). Interactions between Activating Region 3 of the *Escherichia coli* cyclic AMP receptor protein and region 4 of the RNA polymerase σ^{70} subunit: application of suppression genetics. *J Mol Biol* **299**, 311–324.

Savery, N., Belyaeva, T., Wing, H. & Busby, S. (1996). Regulation of promoters by two transcription activators: evidence for a "simultaneous touching" model. *Biochem Soc Trans* **24**, 351–353.

Schultz, S., Shields, G. & Steitz, T. (1991). Crystal structure of a CAP-DNA complex: the DNA is bent by 90°. *Science* **253**, 1001–1007.

Scott, S., Busby, S. & Beacham, I. (1995). Transcriptional coactivation at the *ansB* promoters: involvement of the activating regions of CRP and FNR when bound in tandem. *Mol Microbiol* **18**, 521–532.

Spiro, S. (1994). The FNR family of transcriptional regulators. *Antonie Leeuwenhoek* **66**, 23–36.

Spiro, S. & Guest, J. R. (1987a). Regulation and overexpression of the *fnr* gene of *Escherichia coli*. *J Gen Microbiol* **133**, 3279–3288.

Spiro, S. & Guest, J. R. (1987b). Activation of the *lac* operon of *Escherichia coli* by a mutant FNR protein. *Mol Microbiol* **1**, 53–58.

Spiro, S. & Guest, J. R. (1990). FNR and its role in oxygen-regulated gene expression in *Escherichia coli*. *FEMS Microbiol Rev* **75**, 399–428.

Spiro, S., Gaston, K. L., Bell, A. I., Roberts, R. E., Busby, S. J. W. & Guest, J. R. (1990). Interconversion of the DNA-binding specificities of two transcription regulators, CRP and FNR. *Mol Microbiol* **4**, 1831–1838.

Tyson, K., Bell, A., Cole, J. & Busby, S. (1993). Identification of the nitrite and nitrate response elements at the anaerobically-inducible *Escherichia coli nirB* promoter. *Mol Microbiol* **7**, 151–157.

Tyson, K., Cole, J. & Busby, S. (1994). Nitrite and nitrate regulation at the promoters of two *Escherichia coli* operons encoding nitrite reductase. *Mol Microbiol* **13**, 1045–1046.

Ushida, C. & Aiba, H. (1990). Helical phase dependent action of CRP: effect of the distance between the CRP site and the −35 region on promoter activity. *Nucleic Acids Res* **18**, 6325–6330.

Williams, R., Bell, A., Sims, G. & Busby, S. (1991). The role of two surface exposed loops in transcription activation by the *Escherichia coli* CRP and FNR proteins. *Nucleic Acids Res* **19**, 6705–6712.

Williams, R., Rhodius, V., Bell, A., Kolb, A. & Busby, S. (1996). Orientation of functional Activating Regions in the *Escherichia coli* CRP protein during transcription activation at Class II promoters. *Nucleic Acids Res* **24**, 1112–1118.

Williams, S., Savery, N., Busby, S. & Wing, H. (1997). Transcription activation at Class I FNR-dependent promoters: identification of the activating surface of FNR and the corresponding contact site in the C-terminal domain of the RNA polymerase α subunit. *Nucleic Acids Res* **25**, 4028–4034.

Williams, S., Wing, H. & Busby, S. (1998). Repression of transcription initiation by *Escherichia coli* FNR protein: repression by FNR can be simple. *FEMS Microbiol Lett* **163**, 203–208.

Wing, H., Williams, S. & Busby, S. (1995). Spacing requirements for transcription regulation by *Escherichia coli* FNR protein. *J Bacteriol* **177**, 6704–6710.

Wing, H., Green, J., Guest, J. & Busby, S. (2000). Role of Activating Region 1 of *Escherichia coli* FNR protein in transcription activation at Class II promoters. *J Biol Chem* **275**, 29061–29065.

Wu, H.-C., Tyson, K., Cole, J. & Busby, S. (1998). Regulation of the *E. coli nir* operon by two transcription factors: a new mechanism to account for co-dependence on two activators. *Mol Microbiol* **27**, 493–505.

Zhang, X. & Ebright, R. H. (1990). Substitution of 2 base pairs (1 base pair per half site) within the *Escherichia coli lac* promoter DNA site for catabolite gene activator protein places the *lac* promoter in the FNR regulon. *J Biol Chem* **265**, 12400–12403.

What can be learned from the LacR family of *Escherichia coli*?

Benno Müller-Hill

Institut für Genetik der Universität zu Köln, Weyertal 121, 50931 Köln, Germany

'(a protein structure) emerges one day like Venus from the waves ...'

Max Perutz (2000) *Cell* **101**, 23–24

INTRODUCTION

The model of negative transcriptional control of the *lac* operon and of phage lambda was proposed in 1961 by Jacob & Monod (1961a, b). It was so beautiful that almost everybody believed that this was the general model which was to be found in *Escherichia coli* and all living systems. It could be confirmed. Lac repressor (LacR) could be isolated (Gilbert & Müller-Hill, 1966). It was a protein which bound specifically to *lac* operator DNA *in vitro* (Gilbert & Müller-Hill, 1967). It could be produced in gram amounts (Müller-Hill *et al.*, 1968) to allow the determination of its sequence (Beyreuther *et al.*, 1973). Positive transcriptional control through activators was thought to be ruled out and nonexistent. Therefore people like Ellis Englesberg (Englesberg, 1961; Englesberg *et al.*, 1965) and Maxime Schwartz (Schwartz, 1967) had a hard time when they proposed their views of the positive controls of the arabinose or maltose systems in *E. coli*. It took many years until it was understood that in eukaryotes transcription is repressed by the general formation of nucleosomes, and that it has to be turned on by positive control through activators, which then eventually may be closed down by negative control through repressors (Lemon & Tijan, 2000). It was the overwhelming beauty of the Jacob–Monod model that blinded many researchers for years to appreciate reality.

SGM symposium 61: Signals, switches, regulons and cascades: control of bacterial gene expression.
Editors D. A. Hodgson, C. M. Thomas. Cambridge University Press. ISBN 0 521 81388 3 ©SGM 2002.

SIMILARITIES AND DIFFERENCES IN THE LacR/GalR FAMILY

When in 1982 it was recognized that Gal repressor (GalR) is homologous to LacR (von Wilcken-Bergmann & Müller-Hill, 1982), a more specific question could be asked: do these two homologous repressors work in an identical or similar fashion? Today, where the entire genome of E. coli is sequenced, we know about 15 repressors which are homologous to LacR. Only a few of them have been analysed in detail. The remarkable result is that all of those which have been analysed in detail repress in different ways. The lac system uses just one way to do so. All the other modes which are known today are equally elegant. Thus there is not one single solution to the problem of repression, there are many, and today we know only a few of them.

LacR has a modular structure (Adler et al., 1972; Lewis et al., 1996; Bell & Lewis, 2000, 2001). It consists of a DNA-binding headpiece (residues 1–60), an inducer binding and dimerization core (residues 61–330) and two tetramerizing heptad repeats (residues 330–360). Apparently it is the only member of the family carrying C-terminal heptad repeats. Thus some of the other repressors use other means to become tetramers.

The N-terminal headpiece contains a helix–turn–helix motif (Matthews et al., 1982). The second helix is the recognition helix. Residues 1, 2, 5 and 6 of the recognition helix recognize bases in the major groove of the operator (Lewis et al., 1996). Therefore one can change the specificity of DNA recognition by just changing these residues according to specific rules (Lehming et al., 1990). Some of the other residues of the headpiece are important for positioning the recognition helix appropriately in the major groove of the operator. We find the helix–turn–helix motif not only in the LacR family (Matthews et al., 1982). Has it been reinvented several times or do all helix–turn–helix motifs go back to one founder? We do not know.

The core of the Lac repressor is homologous to several monomeric proteins which are found in the periplasm between the inner and outer membranes (Müller-Hill, 1983). They bind various small molecules, in particular sugars. Some of these proteins have changed their sequence to such an extent that no homology is apparent, and only the three-dimensional structure, as revealed through X-ray crystallography, is similar (Sack et al., 1989; Kang et al., 1992). X-ray crystallographic analysis in addition has shown that CysB protein, an activator of transcription and member of the large LysR family, has a similar core structure to that of the members of the LacR family. However, dimerization is different and no amino acid sequence homology can be seen (Tyrrell et al., 1997).

An operator in E. coli has to consist of about 10–12 bases in order to be unique in the genome. One headpiece can recognize 5–6 bases. Thus repressors have to bind as dimers to their operators in *E. coli*. All the 15 repressors of the LacR family thus should have acquired the ability to form homodimers. All relatives of LacR that have been analysed are indeed dimers or multimers (LacR is a tetramer, GalR is a dimer). The X-ray crystallography structures of LacR (Lewis *et al.*, 1996), RafR (Hars *et al.*, 1998) and PurR (Schumacher *et al.*, 1994, 1995) have been solved. Therefore the residues which form the dimer interface are known. If one compares the residues of the interface amongst the various repressors of the LacR family one finds extremely different sequences (Dong *et al.*, 1999). They are so different that one would never guess that they belong to the same family.

How do LacR and the other members of the family repress the promoters they control? We have a detailed analysis only for four of them: LacR, GalR, CytR and RafR (see Fig. 1). In every case, they repress promoters which are activated by the CAP/CRP protein.

Tetrameric LacR represses a thousandfold in the wild-type situation. It competes with RNA polymerase by binding with two of its four subunits to *O1*, which is situated +11 bp downstream of the transcription start site (Schlax *et al.*, 1995). There is a CAP site at position −61.5. LacR binds with its two other subunits to either *O2* (situated +412 bp downstream of the transcription start site) or *O3* (situated −82 bp upstream of the transcription start site) (Oehler *et al.*, 1990, 1994). This tightens the binding to *O1*. The auxiliary operators *O2* and *O3* thus together increase repression by *O1* by a factor of 70. If only one of them is present, repression is still increased 30-fold. Thus repression of *O1* is strongly increased by the presence of one or both auxiliary operators (Oehler *et al.*, 1990, 1994; Müller-Hill, 1998).

Recently it has been shown (Fried & Daugherty, 2001) that tetrameric Lac repressor forms *in vitro* a complex with two dimers of CAP in the presence of cAMP. Dimeric Lac repressor which lacks the C-terminal four helical bundle does not form such a complex with CAP. This may explain the fact that tetrameric Lac repressor forms a loop between *O1* and *O3* +5 bp upstream in the presence of bound CAP (Fried & Hudson, 1996). The structure of this complex is unclear. One may ask whether there are special conditions which favour the loop *O1–O3* over the loop *O1–O2* and vice versa. This is not known.

In the *gal* system, two molecules of dimeric GalR bind to two operators: *OE* (situated −60.5 bp upstream of the transcription start site) and *OI* (situated +53.5 bp downstream of the transcription start site). Binding of HU protein in between the two GalR dimers in supercoiled DNA makes the two dimers come together and to form a GalR tetramer (Aki & Adhya, 1997). One might have assumed that HU needs specific DNA sequences to bind to the *gal* operator *in vivo*. This is not so. Apparently any sequence

(a) Positions of the sites of control

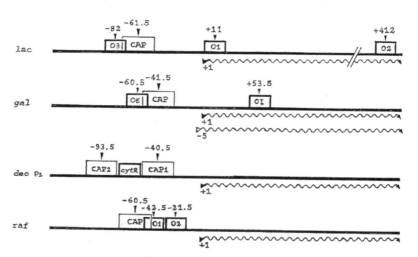

(b) In the absence of repressors   (c) In the presence of repressors

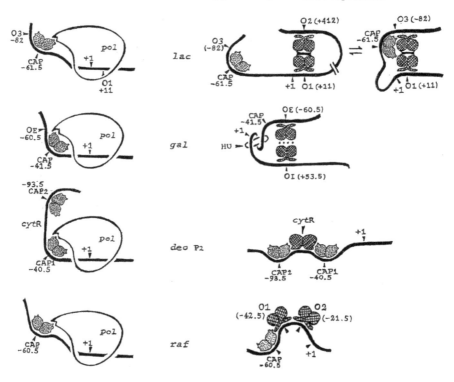

will do (Perez *et al.*, 2000). The HU–GalR complex makes it impossible for the CAP protein to bind to its site situated −41.5 bp upstream of the transcription site. Thereby transcription is repressed about 50-fold. Stronger repression is apparently not necessary. Recently the presence of a third *gal* promoter at position −100 has been demonstrated. It is unclear how it is repressed (Sur *et al.*, 2001).

Even at very high (mg ml⁻¹) concentrations GalR is a dimer. When it represses *in vivo* it forms a tetramer. Why is it not a tetramer like LacR in solution? Tetrameric LacR finds *lacO* as fast as dimeric LacR *in vitro* when *lacO* is placed on a short piece of DNA. However, when *lacO* is positioned within the 49 000 bp of phage lambda DNA, then the association rate of tetrameric LacR is increased 5–10-fold over dimeric LacR (Fickert & Müller-Hill, 1992). We explain this with the Tarzan model: tetrameric LacR binds briefly with one half site to DNA, then searches with the other half site for a *lac* operator and thus finds it faster than dimeric LacR.

The recognition helix of LacR recognizes only one DNA sequence whereas the recognition helix of GalR recognizes several DNA sequences (Lehming *et al.*, 1990). GalR is not as specific as LacR in DNA binding. When we test *in vitro* the association rate of *galO* to LacR modified with the DNA-binding specificity of GalR we find that the association rate is as fast for the dimer as for the tetramer, when *galO* is placed in short DNA. When *galO* is buried in lambda DNA, however, the association rate of tetrameric but not dimeric LacR with GalR specificity is decreased by a factor of 5 to 10! In other words, tetrameric GalR equivalent gets lost in the jungle of very weak binding sites, whereas native LacR is capable of speeding up its reaction! Thus it would be disadvantageous for GalR to be a tetramer (Barker *et al.*, 1998).

Fig. 1. Control of the *lac*, *gal*, *deo*P2 and *raf* promoters. Modified from Müller-Hill (1998). (a) Positions of the repressor-binding and activator-binding sites with respect to the start of transcription (+1), i.e. the positions of the promoters. *O1* is the main operator and *O2* and *O3* are auxiliary operators of the *lac* system. *OE* is the main operator and *OI* is the auxiliary operator in the *gal* system. Cyt repressor (*cytR*) binds only in the presence of CAP. *O1* and *O2* are the two operators of the *raf* system. (b) The same four systems in the absence of their repressors. Note: RNA polymerase binds CAP with one of its two alpha subunits. The presence of RNA polymerase symbolizes frequent transcription. (c) The four systems when repressed. The tetrameric LacR binds to *O1* and can then bind to another operator, *O2* or *O3*, close by. This strengthens the binding to *O1* and prevents binding of RNA polymerase and represses transcription. In the presence of CAP, tetrameric Lac repressor may form a complex between *O1* and a target 5 bp upstream of *O3*. The detailed structure of this complex is unknown. Gal repressor is a dimer which binds to *OE* and *OI* and *HU* binds to the *gal* promoter region (+1) supercoiled DNA, resulting in repression of transcription. CytR repressor binds to the *cytR* site between two bound CAP dimers and shields CAP from the alpha subunit of RNA polymerase. The two Raf repressors bound to *O1* and *O2* allow the binding of CAP but not of RNA polymerase to their sites. Dotted fill, CAP; hatched fill, repressor: LacR, GalR, CytR or RafR.

Fig. 2. Overall structure of the *lac–gal* regulon of *Streptococcus thermophilus*. Adapted from Vaughan *et al.* (2001).

CytR represses the *deo*P2 promoter in a peculiar manner (Kristensen *et al.*, 1996; Kallipolitis *et al.*, 1997). The P2 promoter has two CAP-binding sites: one at the position −40.5 (like the *gal* promoter) and the other at −93.5. Activation occurs mainly through the CAP protein bound at the site positioned −40.5. Now, CytR has a weak binding site in between the two CAP-binding sites. When bound there it shields the surface of the promoter distal subunit of CAP protein bound to the site positioned at −40.5, and the promoter proximal site of the CAP protein bound to the site positioned at −93.5. CytR thereby makes the parts of the CAP surface which normally interact with the alpha subunit of RNA polymerase inaccessible to RNA polymerase. Therefore the CAP protein cannot activate anymore, even when bound to its ideal site.

Finally, in the *raf* system (Muiznieks *et al.*, 1999), there is a CAP site at position −60.5. There are two binding sites for Raf repressor at −42.5 and −21.5. RafR bound to these sites competes with RNA polymerase directly for binding. The effect of the two repressors bound is therefore very strong, even if the CAP site at position −60.5 is not hindered.

So we learn that in every one of these four systems a different strategy is used for repression (Fig. 1). However, all four of them use a weak promoter which is activated by CAP. The positions for the operator or CAP sites are always optimal or almost optimal for the particular complex. The CAP sites are positioned at −40 or −61, the optimal positions (Gaston *et al.*, 1990). *lac* O3 is also positioned at an optimal position as shown by Müller *et al.* (1996). It is truly remarkable that evolution has always selected the best possible positions for its operators. Ideally we would like to know *all* possibilities used by nature. For example, the DNA sequence of the *gal–lac* regulon of *Streptococcus thermophilus* has recently been determined (Vaughan *et al.*, 2001) (see Fig. 2). The structure of the regulon is amazing. We have a Gal repressor gene and the *galK*, *galT* and *galE* genes, which apparently are transcribed as one operon. Please note that the order of the *gal* genes is opposite to that in *E. coli*! Furthermore, directly connected to these *gal* genes is a *lac* operon consisting of a *lacS* (permease) and a *lacZ* (β-galactosidase) gene. Where are the operators? This could not be predicted! Now that the sequence of the *E. coli* chromosome is known, could one predict properly how all the other repressors work?

This is a real challenge. If one fails here, how can one expect to understand the regulation of eukaryotic promoters, which is so much more complicated?

DESIGNING LacR WITH NEW PROPERTIES

The properties of the repressors of the LacR/GalR family have been altered by evolution in *E. coli* and in other bacteria. Various members recognize different DNA sequences and different inducers. They form only homodimers with themselves and do not form heterodimers with others. Some members have become heat-resistant.

Ten years ago (Lehming *et al.*, 1990) it was demonstrated that residues 1, 2 and 6 of the recognition helix of LacR determine in an additive manner recognition of bp 4, 5 and 6 of the *lac* operator

$$6 \; 5 \; 4 \; 3 \; 2 \; 1 \; 1' \; 2' \; 3' \; 4' \; 5' \; 6'$$
$$5' \; G \; T \; G \; A \; G \; C \; G \; C \; T \; C \; A \; C.$$

Residue 5 of the recognition helix interacts with residues 1 and 2 of the recognition helix of LacR. This makes predictions of the interactions with the bases difficult (Sartorius *et al.*, 1991). The rules found for mutant Lac repressors seem to hold also in other examples of the LacR/GalR family (Lehming *et al.*, 1991). So far, this type of analysis has not been tried with any of the other members of the LacR family.

The members of the LacR/GalR family all form homodimers which may then aggregate to higher oligomers. There are many ways by which a homodimer bond can be strengthened (Swint-Kruse *et al.*, 2001). We concentrate here on direct and specific strengthening of the interface between monomers. Intuitively one may think that to exchange two or even more residues of the interface with the equivalent residues found in another member of the family will do. This was tried with no results. Astonishingly, the best mutants with new dimerization specificity are single exchanges like D278L and L251M (Spott *et al.*, 2000). Residues 278 and 251 of LacR are of great importance for the interface according to Miller's mutant analysis of LacR (Suckow *et al.*, 1996). The residues which were put in occur in some members of the LacR/GalR family. The results suggest that a change in dimerization specificity is straightforward and easily achieved in evolution, at least in this family of proteins. We do not know whether this is generally true.

E. coli grows at 43 °C. At higher temperatures it does not grow. Therefore it makes sense that LacR fast loses its activity when heated to 54 °C. One may ask the question, can one stabilize LacR against heat denaturation by single amino acid exchanges? D84L is a candidate since it is more stable than wild-type LacR against denaturation by urea (Nichols & Matthews, 1997). Indeed, this single exchange stabilizes LacR by 40 degrees! LacR mutant D84L is stable at 93 °C and becomes unstable at 94 °C (Pereg-Gerk *et al.*, 2000). The X-ray structure of the K84L mutant has been solved (Bell *et al.*, 2001). It is not known if the mechanism of creating a highly stable lipophilic interaction may occur generally.

The last bastion to be taken is the inducer binding site. So far, all attempts to change the specificity by specific exchanges have failed. This tells us how little we understand in spite of all the knowledge accumulated in the last few years.

CONCLUSIONS

If one analyses a biological system down to a level which has not been reached in any other comparable system, one will be overwhelmed by the beauty, elegance and economy of the system. One may believe that many or even all other systems must function similarly. But then when one analyses such similar systems one suddenly discovers that there are many elegant solutions to the same problem. There is not one solution, there are many solutions depending presumably on small differences in the original DNA sequence. Out of these different sequences originate then the various CAP sites and the operators. It is not just Venus emerging from the waves, to quote Max Perutz, it is a whole universe of goddesses appearing. To participate in finding some of them is indeed the greatest pleasure science can provide in molecular biology. And then to explain how they have arisen. We do not understand it, but we have to understand it.

ACKNOWLEDGEMENTS

This work was supported by Deutsche Forschungsgemeinschaft and Fonds der Chemie.

REFERENCES

Adler, K., Beyreuther, K., Fanning, E., Geisler, N., Gronenborn, B., Klemm, A. & Müller-Hill, B. (1972). How Lac repressor binds to DNA. *Nature* **237**, 322–327.

Aki, T. & Adhya, S. (1997). Repressor induced site specific binding of HU for transcriptional regulation. *EMBO J* **16**, 3666–3674.

Barker, A., Fickert, R., Oehler, S. & Müller-Hill, B. (1998). Operator search by mutant Lac repressors. *J Mol Biol* **278**, 549–568.

Bell, C. E. & Lewis, M. (2000). A closer view of the conformation of the Lac repressor bound to operator. *Nat Struct Biol* **7**, 209–214.

Bell, C. E. & Lewis, M. (2001). The Lac repressor: a second generation of structural and functional studies. *Curr Opin Struct Biol* **11**, 19–25.

Bell, C. E., Barry, J., Matthews, K. S. & Lewis, M. (2001). Structure of a variant of lac repressor with increased thermostability and decreased affinity for operator. *J Mol Biol* **313**, 99–109.

Beyreuther, K., Adler, K., Geisler, N. & Klemm, A. (1973). Amino acid sequence of lac repressor. *Proc Natl Acad Sci U S A* **7**, 3576–3580.

Dong, F., Spott, S., Zimmermann, O., Kisters-Woike, B., Müller-Hill, B. & Barker, A. **(1999)**. Dimerisation mutants of Lac repressor I: a monomeric mutant, L251A, that binds *lac* operator as a dimer. *J Mol Biol* **290**, 653–666.

Englesberg, E. **(1961)**. Discussion of the Jacob & Monod paper. *Cold Spring Harbor Symp Quant Biol* **26**, 209–210.

Englesberg, E., Irr, J., Power, J. & Lee, N. **(1965)**. Positive control of enzyme synthesis by gene C in the L-*arabinose* system. *J Bacteriol* **90**, 946–957.

Fickert, R. & Müller-Hill, B. **(1992)**. How Lac repressor finds *lac* operator *in vitro*. *J Mol Biol* **226**, 59–68.

Fried, M. G. & Daugherty, M. A. **(2001)**. *In vitro* interaction of the *Escherichia coli* cyclic AMP receptor protein with the lactose repressor. *J Biol Chem* **276**, 11226–11229.

Fried, M. G. & Hudson, J. M. **(1996)**. DNA looping and Lac repressor-CAP interaction. *Science* **274**, 1930–1931.

Gaston, K., Bell, A., Kolb, A., Buc, H. & Busby, S. **(1990)**. Stringent spacing requirements for transcription activation by CRP. *Cell* **62**, 733–743.

Gilbert, W. & Müller-Hill, B. **(1966)**. Isolation of the lac repressor. *Proc Natl Acad Sci U S A* **56**, 1891–1898.

Gilbert, W. & Müller-Hill, B. **(1967)**. The *lac* operator is DNA. *Proc Natl Acad Sci U S A* **58**, 2415–2421.

Hars, U., Horlacher, R., Boos, W., Welte, W. & Diederichs, K. **(1998)**. Crystal structure of the effector-binding domain of the trehalose-repressor of *Escherichia coli*, a member of the LacI family, in its complexes with inducer trehalose-6-phosphate and noninducer trehalose. *Protein Sci* **7**, 2511–2521.

Jacob, F. & Monod, J. **(1961a)**. Genetic regulatory mechanisms in the synthesis of proteins. *J Mol Biol* **3**, 318–356.

Jacob, F. & Monod, J. **(1961b)**. On the regulation of gene activity. *Cold Spring Harbor Symp Quant Biol* **26**, 193–211.

Kallipolitis, B. H., Nørregaard-Madsen, M. & Valentin-Hansen, P. **(1997)**. Protein-protein communication: structural model of the repression complex formed by CytR and the global regulator CRP. *Cell* **89**, 1101–1109.

Kang, C. H., Shin, W. C., Yamagata, Y., Gokcen, S., Ames, G. F. L. & Kim, S. H. **(1992)**. Crystal structure of the lysine-, arginine-, ornithine-binding protein (LAO) from *Salmonella typhimurium* at 2.7 Å resolution. *J Biol Chem* **266**, 23893–23899.

Kristensen, H. H., Valentin-Hansen, P. V. & Søgaard-Andersen, L. **(1996)**. CytR/cAMP-CRP nucleoprotein formation in *E. coli*: the CytR repressor binds its operator as a stable dimer in a ternary complex with cAMP-CRP. *J Mol Biol* **260**, 113–119.

Lehming, N., Sartorius, J., Kisters-Woike, B., von Wilcken-Bergmann, B. & Müller-Hill, B. **(1990)**. Mutant *lac* repressors with new specificities hint at rules for protein-DNA recognition. *EMBO J* **9**, 615–621.

Lehming, N., Sartorius, J., Kisters-Woike, B., von Wilcken-Bergmann, B. & Müller-Hill, B. **(1991)**. Rules for protein DNA recognition for a family of HTH proteins. In *Nucleic Acids and Molecular Biology*, vol. 5, pp. 114–125. Edited by F. Eckstein & D. M. J. Lilley. Berlin & Heidelberg: Springer.

Lemon, B. & Tijan, R. **(2000)**. Orchestrated responses: a symphony of transcription factors for gene control. *Genes Dev* **14**, 2551–2569.

Lewis, M., Chang, G., Horton, N. C., Kercher, M. A., Pace, H. C., Schumacher, M. A., Brennan, R. G. & Lu, P. **(1996)**. Crystal structure of the lactose operon repressor and its complexes with DNA and inducer. *Science* **271**, 1247–1254.

Matthews, B. W., Ohlendorf, D. H., Anderson, W. F. & Takeda, Y. **(1982)**. Structure of the DNA binding region of *lac* repressor inferred from the homology with *cro* repressor. *Proc Natl Acad Sci U S A* **79**, 1428–1432.

Muiznieks, I., Rostocks, N. & Schmitt, R. (1999). Efficient control of *raf* gene expression by CAP and two Raf repressors that bend DNA in opposite directions. *Biol Chem* **380**, 19–29.

Müller, J., Oehler, S. & Müller-Hill, B. (1996). Repression of *lac* promoter as a function of distance, phase and quality of an auxiliary operator. *J Mol Biol* **257**, 21–29.

Müller-Hill, B. (1983). Sequence homology between Lac and Gal repressors and three sugar-binding periplasmic proteins. *Nature* **302**, 163–164.

Müller-Hill, B. (1998). The function of auxiliary operators. *Mol Microbiol* **29**, 13–18.

Müller-Hill, B., Crapo, L. & Gilbert, W. (1968). Mutants which make more Lac repressor. *Proc Natl Acad Sci U S A* **59**, 1259–1264.

Nichols, J. C. & Matthews, K. S. (1997). Combinatorial mutations of *lac* repressor. *J Biol Chem* **272**, 18550–18557.

Oehler, S., Eismann, E. R., Krämer, H. & Müller-Hill, B. (1990). The three operators of the *lac* operon cooperate in repression. *EMBO J* **9**, 973–979.

Oehler, S., Amouyal, M., Kolkhof, P., von Wilcken-Bergmann, B. & Müller-Hill, B. (1994). Quality and position of the three *lac* operators in *E. coli* define efficiency of repression. *EMBO J* **13**, 3348–3355.

Pereg-Gerk, L., Leven, O. & Müller-Hill, B. (2000). Increasing the thermostability of Lac repressor by 40 °C. *J Mol Biol* **299**, 805–812.

Perez, N., Renault, M. & Amouyal, M. (2000). A functional assay in *Escherichia coli* to detect non-assisted interaction between galactose repressor dimers. *Nucleic Acids Res* **28**, 3600–3604.

Sack, J. S., Trakhanov, S. D., Tsigannik, I. H. & Quiocho, F. A. (1989). Structure of L-leucine-binding protein refined at 2.4 Å resolution and comparison with the Leu/Ile/Val-binding protein structure. *J Mol Biol* **206**, 193–207.

Sartorius, J., Lehming, N., Kisters-Woike, B., von Wilcken-Bergmann, B. & Müller-Hill, B. (1991). The roles of residues 5 and 9 of the recognition helix of Lac repressor in *lac* operator binding. *J Mol Biol* **218**, 313–321.

Schlax, P. J., Capp, M. W. & Record, T. M. (1995). Inhibition of transcription initiation by *lac* repressor. *J Mol Biol* **245**, 331–350.

Schumacher, M. A., Choi, K. Y., Zalkin, H. & Brennan, R. G. (1994). Crystal structure of LacI member, PurR, bound to DNA: minor groove binding by α helices. *Science* **266**, 763–770.

Schumacher, M. A., Choi, K. Y., Lu, F., Zalkin, H. & Brennan, R. G. (1995). Mechanism of corepressor-mediated specific DNA binding by the purine repressor. *Cell* **83**, 147–155.

Schwartz, M. (1967). Expression phénotypique et localisation génétique de mutations affectant le métabolism du maltose chez *Escherichia coli* K12. *Ann Inst Pasteur* **112**, 673–702.

Spott, S., Dong, F., Kisters-Woike, B. & Müller-Hill, B. (2000). Dimerisation mutants of Lac repressor II: a single amino acid substitution, D278L, changes the specificity of dimerisation. *J Mol Biol* **296**, 673–684.

Suckow, J., Markiewicz, P., Kleina, L. G., Miller, J., Kisters-Woike, B. & Müller-Hill, B. (1996). Genetic studies of the Lac repressor XV: 4000 single amino acid substitutions and analysis of the resulting phenotypes on the basis of the protein structures. *J Mol Biol* **261**, 509–523.

Sur, R., Debnath, D., Mukhopadhyay, J. & Parrack, P. (2001). A novel RNA polymerase binding site upstream of the galactose promoter in *Escherichia coli* exhibits promoter-like activity. *Eur J Biochem* **268**, 2344–2350.

Swint-Kruse, L., Elam, C. R., Lin, J. W., Wycuff, D. R. & Matthews, K. S. (2001). Plasticity of quaternary structure: twenty-two ways to form a LacI dimer. *Protein Sci* **10**, 262–276.

Tyrrell, R., Verschueren, K. H. G., Dodsen, E. J., Murshudov, G. N., Addy, C. & Wilkinson, A. J. (1997). The structure of the cofactor-binding fragment of the LysR family member, CysB: a familiar fold with a surprising subunit arrangement. *Structure* **5**, 1017–1022.

Vaughan, E. E., van den Bogard, T. C., Catzeddu, P., Kuipers, O. P. & de Vos, W. M. (2001). Activation of silent *gal* genes in the *lac-gal* regulon of *Streptococcus thermophilus*. *J Bacteriol* **183**, 1184–1194.

von Wilcken-Bergmann, B. & Müller-Hill, B. (1982). Sequence of *galR* gene indicates a common evolutionary origin of the *lac* and *gal* repressor in *Escherichia coli*. *Proc Natl Acad Sci U S A* **79**, 2427–2431.

Regulation of the L-arabinose operon in *Escherichia coli*

Robert Schleif

Biology Department, Johns Hopkins University, 3400 N. Charles St, Baltimore, MD 21218, USA

GENERAL BACKGROUND

Escherichia coli can grow on L-arabinose as a source of carbon and energy (Fig. 1) (Schleif, 1996). Not surprisingly, because the amounts of enzymes required to convert arabinose into a component of the pentose phosphate shunt are significant, the levels of the enzymes are regulated. In the absence of arabinose, the uninduced or basal level of expression of the proteins is about 1/300 the induced level. It is a product of one of the genes of the arabinose operon, the AraC protein, that controls the expression level and hence is a sensor of the presence of arabinose. AraC protein transduces information about the sugar's presence into an induced synthesis rate of the AraE, AraF and AraG proteins, which are required for arabinose uptake, as well as the AraB, AraA and AraD enzymes, which are required for the catabolism of arabinose.

Genetic analysis of the arabinose operon was begun more than 40 years ago by Ellis Englesberg (Gross & Englesberg, 1959). In a series of genetic experiments of increasing rigour and elegance, Englesberg and his collaborators then provided rather convincing evidence that the primary activity of AraC protein was in inducing the expression of the other arabinose specific proteins (Sheppard & Englesberg, 1967; Englesberg *et al.*, 1965, 1969b; Gielow *et al.*, 1971); that is, AraC acted positively to turn on expression rather than acting negatively to turn off expression like the *lac* repressor turns off expression of the *lac* operon. That means that the intrinsic set state of the *ara*-specific promoters is off and AraC turns them on, whereas the set state of the *lac* operon promoter is on and the *lac* repressor turns it off. Subsequently, definitive biochemical

SGM symposium 61: Signals, switches, regulons and cascades: control of bacterial gene expression.
Editors D. A. Hodgson, C. M. Thomas. Cambridge University Press. ISBN 0 521 81388 3 ©SGM 2002.

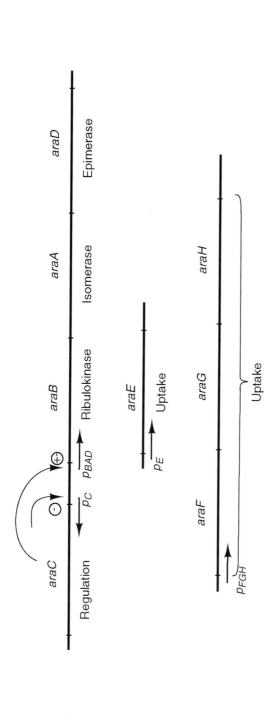

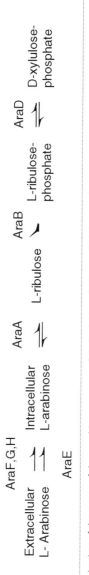

Fig. 1. Organization of the genes of the L-arabinose operon of *Escherichia coli*. The *araCBAD* gene cluster is at 1.45 min on the genetic map, *araE* is at 64.2 min, and *araFGH* is at 42.7 min. Also shown is the metabolic pathway for conversion of L-arabinose to D-xylulose 5-phosphate.

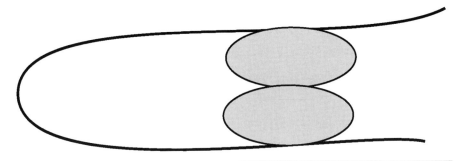

Fig. 2. Schematic diagram of DNA looping where a protein or proteins simultaneously binds to two well-separated sites on DNA and generate a DNA loop.

experiments proved that, indeed, AraC acted positively to turn on expression of the promoter that serves the *araB*, *araA* and *araD* genes, p_{BAD} (Greenblatt & Schleif, 1971).

Englesberg's genetic studies also revealed a secondary activity of AraC protein, that of acting negatively to repress the p_{BAD} promoter (Englesberg *et al.*, 1969a). Due to the implausibility of this negative activity, which by Englesberg's experiments appeared to act from upstream of the promoter, an extensive deletion screen was carried out to confirm the existence of the repressive phenomenon (Schleif, 1972; Schleif & Lis, 1975). This work indicated not only that repression existed, but that the site required for repression lay at least several hundred nucleotides upstream from the promoter. These experiments to verify the existence of the repression from upstream then led to the discovery of DNA looping (Dunn *et al.*, 1984). In looping, AraC protein simultaneously binds to two half-sites that are separated on the DNA by more than 200 bp (Fig. 2). Consequently, a loop is formed in the DNA. The discovery and demonstration of DNA looping came just in time to explain the puzzling properties posed by eukaryotic enhancer elements whose action at a distance properties had just been discovered. DNA looping provided a simple mechanism by which enhancers could act from sites located hundreds or thousands of nucleotides away to influence transcription from a promoter.

MECHANISMS OF REGULATION, DNA LOOPING AND LIGHT SWITCH RESPONSE TO ARABINOSE

Much work has gone into the study of DNA looping and into the mechanism by which AraC protein activates transcription from the *ara* p_{BAD} promoter. A few of the highlights of the earlier work on DNA looping were the development of *in vivo* footprinting and the demonstration of cooperativity in AraC protein binding to the two well-separated DNA-binding sites (Martin *et al.*, 1986), the subsequent discovery of DNA looping in other prokaryotic systems (Dandanell & Hammer, 1985; Eismann *et al.*,

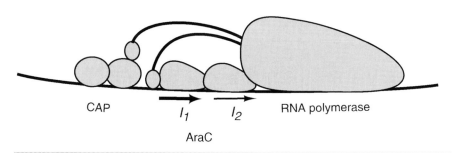

Fig. 3. Interactions thought to occur at the *ara* p_{BAD} promoter between RNA polymerase and AraC protein and CAP protein.

1987), the measurement of the *in vivo* helical pitch of DNA (Lee & Schleif, 1989), where it was found to be a little over 11 bp per turn, size limits on the DNA loop in the *ara* system (Lee & Schleif, 1989), and the construction of an *in vitro* system displaying DNA looping and unlooping caused by the presence of arabinose (Lobell & Schleif, 1990). More recently, AraC protein has been shown, by a gel electrophoretic assay that was developed for the study of multi-protein transcription complexes (Zhang *et al.*, 1996), to stimulate transcription by both assisting the binding of RNA polymerase to the p_{BAD} promoter and speeding the isomerization of RNA polymerase to an open complex. Also, the energetics of AraC protein binding to its half-sites of slightly different sequences and its energetic preference for looping DNA when there is no arabinose present, and its preference for binding to two adjacent sites when arabinose is present have been described thermodynamically (Seabold & Schleif, 1998). One of the two C-terminal domains on the two alpha subunits of RNA polymerase appears to bind to the polymerase distal subunit of AraC and the second likely binds to the cAMP receptor protein (Fig. 3). Mutations have been found in CAP that identify the contact region used by CAP when stimulating p_{BAD} as well as the promoter that is activated by the AraC homologue RhaR (Zhang & Schleif, 1998).

Much of our current understanding of the mechanism of regulation of the arabinose operon in *E. coli* is summarized in Fig. 4, which shows the light switch mechanism that lies at the heart of the protein's arabinose response (Saviola *et al.*, 1998). A monomer of the homodimeric AraC protein consists of a dimerization domain that also binds arabinose. This is loosely connected to a DNA-binding domain that also interacts with RNA polymerase to activate transcription. In the absence of arabinose, AraC protein loops the DNA by binding to the I_1 and O_2 DNA half-sites that are located 210 bp apart. This loop both restricts RNA polymerase access to the p_C and p_{BAD} promoters as well as keeps a DNA-binding domain of AraC from binding to the I_2 half-site. When AraC is bound in this state, the p_{BAD} is off. When arabinose is present, however, p_{BAD} is on.

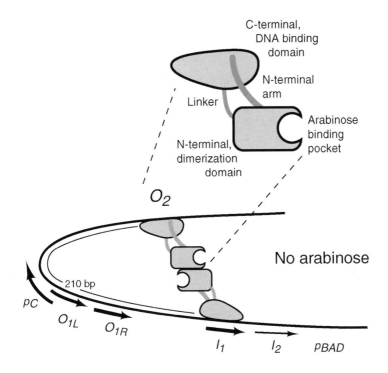

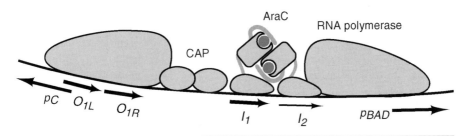

Fig. 4. Light switch mechanism of AraC protein action. In the absence of arabinose, the N-terminal arm interacts with the DNA-binding domains and holds the latter such that the protein loops between $araO_2$ and $araI_1$. This represses both p_C and p_{BAD}. In the presence of arabinose, interactions between arabinose and the N-terminal arms hold the latter over the arabinose, thus freeing the DNA-binding domains. These domains are then free to bind to I_1 and I_2, which then leads to induction of the p_{BAD} promoter. The binding of AraC to O_{1L} and O_{1R}, left and right, directly represses the p_C.

AraC then prefers to bind to the adjacent I_1 and I_2 half-sites. Binding to the I_2 half-site probably results in direct interactions with RNA polymerase, whereas RNA polymerase also contacts the DNA-binding domain bound at I_1 by reaching over the polymerase proximal subunit of AraC. Hence induction ensues because AraC is no longer looping the DNA and because AraC actively promotes transcription by its interactions with RNA polymerase.

AraC protein prefers to loop in the absence of arabinose because an N-terminal arm of about 15 amino acids extends from the dimerization domain and binds to the DNA-binding domain. The combination of the direct connection between the dimerization domains and the DNA-binding domains and the noncovalent arm-mediated connection holds the DNA-binding domains in a relative orientation that favours DNA looping between the O_2 and I_1 half-sites (Seabold & Schleif, 1998). In order for AraC to bind to the adjacent I_1 and I_2 half-sites, AraC would have to be significantly distorted or at least one of its two arms would have to be removed from a DNA-binding domain. Hence the arm–DNA-binding domain interactions make it energetically more favourable for the protein to engage in DNA-looping interactions between the distally located half-sites rather than utilize the adjacent half-sites. When arabinose is present, however, the arms find it energetically more favourable to bind over the arabinose that is bound to the dimerization domains than to bind to the DNA-binding domains. Thus the arms shift position from the DNA-binding domains to the dimerization domains. As a result of the loss of the interactions with the arms, the DNA-binding domains are no longer constrained in their relative orientations. As a consequence of this, the AraC dimer now finds it energetically more favourable to bind to the adjacent half-sites I_1 and I_2 than to bind to I_1 and O_2.

How does the arm determine that arabinose is present on the dimerization domain if it is bound to the DNA-binding domain, and how is the DNA-binding domain at O_2 directed to the I_2 half-site? These two questions are generated due to the static nature of the representation in Fig. 4. In reality, part of the time the arm is bound to the DNA-binding domain, but some of the time it must be bound to the dimerization domain and part of the time it must be free in solution. The fact that its lowest energy state occurs when it is bound to the DNA-binding domain merely means that it spends most of the time at that position. It is in equilibrium with the other states, and likely shifts from one to another on a millisecond to microsecond timescale. Hence, despite the fact that most of the time the arm is bound to the DNA-binding domain, the arm samples arabinose occupancy of the dimerization domain, and can almost instantaneously respond to the presence of arabinose. This same diffusional motion is also the reason that the DNA-binding domain that occupies the O_2 site in the absence of arabinose does not have to be explicitly directed to the I_2 half-site when it is freed from the constraint provided by the arm. Very quickly, probably in the order of milliseconds, the DNA-binding domain can shift from O_2 to I_2.

What binds the arm to the dimerization domain in the presence of arabinose? The binding of arabinose to the dimerization domain could bind the arm to the dimerization domain by either of two mechanisms. Arabinose could change the structure of the dimerization domain and the arm could bind to this altered structure. Alternatively, direct interactions between the arm and arabinose could hold the arm in place. The structures of the dimerization domain that were determined by X-ray crystallography of crystals grown in both the presence and absence of arabinose provide some information on this question, but less than might have been hoped (Soisson *et al.*, 1997). While there is virtually no change in the structure of the dimerization domain in the presence and absence of arabinose, when arabinose is absent, pairs of subunits engage in a French kiss, with each subunit inserting a tyrosine from near the base of the arm into the arabinose-binding pocket of its partner. Conceivably, the tyrosine also generates the same conformational change that is induced by arabinose. More likely, however, is the possibility that arabinose (and tyrosine) does not induce a significant change in the structure of the dimerization. If this is the case, the main determinant of arm position would then be direct arm–arabinose interactions. The interaction energies of such interactions calculated from molecular dynamics simulations of the dimerization domain are consistent with this notion.

EVIDENCE FOR THE LIGHT SWITCH MECHANISM

A number of lines of evidence are consistent with the light switch mechanism for AraC action. The simplest is the effect of deleting the N-terminal arm from AraC. The light switch model predicts that in the absence of the arm, the DNA-binding domains of AraC will behave the same as they do when arabinose is present; that is, armless AraC should make the arabinose operon constitutive. Indeed, that is what was found (Saviola *et al.*, 1998). Deleting the first five amino acids from the arm did nothing, but deleting more residues made the protein induce p_{BAD} in the absence of arabinose. It was concluded that the first four or five amino acids of AraC do not make important contacts with the DNA-binding domain. These same residues were also not structured in the crystals of the dimerization domain, and hence they appear not to play an important role in the function of AraC.

Several lines of evidence whose common theme was rigidity support the light switch model. The simplest was the effect of reversing the orientation of the O_2 half-site. When this was done, taking care to retain its binding face on the same side of the DNA, AraC could no longer form a DNA loop between I_1 and the reversed O_2 half-site (Seabold &

Schleif, 1998). This indicated that the AraC protein itself is rigid in the absence of arabinose and cannot adapt to the reoriented O_2 half-site.

In vivo, AraC was seen to shift from forming a DNA loop to binding to adjacent half-sites when arabinose was added (Martin *et al.*, 1986). *In vitro*, the consequence of adding arabinose was that AraC bound about 30 times more tightly to adjacent half-sites (Hendrickson & Schleif, 1984). This difference reflects the energetic costs of bending AraC with the arms binding to the DNA-binding domain or of releasing at least one of the arms from the DNA-binding domain. If DNA were provided whose half-sites could be freely positioned and orientated just the way AraC would prefer them, both in the presence and absence of arabinose, this energetic difference should be eliminated. Consequently, AraC would bind to such DNA with the same affinity in the presence and absence of arabinose. Such DNA was constructed by connecting double-stranded regions constituting the half-sites with single-stranded, and hence highly flexible, DNA. The DNA for this experiment was constructed by hybridizing two 25 base oligonucleotides to each end of a 75 base oligonucleotide, leaving a long stretch of single-stranded DNA in the middle connecting two double-stranded I_1 half-sites. AraC dissociated from this DNA at a rate independent of the presence of arabinose (Fig. 5) (Harmer *et al.*, 2001). In a parallel experiment, AraC dissociated nearly 30 times faster from normal double-stranded DNA in the absence of arabinose than in the presence of arabinose. Not only was this experimental result consistent with the light switch mechanism, it ruled out the possibility that the arm changes the intrinsic affinity of the DNA-binding domains for DNA or for special sequences of DNA.

According to the light switch mechanism, the two DNA-binding domains of AraC are comparatively free to move and orient themselves in the presence of arabinose. Hence if the two DNA-binding domains of AraC were connected not by dimerization domains, but merely by a peptide linker so that the single polypeptide chain included two DNA-binding domains and a peptide linker, the resulting protein ought to behave like AraC in the presence of arabinose. Indeed, the double DNA-binding domain protein acted *in vivo* to induce p_{BAD} and did not detectably loop between I_1 and O_2 (Harmer *et al.*, 2001). Another way to look at this result is that a signal must be sent from the dimerization domain of AraC to the DNA-binding domains to inform the DNA-binding domains of the presence or absence of arabinose. Formally, one signal could be sent in the presence of arabinose, and a different signal sent in the absence of arabinose. More likely, however, is the possibility that no signal is sent in one situation, and signal is sent in the other situation. The results with double DNA-binding domain protein showed that the no signal state corresponds to the plus arabinose situation.

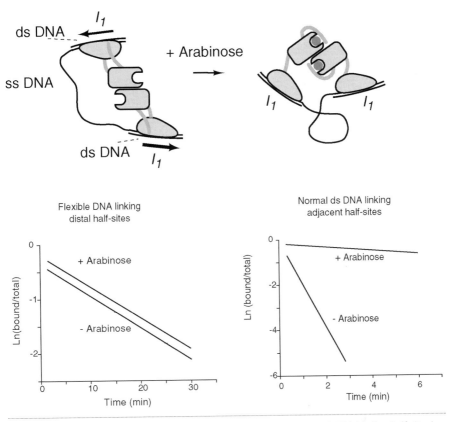

Fig. 5. Dissociation rate of AraC from DNA consisting of flexibly connected DNA binding half-sites is independent of arabinose. The dissociation rate of AraC from normal double-stranded DNA shows a 30-fold arabinose dependence.

Although they were developed for the study of the mechanism of the ligand response of AraC, the two methods of eliminating rigidity in the DNA or in the protein are generally applicable to the study of DNA-binding proteins that respond to ligands. They can determine whether a ligand modulates a protein's DNA-binding affinity by changing the intrinsic affinity of a DNA-binding domain for DNA or whether the ligand changes the relative positioning of two DNA-binding domains with respect to each other. By changing the positioning of DNA subunits, DNA-binding affinity is altered by changing the DNA-binding cooperativity between the two domains. This results from the fact that when one domain binds, the other domain may then be properly positioned for binding itself (positive cooperativity) or improperly positioned (negative cooperativity). Elimination of a ligand-binding domain allows identification of a 'no signal' state of the protein.

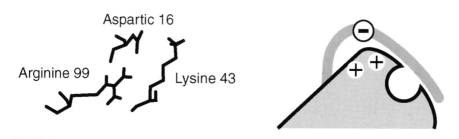

Fig. 6. Schematic drawing of the locations of the aspartic acid residue at position 16 in the N-terminal arm and the charged arginine and lysine residues that lie nearby when the arm binds over the arabinose-binding pocket in the dimerization domain.

The light switch mechanism predicted the existence of a number of different classes of mutations that affect the regulatory properties of AraC. Since the N-terminal arm was postulated to bind to two different sites, it was possible to imagine mutations that either strengthened or weakened its binding to either of these two sites. Furthermore, the mutations could be imagined to lie in either the binding sites on the domains, or in the arm itself. Hence there are eight classes of mutations that might be found that would directly affect arm positioning and which should have easily observable effects on regulation by AraC. Several of these expected classes have been identified (Saviola *et al.*, 1998; Reed & Schleif, 1999).

A mutation in the dimerization domain that binds the N-terminal arm more tightly would tend to make AraC protein constitutive; that is, not to require arabinose for it to be in the state where it binds *cis* and induces transcription from p_{BAD}. The mutation in the arm from asparagine at position 16 to aspartic acid appeared to be such a mutation (Wu & Schleif, 2001). The change made AraC constitutive. Structurally, it was easy to see how the change would lead to the arm binding more tightly to the dimerization domain. In the plus arabinose state, asparagine 16 lies very near the positively charged lysine 43 and arginine 99 residues (Fig. 6). Hence the arm likely is held more tightly to the dimerization domain by electrostatic interactions. Consistent with this idea is the fact that changing any one of the three charged amino acids to uncharged alanine significantly reduced the constitutivity and moved the behaviour of the protein back towards wild-type in that it requires arabinose for induction.

Mutations with a similar effect of binding the arm more tightly have also been identified on the DNA-binding domain (Saviola *et al.*, 1998). A number of these mutations left AraC unable to respond to arabinose, presumably because the arm cannot be pulled off the DNA-binding domain. Many of these introduced a positively charged amino acid, implying that a new electrostatic interaction was created. The arm contains two negatively charged amino acids, aspartic acid at positions 3 and 7. Changing aspartate

7 to alanine relieved the effect of introducing a positive change on the DNA-binding domain. Thus this second site suppressor indicated the existence of a direct interaction between the N-terminal arm and the DNA-binding domain.

The two classes of mutations discussed above both strengthened arm–domain interactions. Mutations that apparently weaken the interactions have also been identified. An example would be a mutation in the DNA-binding domain that leads AraC not to repress by looping DNA, but instead to induce to some degree. Such mutations have also been found (Wu, 2000). Once their existence was shown, an alanine and glutamate scan of the surface of the DNA-binding domain was performed. The scan identified ten different residues whose side chains were required for the full repressive abilities of AraC protein. These residues lie in a path across the side of the domain opposite to the DNA-binding side and suggest that the arm binds along the path.

Surface residues of the DNA-binding domain could be targeted for the scan because two homologues of the DNA-binding domain of AraC, MarA and Rob have been crystallized in the presence of their DNA-binding sites and their structures determined (Rhee *et al.*, 1998; Kwon *et al.*, 2000). The amino acid sequence of AraC could be threaded through these homologous structures to provide an approximate structure of the DNA-binding domain of AraC.

Alanine and glutamic acid were chosen as residues for the scan for two reasons. Alanine possesses a minimal side chain and its introduction rarely disrupts proteins. It might be thought that substitution of alanine for a residue whose side chain is involved in an interaction would weaken the stability of a structure; this is often not the case. The increased flexibility resulting from the absence of a side chain leads to a phenomenon called entropy–enthalpy compensation (Dunitz, 1995), and the resulting free energy differences can be surprisingly small. As a result, residues that are directly involved in interactions may not be identified in alanine scans. The substitution of the large and charged glutamic acid for such a residue almost surely will interfere with an interaction, however. The combination of alanine and glutamate is advantageous because in oligonucleotide-directed mutagenesis, a mixed base oligonucleotide may be used that will introduce alanine or glutamic acid, but no other residues. Then, following mutagenesis, the absence of transformants with altered phenotype means that neither alanine nor glutamic acid at that position generates a change. Any candidates with altered phenotype can be sequenced to determine whether the alteration is due to alanine or glutamic acid.

Finally, direct physical evidence for the light switch mechanism was obtained by observing binding between a peptide with a sequence of the N-terminal arm, and a dimerization domain that was deleted of the arm. These experiments were carried out with a Biacore machine, which uses plasmon resonance to measure very small concentration changes within about 100 Å of a surface. Peptide with a sequence of the N-terminal arm of AraC was immobilized in the surface, and solution containing the arm-deleted dimerization domain of AraC was flowed past (Ghosh & Schleif, 2001). Binding of the dimerization domain to the immobilized arm was seen only when arabinose was also present. No binding was seen when glucose was substituted for arabinose, and intermediate binding was seen when fucose, an analogue of arabinose, was provided. Similarly, using an arm peptide containing the asparagine to aspartate change at position 16 showed tighter binding between the peptide and the domain. A mutation, phenylalanine 15 to leucine, is uninducible *in vivo* and likely is a result of weaker arm–arabinose binding. Dimerization domain did not detectably bind to this peptide.

SUMMARY

In summary, two important and general biochemical mechanisms have been discovered in the studies of the arabinose system. These are the phenomenon of DNA looping, and the light switch mechanism for regulation of protein activity. The consequences of DNA looping have been extensively discussed (Schleif, 1988, 1992) and will not be further addressed here. More should be said about the light switch mechanism, however. The mechanism is a special case of arm–domain interactions. While such interactions in proteins have not been given much attention, they are being found in an increasing number of systems. Such interactions provide a simple way for evolution to add interactions with one protein to another protein or to a whole class of proteins (Schleif, 1999). An arm sequence that binds to one protein might be added to many proteins, thus enabling each of them to bind to the first protein. Several examples for this phenomenon are that the proteins in the prokaryotic and in the eukaryotic DNA replication apparatus bind to one another via arm–domain interactions. Cell sorting and transport proteins also use this interaction mechanism.

Arm–domain interactions are also attractive for the engineering of new protein–protein interactions. While it is difficult to imagine engineering the surfaces of two proteins so that they will now bind to one another, it is simple to imagine finding a peptide that will bind to one of the protein's surfaces. This peptide could then be added to the other protein. Such peptides may prove to be readily identified through the use of phage display libraries. It is even possible to imagine designing an allosteric regulation mechanism based on the light switch mechanism. If part of the peptide arm of AraC activated or inhibited an enzyme to which it bound, then the addition of arabinose would change the enzyme's activity by pulling the arm off the enzyme and onto the dimerization domain of AraC.

REFERENCES

Dandanell, G. & Hammer, K. (1985). Two operator sites separated by 599 base pairs are required for *deoR* repression of the *deo* operon of *Escherichia coli. EMBO J* **4**, 3333–3338.

Dunitz, J. D. (1995). Win some, lose some: enthalpy-entropy compensation in weak intermolecular interactions. *Chem Biol* **2**, 709–712.

Dunn, T., Hahn, S., Ogden, S. & Schleif, R. (1984). An operator at −280 base pairs that is required for repression of *araBAD* operon promoter: addition of DNA helical turns between the operator and promoter cyclically hinders repression. *Proc Natl Acad Sci U S A* **81**, 5017–5020.

Eismann, E., von Wilcken-Bergmann, B. & Müller-Hill, B. (1987). Specific destruction of the second *lac* operator decreases repression of the *lac* operon in *Escherichia coli* five-fold. *J Mol Biol* **195**, 949–952.

Englesberg, E., Irr, J., Power, J. & Lee, N. (1965). Positive control of enzyme synthesis by gene C in the L-arabinose system. *J Bacteriol* **90**, 946–957.

Englesberg, E., Squires, C. & Meronk, F. (1969a). The L-arabinose operon in *Escherichia coli* B/r: a genetic demonstration of two functional states of the product of a regulator gene. *Proc Natl Acad Sci U S A* **62**, 1100–1107.

Englesberg, E., Sheppard, D., Squires, C. & Meronk, F., Jr (1969b). An analysis of "revertants" of a deletion mutant in the C gene of the L-arabinose gene complex in *Escherichia coli* B/r: isolation of initiator constitutive mutants (Ic). *J Mol Biol* **43**, 281–298.

Ghosh, M. & Schleif, R. (2001). Biophysical evidence of arm-domain interactions in AraC. *Anal Biochem* **295**, 107–112.

Gielow, L., Largen, M. & Englesberg, E. (1971). Initiator constitutive mutants of the L-arabinose operon (OIBAD) of *Escherichia coli* B/r. *Genetics* **69**, 289–302.

Greenblatt, J. & Schleif, R. (1971). Regulation of the arabinose operon *in vitro. Nat New Biol* **233**, 166–170.

Gross, J. & Englesberg, E. (1959). Determination of the order of mutational sites governing L-arabinose utilization in *Escherichia coli* B/r by transduction with phage P1bt. *Virology* **9**, 314–331.

Harmer, T., Wu, M. & Schleif, R. (2001). The role of rigidity in DNA looping-unlooping by AraC. *Proc Natl Acad Sci U S A* **98**, 427–431.

Hendrickson, W. & Schleif, R. (1984). Regulation of the *Escherichia coli* L-arabinose operon studied by gel electrophoresis DNA binding assay. *J Mol Biol* **178**, 611–628.

Kwon, H. J., Bennik, M. H., Demple, B. & Ellenberger, T. (2000). Crystal structure of the *Escherichia coli* Rob transcription factor in complex with DNA. *Nat Struct Biol* **7**, 424–430.

Lee, D. & Schleif, R. (1989). *In vitro* DNA loops in *araCBAD*: size limits and helical repeat. *Proc Natl Acad Sci U S A* **86**, 476–480.

Lobell, R. & Schleif, R. (1990). DNA looping and unlooping by AraC protein. *Science* **250**, 528–532.

Martin, K., Huo, L. & Schleif, R. (1986). The DNA loop model for *ara* repression: AraC protein occupies the proposed loop sites in vivo and repression-negative mutations lie in these same sites. *Proc Natl Acad Sci U S A* **83**, 3654–3658.

Reed, W. & Schleif, R. (1999). Hemiplegic mutations in AraC protein. *J Mol Biol* **294**, 417–425.

Rhee, S., Martin, R. G., Rosner, J. L. & Davies, D. R. (1998). A novel DNA-binding motif in

MarA: the first structure for an AraC family transcriptional activator. *Proc Natl Acad Sci U S A* **95**, 10413–10418.

Saviola, B., Seabold, R. & Schleif, R. (1998). Arm-domain interactions in AraC. *J Mol Biol* **278**, 539–548.

Schleif, R. (1972). Fine-structure deletion map of the *Escherichia coli* L-arabinose operon. *Proc Natl Acad Sci U S A* **69**, 3479–3484.

Schleif, R. (1988). DNA looping. *Science* **240**, 127–128.

Schleif, R. (1992). DNA looping. *Annu Rev Biochem* **61**, 199–223.

Schleif, R. (1996). Two positively regulated systems, *ara* and *mal*. In *Escherichia coli and Salmonella: Cellular and Molecular Biology*, pp. 1300–1309. Edited by F. C. Neidhardt, R. Curtiss, III, J. L. Ingraham, E. C. C. Lin, K. B. Low, B. Magasanik, W. S. Reznikoff, M. Riley, M. Schaechter & H. E. Umbarger. Washington, DC: American Society for Microbiology.

Schleif, R. (1999). Arm-domain interactions in proteins: a review. *Proteins* **34**, 1–3.

Schleif, R. & Lis, J. T. (1975). The regulatory region of the L-arabinose operon: a physical, genetic and physiological study. *J Mol Biol* **95**, 417–431.

Seabold, R. & Schleif, R. (1998). Apo-AraC actively seeks to loop. *J Mol Biol* **278**, 529–538.

Sheppard, D. & Englesberg, E. (1967). Further evidence for positive control of the L-arabinose system by gene *araC. J Mol Biol* **25**, 443–454.

Soisson, S., MacDougall-Shackleton, B., Schleif, R. & Wolberger, C. (1997). Structural basis for ligand-regulated oligomerization of AraC. *Science* **276**, 421–425.

Wu, M. (2000). *Arm domain interactions in AraC*. PhD thesis, Johns Hopkins University.

Wu, M. & Schleif, R. (2001). Strengthened arm-dimerization domain interactions in AraC. *J Biol Chem* **276**, 2562–2564.

Zhang, X. & Schleif, R. (1998). Catabolite gene activator protein mutations affecting activity of the *araBAD* promoter. *J Bacteriol* **180**, 195–200.

Zhang, X., Reeder, T. & Schleif, R. (1996). Transcription activation parameters at *ara* p_{BAD}. *J Mol Biol* **258**, 14–24.

Transcription termination control in bacteria

Tina M. Henkin

Department of Microbiology, The Ohio State University, 484 W. 12th Avenue, Columbus, OH 43210, USA

INTRODUCTION

A variety of systems for control of gene expression at the level of premature termination of transcription have been identified in bacteria. Genes regulated by mechanisms of this type are characterized by the presence of signals for transcription termination located in the leader region of the transcript, upstream of the start of the regulated coding sequences. The activity of the leader region transcriptional terminator can be modulated in response to a regulatory signal, so that readthrough, and synthesis of the full-length transcript, occurs only under appropriate conditions. Mechanisms for regulation of terminator function include modification of the processivity of RNA polymerase (RNAP), such that termination signals are no longer recognized, or alterations in the leader RNA, so that termination signals are unable to form or are inaccessible to the required factors. This chapter will review the best-characterized termination control systems, with a focus on how subtle variations in the molecular mechanism can be exploited to allow responses to a variety of physiological signals. Emphasis will be placed on the most recent literature; additional references are available in previous reviews (Henkin, 1996, 2000; Weisberg & Gottesman, 1999; Landick et al., 1996).

TRANSCRIPTION TERMINATION AND ITS CONTROL

Recent advances in the analysis of the structure and function of RNAP are providing new insight into how transcription termination occurs, and how it can be modulated. While the molecular details of transcription elongation and termination are the subject of active investigation (Richardson & Greenblatt, 1996; Landick, 1999; Korzheva et

SGM symposium 61: Signals, switches, regulons and cascades: control of bacterial gene expression.
Editors D. A. Hodgson, C. M. Thomas. Cambridge University Press. ISBN 0 521 81388 3 ©SGM 2002.

al., 2000), many features of the process are reasonably clear. Two classes of transcriptional termination sites have been described, both of which require pausing of RNAP in response to signals in the nascent transcript. Intrinsic terminators consist of a GC-rich helical region followed by a run of U residues in the nascent transcript; formation of the helix results in destabilization of the elongation complex, followed by release of RNAP from the template DNA and the nascent transcript. In contrast, factor-dependent terminators are dependent on interaction of a protein factor (designated Rho) with a *rut* (Rho utilization) site on the nascent transcript; Rho then contacts the elongation complex and triggers transcription termination.

The activity of both intrinsic and Rho-dependent terminators can be affected by alterations in the processivity of RNAP (Weisberg & Gottesman, 1999). Intrinsic terminators can also be controlled by alternative folding of the nascent transcript, to promote or prevent formation of the terminator helix, while Rho-dependent terminators can be controlled by blocking access of Rho protein to the *rut* site (Yanofsky, 2000). Modulation of RNAP processivity and leader RNA structure can be conferred by *trans*-acting proteins, *cis*-acting RNA elements, *trans*-acting RNAs, and translation of the leader RNA, which in turn allows participation by the nascent peptide. Each of these regulatory factors can in principle act either positively to promote readthrough of the leader region terminator or negatively to promote termination. In addition, many of these regulatory factors permit responses to physiological signals, by interactions with effector molecules or additional regulatory proteins. Classes of termination control mechanisms, and well-characterized examples, are summarized in Fig. 1. These examples, and variations on the themes they illustrate, are discussed below.

Protein-directed antitermination

Phage λ N and Q. Bacteriophage λ utilizes sequential readthrough of a series of transcriptional terminators to provide temporal control of gene expression during the lytic cycle. The first of these events is dependent on the phage-encoded N protein, which binds to two RNA sites (*nut*$_L$ and *nut*$_R$) to direct assembly of a complex of host-encoded proteins (NusA, NusB, NusG and ribosomal protein S10) that interact with RNAP to convert it to a highly processive form, resistant to downstream pause and termination signals (Friedman & Court, 2001; Fig. 1a). An interesting variation on the λ system is found in the related phage H-19B, which uses a different type of *nut* site, and does not require NusB or S10 (Neely & Friedman, 2000); this suggests that the basic type of mechanism for which λ N is the paradigm may be subject to multiple variations.

The transition to late gene expression in phage λ requires a second antitermination event, mediated by the Q protein. Association of Q with RNAP requires transcriptional pausing early in the transcript, which is dependent on the σ subunit of RNAP (Ko *et al.*, 1998; Yarnell & Roberts, 1999). Unlike N-dependent antitermination, Q utilizes only a single host-encoded factor, NusA. Analysis of both the N and Q systems has provided important insight into the biochemistry of transcription elongation and termination.

The RfaH system. Expression of a number of genes in *Salmonella typhimurium* and *Escherichia coli* is controlled at the level of transcription termination by the RfaH protein. These genes are generally involved in virulence or plasmid transfer, and contain a common sequence, designated *ops* (operon polarity suppressor), which acts as a pause signal for RNAP (Artsimovitch & Landick, 2000). Interaction of RfaH with the *ops* sequence results in persistent readthrough of both intrinsic and Rho-dependent transcriptional terminators, suggesting that RNAP is modified into a termination-resistant form (Bailey *et al.*, 2000). RfaH is related to the NusG protein, a component of the normal transcriptional machinery (Bailey *et al.*, 1997). It therefore appears that the RfaH system represents a bacterial system functionally analogous to the antitermination systems of the lambdoid phages, and analysis of this system is likely to provide further insight into the biochemistry of transcription elongation and termination.

Csp proteins. The *E. coli* cold-shock proteins CspA, CspE and CspC are RNA-binding proteins that promote translation by disrupting RNA secondary structure. These proteins also cause readthrough of intrinsic terminators, both *in vivo* and *in vitro* (Bae *et al.*, 2000b). Cold-shock conditions result in increased levels of the Csp proteins, suggesting that regulation of target genes is mediated directly by Csp protein availability. Although the mechanism of antitermination is unknown, destabilization of the terminator helix is a reasonable model.

The BglG system. The *E. coli bgl* operon is the paradigm for a set of sugar utilization operons in a variety of bacteria regulated by a common mechanism. The leader regions of these genes contain alternative terminator and antiterminator structures; the terminator represents the more stable structure, so that transcription terminates in the absence of the regulatory signal, conferred by the presence of the sugar substrate (Amster-Choder & Wright, 1993). Readthrough is dependent on binding of an antiterminator protein (BglG and its homologues) to the leader RNA. Binding of BglG stabilizes the antiterminator form of the leader, preventing terminator formation (Fig. 1a). The availability of the substrate sugar is monitored by the transport system for that sugar (Amster-Choder & Wright, 1997; Chen *et al.*, 2000). In the absence of the substrate, the sugar-specific phosphotransferase protein (BglF) phosphorylates the antiterminator protein (BglG), inactivating the antiterminator protein. In the presence of the sugar, BglF phosphorylates the sugar and dephosphorylates BglG, allowing dimerization and antitermination function. Although the antitermination activity of BglG can be

Modification of RNAP

Alternative folding of RNA

(a) Protein-directed antitermination

λ N

E. coli bgl

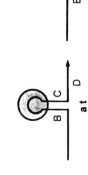

(b) Protein-directed termination

HK022 Nun

B. subtilis trp

(c) RNA-directed antitermination

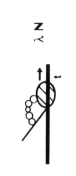

HK022 *put*

B. subtilis tyrS
(T Box)

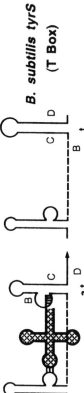

(d) RNA-directed termination

??

pT181 *repC*

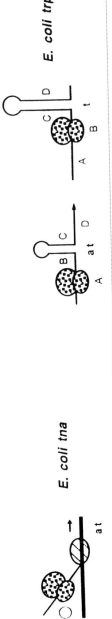

(e) Translation-directed antitermination

E. coli tna

E. coli trp

Fig. 1. Mechanisms for transcription termination control. (a) Protein-directed antitermination can operate by modification of the transcription elongation complex by binding of proteins such as λ N and N-recruited host factors (small circles) to the nascent transcript (diagonal line) and RNAP (hatched ellipse) to direct readthrough (arrow) of termination signals (t). RNA-binding proteins like *E. coli* BglG (shaded circle) can also direct readthrough by stabilizing the antiterminator form (at; B : C pairing) of the RNA, blocking formation of the competing terminator form (t; C : D pairing). (b) Protein-directed termination can operate by modification of RNAP (hatched ellipse) by HK022 Nun and host factors (small circles). Proteins can also control the leader RNA structure, as in the *B. subtilis trp* operon, where binding of TRAP protein (shaded circle) destabilizes the antiterminator form (at; B : C pairing), allowing formation of the terminator (t; C : D pairing). (c) RNA-directed termination can operate by a direct interaction of the nascent transcript (curved line) with RNAP (hatched ellipse), converting RNAP into a form resistant to termination sites (t), as in the HK022 *put* system. Alternatively, an interaction between the leader RNA and a second RNA molecule (uncharged tRNA for the T box system) can direct folding of the leader RNA into an antiterminator form (at; B : C pairing) instead of the competing terminator form (t; C : D pairing). (d) RNA-directed termination, such as the pT181 *repC* antisense control system, utilizes a *trans*-acting RNA (dotted line) to block formation of the antiterminator form (at; B : C pairing), allowing formation of the competing terminator (t; C : D pairing); RNA-directed modification of RNAP to a termination-prone form has not yet been demonstrated. (e) Translation of the nascent RNA allows the translating ribosome (dotted ovals) to affect the behaviour of RNAP. In the *E. coli tna* operon, the ribosome, stalled by the nascent peptide, blocks access of transcription termination factor Rho (small circle). In the *E. coli trp* operon, pausing of the ribosome at region A, due to low availability of charged tRNA^Trp, results in antiterminator formation (at; B : C pairing) rather than terminator formation (t; C : D pairing).

explained on the basis of alternative folding of the leader RNA, an interaction between BglG and the β' subunit of RNAP has also been reported, suggesting additional complexities in the regulatory mechanism (Nussbaum-Schochat & Amster-Choder, 1999). A number of systems that use analogous proteins and binding sites, such as the *Bacillus subtilis sac* genes (Declerck *et al.*, 1999), have also been described. The *B. subtilis glp* system, involved in glycerol utilization, shares some of the features of these systems, although the antiterminator protein GlpP is unrelated to the BglG family (Glatz *et al.*, 1998).

The *nas* operon. The genes for nitrate assimilation in *Klebsiella oxytoca* are regulated by transcription termination using a protein factor, NasR, that binds to a stem–loop structure in the leader region of the *nasF* operon to promote readthrough of an intrinsic terminator (Chai & Stewart, 1998, 1999). In this case, there is no evidence for competition between the NasR-binding region and the terminator region of the leader, and the mechanism by which NasR binding blocks terminator function is unknown.

The amidase system. The *Pseudomonas aeruginosa* amidase operon is similar to the *K. oxytoca nas* operon in that binding of a regulatory protein, AmiR, to the leader is required for antitermination, although the mechanism of antitermination is unknown. This system has the added element of a second regulatory protein, AmiC, which acts to block AmiR activity (O'Hara *et al.*, 1999; Norman *et al.*, 2000). In the presence of the inducer acetamide, AmiR is active in RNA binding and antitermination. The corepressor butyramide instead triggers formation of an AmiC–AmiR complex that is inactive in antitermination. The regulation of AmiR function by a second protein, which is responsible for detection of the effector, is similar to the regulation of BglG activity by BglF; however, AmiC regulates AmiR function by formation of a stable complex rather than by phosphorylation.

Protein-directed transcription termination

Phage HK022 Nun. Phage HK022, which is related to λ, encodes a protein, designated Nun, in place of N protein. Nun promotes transcription termination by other lambdoid phages, by binding both to the *nut* site on the nascent transcript and also to the template DNA (Watnick & Gottesman, 1999). Nun binding is dependent on NusA, one of the host factors required for N-dependent antitermination (Watnick & Gottesman, 1998; Fig. 1b). Nun provides a competitive advantage for the phage, since HK022 uses an N-independent antitermination mechanism for its own gene expression (see below), and Nun is expressed even in HK022 lysogens, preventing superinfection by other lambdoid phages (King *et al.*, 2000).

TRAP. The *B. subtilis trp* genes are regulated by transcription termination using an RNA-binding protein, designated TRAP, that binds to the leader RNA to prevent formation of a stable antiterminator structure, thereby allowing formation of the less stable terminator (Babitzke, 1997; Fig. 1b). TRAP, in the presence of the corepressor tryptophan, forms an 11-mer ring, each subunit of which interacts with one of 11 closely spaced GAG or AAG trimers in the leader (Antson *et al.*, 1999); binding of the TRAP–tryptophan complex to the leader RNA is further stimulated by an additional stem–loop at the 5′ end of the transcript (Du *et al.*, 2000). TRAP-dependent repression is further modulated by the *yczA–ycbK* operon, expression of which is induced by uncharged tRNA^{Trp}, via the T box transcription antitermination mechanism (see below), so that *trp* gene expression in *B. subtilis* responds to both tryptophan and tRNA^{Trp} charging (Sarsero *et al.*, 2000). A dual response is also observed for *trp* gene regulation in *E. coli*, although in that case the direct response to tryptophan is at the level of transcription initiation while the response to tRNA^{Trp} is at the level of transcription termination, but by a mechanism different from that observed in *B. subtilis* (Landick *et al.*, 1996; see below).

PyrR. The *B. subtilis* pyrimidine biosynthesis genes are regulated by a system similar to the TRAP system in that binding of a regulatory protein to the leader region of the *pyr* operon mRNA promotes transcription termination. In this case, the regulatory protein, PyrR, in the presence of its corepressor, binds to an RNA element that serves as an anti-antiterminator, since it competes with the antiterminator (Switzer *et al.*, 1999). Stabilization of the anti-antiterminator therefore prevents antiterminator formation, allowing the terminator to form.

The S box regulon. A conserved set of leader sequence and structural elements was identified upstream of multiple genes involved in methionine metabolism in certain Gram-positive organisms, including *Bacillus*, *Clostridium* and *Staphylococcus* species (Grundy & Henkin, 1998). These leaders include an intrinsic terminator and competing antiterminator, as well as an additional structure that competes with the antiterminator and appears to function as an anti-antiterminator. The leader region terminator is active when cells are grown in the presence of methionine, and readthrough is induced by methionine starvation. The anti-antiterminator is postulated to serve as the binding site for an unknown repressor protein, since mutations that disrupt this element result in constitutive expression. The mechanism by which methionine is sensed and the leader region structure is modulated is not yet known. This system is most like the *B. subtilis pyr* system in that stabilization of an anti-antiterminator is required for premature termination of transcription.

RNA-directed antitermination

Phage HK022 *put* RNA. λ N function in directing the transition from early to delayed early gene expression is replaced in phage HK022 by an RNA-directed transcription antitermination system (Fig. 1c). A *put* (polymerase utilization) sequence in the nascent transcript is required for readthrough of subsequent terminators, but no phage- or host-encoded proteins appear to be required (Banik-Maiti *et al.*, 1997). Specific mutations in the β' subunit of RNAP interfere with antitermination, and a direct interaction between RNAP and the *put* RNA that results in persistent antitermination through downstream termination sites has been postulated (Clerget *et al.*, 1995; Sen *et al.*, 2001).

T box system. Many aminoacyl-tRNA synthetase, amino acid biosynthesis and transporter genes in Gram-positive bacteria are regulated by a common transcription termination control system, in which each gene responds specifically to the charging ratio of the cognate tRNA. Multiple genes in the same organism can be regulated by this mechanism, with the specificity of the regulatory response determined primarily by the presence of a single codon (the specifier sequence) embedded at a precise position within the leader region structure (Grundy & Henkin, 1993; Pelchat & Lapointe, 1999). This codon interacts with the anticodon of the cognate tRNA. A second interaction between the acceptor end of uncharged tRNA and a bulge in the antiterminator is proposed to stabilize the antiterminator, preventing formation of the competing intrinsic terminator (Grundy *et al.*, 1994; Fig. 1c). Genes regulated by this mechanism can be recognized by conservation of primary sequence and structural elements within the leader, in particular the T box sequence, a highly conserved 14 nucleotide element that forms the 5' side of the antiterminator. Mutational analysis of the *B. subtilis tyrS* leader demonstrated that all of the conserved elements are required for efficient antitermination (Rollins *et al.*, 1997), while much of the tRNA^{Tyr} sequence could be varied if the tertiary structure is maintained (Grundy *et al.*, 2000). Many of the features demonstrated for the *tyrS* system have been confirmed in other T box genes (for example, van de Guchte *et al.*, 1998; Luo *et al.*, 1997). The role of most of the conserved leader elements and possible requirements for host factors in addition to the tRNA remain to be explored.

RNA-directed transcription termination

Plasmid control systems. A number of plasmids utilize regulatory RNAs to control key functions; some of these RNAs have been demonstrated to operate by stimulating transcriptional termination, thereby downregulating expression of regulatory determinants. For example, binding of the antisense RNA of plasmid pT181 to the nascent transcript of the *repC* gene is postulated to confer alternative folding of the transcript, resulting in formation of an intrinsic terminator and reduction in synthesis of RepC protein, the limiting factor in pT181 replication (Brantl & Wagner, 2000; Fig. 1d). The response of transfer functions of *Enterococcus faecalis* plasmids pAD1 and pCF10 to pheromone is also dependent on regulatory RNAs. Overexpression of the mD small RNA of pAD1 results in increased termination at a target intrinsic terminator, blocking expression of downstream *tra* genes required for conjugation (Tomita & Clewell, 2000). The mechanism of action of the mD RNA is not yet understood. A similar role for the Qa RNA has been proposed for pCF10, although in this case interaction with the PrgX regulatory protein appears to be required for downregulation of *tra* gene expression (Bae *et al.*, 2000a). No examples of RNAs promoting persistent antitermination by modification of RNAP have yet been described.

Translation of the nascent transcript

E. coli trp **attenuation.** Many amino acid biosynthesis operons in enteric bacteria are controlled at the level of transcription termination, using a mechanism in which coupling of transcription with translation of the nascent RNA is a key component (Landick *et al.*, 1996). These systems are characterized by the presence in the leader region of a short open reading frame which contains several codons for the amino acid to which expression of the operon responds. When the amino acid is abundant, the charging ratio of the cognate tRNA is high and leader peptide translation is efficient; when the regulatory amino acid is limiting, the ribosome stalls at these codons due to low availability of charged tRNA. The position of the ribosome in the leader controls the folding of the leader RNA so that either an intrinsic terminator or an alternative antiterminator will form; the terminator forms when the amino acid is abundant, while stalling of the ribosome allows formation of the antiterminator (Fig. 1e). The *E. coli trp* and *Salmonella typhimurium his* operons are the paradigm examples of this type of system, but it is also used for many other amino acid and aminoacyl-tRNA synthetase genes in Gram-negative bacteria. It is interesting to note that this type of system appears to be rare in Gram-positive bacteria, and the presence of the expected sequence features does not always predict regulatory function (Craster *et al.*, 1999). The T box system, which is used for the same classes of genes in Gram-positive organisms, also senses tRNA charging, but by a direct interaction between the leader RNA and uncharged tRNA, in the absence of translation.

tna. The *E. coli tna* operon, the product of which is responsible for degradation of tryptophan, is induced during growth in the presence of high concentrations of tryptophan. Expression of the *tnaA* gene, encoding tryptophanase, is dependent on readthrough of a leader region Rho-dependent terminator. Tryptophan-induced readthrough requires synthesis of a leader peptide, TnaC, which acts *in cis* to stall the ribosome at the *tnaC* stop codon; the stalled ribosome blocks access of Rho to the *rut* site, so that termination is prevented and transcription elongation continues, allowing synthesis of the full-length mRNA (Konan & Yanofsky, 2000; Gong & Yanofsky, 2001; Fig. 1e). The molecular basis for the response to tryptophan availability is not yet understood. The *tna* regulatory system is similar to the mechanism used for *E. coli trp* biosynthesis genes in that leader peptide translation is a key component, but the molecular mechanism for transcription termination control is very different in the two cases.

CONCLUSION

Although regulation at the level of transcription initiation is generally perceived as the dominant feature of gene regulation, it is apparent that regulation at other levels, including premature termination of transcription, is of major importance in a variety of systems. It is especially notable in the examples described above that transcription termination control is of great importance in Gram-positive bacteria; it is highly likely that additional systems will be uncovered in the near future. This effort is greatly facilitated by the availability of genomic sequence data, since transcriptional terminators can often be predicted in raw sequence data. The S box system is an example of a global termination control system initially uncovered on the basis of conserved sequence and structural elements. It is also evident that investigation of a variety of experimental systems will reveal new variations on this general regulatory theme.

ACKNOWLEDGEMENTS

Frank J. Grundy is thanked for helpful discussions and critical reading of the manuscript. Work in the author's laboratory is supported by National Institutes of Health grant GM47823.

REFERENCES

Amster-Choder, O. & Wright, A. (1993). Transcriptional regulation of the *bgl* operon of *Escherichia coli* involves phosphotransferase system-mediated phosphorylation of a transcriptional antiterminator. *J Cell Biochem* **51**, 83–90.

Amster-Choder, O. & Wright, A. (1997). BglG, the response regulator of the *Escherichia coli bgl* operon, is phosphorylated on a histidine residue. *J Bacteriol* **179**, 5621–5624.

Antson, A. A., Dodson, E. J., Dodson, G., Greaves, R. B., Chen, X. & Gollnick, P. (1999). Structure of the *trp* RNA binding attenuation protein, TRAP, bound to RNA. *Nature* **401**, 235–242.

Artsimovitch, I. & Landick, R. (2000). Pausing by bacterial RNA polymerase is mediated by mechanistically distinct classes of signals. *Proc Natl Acad Sci U S A* **97**, 7090–7095.

Babitzke, P. (1997). Trp-ing the TRAP or how *Bacillus subtilis* reinvented the wheel. *Mol Microbiol* **26**, 1–9.

Bae, T., Clerc-Bardin, S. & Dunny, G. M. (2000a). Analysis of expression of *prgX*, a key negative regulator of the transfer of the *Enterococcus faecalis* pheromone-inducible plasmid pCF10. *J Mol Biol* **297**, 861–875.

Bae, W., Xia, B., Inouye, M. & Severinov, K. (2000b). *Escherichia coli* CspA-family RNA chaperones are transcription antiterminators. *Proc Natl Acad Sci U S A* **97**, 7784–7789.

Bailey, M. J., Hughes, C. & Koronakis, V. (1997). RfaH and the *ops* element, components of a novel system controlling bacterial transcription elongation. *Mol Microbiol* **26**, 845–851.

Bailey, M. J., Hughes, C. & Koronakis, V. (2000). *In vitro* recruitment of the RfaH regulatory protein into a specialized transcription complex, directed by the nucleic acid *ops* element. *Mol Gen Genet* **262**, 1052–1059.

Banik-Maiti, S., King, R. A. & Weisberg, R. A. (1997). The antiterminator RNA of phage HK022. *J Mol Biol* **272**, 677–687.

Brantl, S. & Wagner, E. G. H. (2000). Antisense RNA-mediated transcriptional attenuation: an *in vitro* study of plasmid pT181. *Mol Microbiol* **35**, 1469–1482.

Chai, W. & Stewart, V. (1998). NasR, a novel RNA-binding protein, mediates nitrate-responsive transcription antitermination of the *Klebsiella oxytoca* M5al *nasF* operon leader *in vitro*. *J Mol Biol* **283**, 339–351.

Chai, W. & Stewart, V. (1999). RNA sequence requirements for NasR-mediated, nitrate responsive transcription antitermination of the *Klebsiella oxytoca* M5al *nasF* operon leader. *J Mol Biol* **292**, 203–216.

Chen, Q., Postma, P. W. & Amster-Choder, O. (2000). Dephosphorylation of the *Escherichia coli* transcriptional antiterminator BglG by the sugar sensor BglF is the reversal of its phosphorylation. *J Bacteriol* **182**, 2033–2036.

Clerget, M., Hin, D. J. & Weisberg, R. A. (1995). A zinc binding region in the β' subunit of RNA polymerase is involved in antitermination of early transcription of phage HK022. *J Mol Biol* **248**, 768–780.

Craster, H. L., Potter, C. A. & Baumberg, S. (1999). End-product control of expression of branched-chain amino acid biosynthesis genes in *Streptomyces coelicolor* A3(2): paradoxical relationships between DNA sequence and regulatory phenotype. *Microbiology* **145**, 2375–2384.

Declerck, N., Vincent, F., Hoh, F., Aymerich, S. & van Tilbeurgh, H. (1999). RNA recognition by transcriptional antiterminators of the BglG/SacY family: functional and structural comparison of the CAT domain from SacY and LicT. *J Mol Biol* **294**, 389–402.

Du, H., Yakhnin, A. V., Dharmaraj, S. & Babitzke, P. (2000). *trp* RNA-binding attenuation protein-5′ stem-loop RNA interaction is required for proper transcription attenuation control of the *Bacillus subtilis trpEDCFBA* operon. *J Bacteriol* **182**, 1819–1827.

Friedman, D. I. & Court, D. L. (2001). Bacteriophage lambda: alive and well and still doing its thing. *Curr Opin Microbiol* **4**, 201–207.

Glatz, E., Persson, M. & Rutberg, B. (1998). Antiterminator protein GlpP of *Bacillus subtilis* binds to *glpD* leader RNA. *Microbiology* **144**, 449–456.

Gong, F. & Yanofsky, C. (2001). Reproducing *tna* operon regulation *in vitro* in an S-30 system. Tryptophan induction inhibits cleavage of TnaC peptidyl-tRNA. *J Biol Chem* **276**, 1974–1983.

Grundy, F. J. & Henkin, T. M. (1993). tRNA as a positive regulator of transcription antitermination in *B. subtilis*. *Cell* **74**, 475–482.

Grundy, F. J. & Henkin, T. M. (1998). The S box regulon: a new global transcription termination control system for methionine and cysteine biosynthesis genes in Gram-positive bacteria. *Mol Microbiol* **30**, 737–749.

Grundy, F. J., Rollins, S. M. & Henkin, T. M. (1994). Interaction between the acceptor end of tRNA and the T box stimulates antitermination in the *Bacillus subtilis tyrS* gene: a new role for the discriminator base. *J Bacteriol* **176**, 4518–4526.

Grundy, F. J., Collins, J. A., Rollins, S. M. & Henkin, T. M. (2000). tRNA determinants for transcription antitermination of the *Bacillus subtilis tyrS* gene. *RNA* **6**, 1131–1141.

van de Guchte, M., Ehrlich, S. D. & Chopin, A. (1998). tRNA^Trp as a key element of antitermination in the *Lactococcus lactis trp* operon. *Mol Microbiol* **29**, 61–74.

Henkin, T. M. (1996). Control of transcription termination in prokaryotes. *Annu Rev Genet* **30**, 35–57.

Henkin, T. M. (2000). Transcription termination control in bacteria. *Curr Opin Microbiol* **3**, 149–153.

King, R. A., Madsen, P. L. & Weisberg, R. A. (2000). Constitutive expression of a transcription termination factor by a repressed prophage: promoters for transcribing the phage HK022 *nun* gene. *J Bacteriol* **182**, 456–462.

Ko, D. C., Marr, M. T., Guo, J. & Roberts, J. W. (1998). A surface of *Escherichia coli* σ^{70} required for promoter function and antitermination by phage λ Q protein. *Genes Dev* **12**, 3276–3285.

Konan, K. V. & Yanofsky, C. (2000). Rho-dependent transcription termination in the *tna* operon of *Escherichia coli*: roles of the *boxA* sequence and the *rut* site. *J Bacteriol* **182**, 3981–3988.

Korzheva, N., Mustaev, A., Kozlov, M., Malhotra, A., Nikiforov, V., Goldfarb, A. & Darst, S. A. (2000). A structural model of transcription elongation. *Science* **289**, 619–625.

Landick, R. (1999). Shifting RNA polymerase into overdrive. *Science* **284**, 598–599.

Landick, R., Turnbough, C. L., Jr & Yanofsky, C. (1996). Transcription attenuation. In *Escherichia coli and Salmonella: Cellular and Molecular Biology*, pp. 1263–1286. Edited by F. C. Neidhardt and others. Washington, DC: American Society for Microbiology.

Luo, D., Leautey, J., Grunberg-Manago, M. & Putzer, H. (1997). Structure and regulation of expression of the *Bacillus subtilis* valyl-tRNA synthetase gene. *J Bacteriol* **179**, 2472–2478.

Neely, M. N. & Friedman, D. I. (2000). N-mediated transcription antitermination in lambdoid phage H-19B is characterized by alternative NUT RNA structures and a reduced requirement for host factors. *Mol Microbiol* **38**, 1074–1085.

Norman, R. A., Poh, C. L., Pearl, L. H., O'Hara, B. P. & Drew, R. E. (2000). Steric hindrance regulation of the *Pseudomonas aeruginosa* amidase operon. *J Biol Chem* **275**, 30660–30667.

Nussbaum-Schochat, A. & Amster-Choder, O. (1999). BglG, the transcriptional antiterminator of the *bgl* system, interacts with the β' subunit of the *Escherichia coli* RNA polymerase. *Proc Natl Acad Sci U S A* **96**, 4336–4341.

O'Hara, B. P., Norman, R. A., Wan, P. T. C., Roe, S. M., Barrett, T. E., Drew, R. E. & Pearl, L. H. (1999). Crystal structure and induction mechanism of AmiC-AmiR: a ligand-regulated transcription antitermination complex. *EMBO J* **18**, 5175–5186.

Pelchat, M. & Lapointe, J. (1999). Aminoacyl-tRNA synthetase genes of *Bacillus subtilis*: organization and regulation. *Biochem Cell Biol* **77**, 343–347.

Richardson, J. & Greenblatt, J. (1996). Control of RNA chain elongation and termination. In *Escherichia coli and Salmonella: Cellular and Molecular Biology*, pp. 822–848. Edited by F. C. Neidhardt and others. Washington, DC: American Society for Microbiology.

Rollins, S. M., Grundy, F. J. & Henkin, T. M. (1997). Analysis of *cis*-acting sequence and structural elements required for antitermination of the *Bacillus subtilis tyrS* gene. *Mol Microbiol* **25**, 411–421.

Sarsero, J. P., Merino, E. & Yanofsky, C. (2000). A *Bacillus subtilis* operon containing genes of unknown function senses tRNA[Trp] charging and regulates expression of the genes of tryptophan biosynthesis. *Proc Natl Acad Sci U S A* **97**, 2656–2661.

Sen, R., King, R. A. & Weisberg, R. A. (2001). Modification of the properties of elongating RNA polymerase by persistent association with nascent antiterminator RNA. *Mol Cell* **7**, 993–1001.

Switzer, R. L., Turner, R. J. & Lu, Y. (1999). Regulation of the *Bacillus subtilis* pyrimidine bio-synthetic operon by transcriptional attenuation: control of gene expression by an mRNA-binding protein. *Prog Nucleic Acid Res Mol Biol* **62**, 329–367.

Tomita, H. & Clewell, D. B. (2000). A pAD1-encoded small RNA molecule, mD, negatively regulates *Enterococcus faecalis* pheromone response by enhancing transcription ter-mination. *J Bacteriol* **182**, 1062–1073.

Watnick, R. W. & Gottesman, M. E. (1998). *Escherichia coli* NusA is required for efficient RNA binding of phage HK022 Nun protein. *Proc Natl Acad Sci U S A* **95**, 1546–1551.

Watnick, R. W. & Gottesman, M. E. (1999). Binding of transcription termination protein Nun to nascent RNA and template DNA. *Science* **286**, 2337–2339.

Weisberg, R. A. & Gottesman, M. E. (1999). Processive antitermination. *J Bacteriol* **181**, 359–367.

Yanofsky, C. (2000). Transcription attenuation: once viewed as a novel regulatory strategy. *J Bacteriol* **182**, 1–8.

Yarnell, W. W. & Roberts, J. W. (1999). Mechanism of intrinsic transcription termination and antitermination. *Science* **284**, 611–615.

Antisense RNAs in programmed cell death

Kenn Gerdes and Jakob Møller-Jensen

Department of Biochemistry and Molecular Biology, University of Southern Denmark, DK-5230 Odense M, Denmark

INTRODUCTION

Antisense RNAs are small regulatory transcripts typically 50–150 nucleotides long. In most cases, antisense RNAs are fully complementary to their target RNAs. In such cases, the antisense RNAs are encoded by the opposite strand relative to that of their target RNA; they are said to be *cis*-encoded. Usually, *cis*-encoded antisense RNAs inhibit the function of their target RNAs by direct RNA–RNA interaction. RNA I of plasmid ColE1 is a *cis*-encoded antisense RNA that regulates the replication frequency of ColE1 by inhibiting the primer function of RNA II (Eguchi *et al.*, 1991). Another antisense RNA, CopA-RNA of plasmid R1, regulates the replication frequency of R1. CopA-RNA is complementary to the leader region of *repA* mRNA and regulates replication by inhibiting translation of the *repA* mRNA that encodes the RepA replication initiation protein (Wagner & Simons, 1994). However, antisense RNAs need not be fully complementary to their target RNAs. In these cases, the genes encoding the antisense RNAs are usually not linked to the genes encoding the target RNAs – such antisense RNAs are said to be *trans*-encoded (Delihas, 1995). MicF-RNA of *Escherichia coli* is an example of a *trans*-encoded antisense RNA. MicF-RNA is partially complementary to the *ompF* mRNA leader and inhibits translation of *ompF* mRNA by direct RNA–RNA interaction. The *ompF* gene encodes outer-membrane porin OmpF. *micF*-RNA down-regulates OmpF expression in response to temperature increases and other cellular stresses (Delihas, 1995). The *micF* gene is located upstream of *ompC*, which also encodes an outer-membrane porin. The *micF* and *ompC* promoters are adjacent but transcribed divergently and regulated such that MicF-RNA expression is activated concomitantly with transcription of *ompC* (Takayanagi *et al.*, 1991; Coyer *et al.*,

SGM symposium 61: Signals, switches, regulons and cascades: control of bacterial gene expression.
Editors D. A. Hodgson, C. M. Thomas. Cambridge University Press. ISBN 0 521 81388 3 ©SGM 2002.

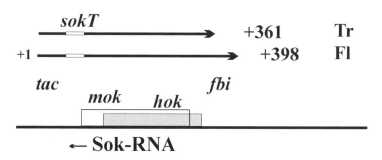

Fig. 1. Genetic structure and regulatory elements of the *hok/sok* system from plasmid R1. Full-length (FI) and truncated (Tr) *hok* mRNAs are 398 and 361 nt long, respectively. Nomenclature: *fbi*, fold-back-inhibition element; *tac*, translational activator element; *mok*, modulation of killing; *hok*, host killing gene; *sok*, suppression of killing; *sokT*, Sok antisense RNA target region.

1990). Thus when *ompC* transcription increases, *micF* transcription also increases. The resulting increase in MicF-RNA concentration confers inhibition of *ompF* mRNA translation and the total amount of OmpC plus OmpF in the outer membrane is by this mechanism kept constant (Delihas, 1995). Recently, several other interesting *trans*-encoded antisense RNAs have been discovered in *E. coli* (Altuvia *et al.*, 1997; Majdalani *et al.*, 1998; Lease *et al.*, 1998). In several such cases, the antisense RNAs are 'promiscuous', meaning that they have multiple target RNAs, and it has been suggested that they function to integrate complex physiological responses to, for example, environmental stresses. Promiscuous antisense RNAs have been described in recent reviews (Wassarman *et al.*, 1999; Lease & Belfort, 2000; Franch & Gerdes, 2000).

ANTISENSE RNA REGULATED PROGRAMMED CELL DEATH: A COMPLEX POST-TRANSCRIPTIONAL CONTROL MECHANISM

Components of the *hok/sok* locus from plasmid R1

The *hok/sok* locus from plasmid R1 was discovered due to its capability of stabilizing the inheritance of plasmids replicating in *E. coli* (Gerdes *et al.*, 1985). The stabilization was very efficient and replicon-independent (Gerdes *et al.*, 1985; Boe *et al.*, 1987; Gerdes, 1988). The genetic organization of the *hok/sok* locus is shown in Fig. 1. The locus encodes three genes denoted *hok*, *sok* and *mok*. The *hok* (host killing) gene specifies a membrane-associated toxin of 52 aa (Gerdes *et al.*, 1986a). Production of the toxin causes irreversible damage to the cell membrane and is thus lethal to host cells (Gerdes *et al.*, 1986b). The *sok* gene (suppression of killing) encodes an antisense RNA of 64 nucleotides (nt) which is complementary to the *hok* mRNA leader region (Gerdes *et al.*, 1988, 1990). Thus Sok-RNA is an example of a *cis*-encoded antisense RNA. Sok-RNA is unstable, but constitutively expressed from a relatively strong promoter. In contrast, *hok* mRNA is very stable and constitutively expressed from a relatively weak promoter (Gerdes *et al.*, 1988, 1990). The *mok* (modulation of killing) reading frame overlaps extensively with *hok*, and is required for expression and regulation of *hok* translation (Thisted & Gerdes, 1992). Genetic analyses showed that Sok-RNA inhibits translation of the *mok* reading frame and that translation of *hok* is coupled to translation of *mok*. Consequently, Sok-RNA regulates translation of *hok* indirectly via *mok*. The secondary structure of the 64 nt Sok-RNA is shown in Fig. 2(a).

Programmed cell death confers plasmid stabilization

Cells cured of a plasmid carrying *hok/sok* were rapidly killed (Gerdes *et al.*, 1986a). The selective cell killing prevented the proliferation of plasmid-free progeny. Phenotypically, this led to plasmid stabilization in growing bacterial cultures. The phenomenon was coined post-segregational killing or PSK (Gerdes *et al.*, 1986a). The discovery of the PSK principle behind the plasmid stabilization phenotype was exciting: the killing appeared paradoxical, since Hok activity was triggered in cells devoid of the *hok* gene, and furthermore, the killing could be regarded as a terminal differentiation event initiated by plasmid-loss. Other examples of programmed cell death in bacteria have been described, and were recently reviewed (Yarmolinsky, 1995; Engelberg-Kulka & Glaser, 1999; Gerdes, 2000).

Programmed cell death by *hok/sok* relies on differential RNA decay

hok mRNA is exceptionally stable (half-life in the order of 30–40 min) and Sok-RNA is labile (half-life in the order of 30 seconds) (Gerdes *et al.*, 1988, 1990; Franch *et al.*, 1997; Mikkelsen & Gerdes, 1997). These basic observations led to the proposal that induction of Hok activity in plasmid-free cells relies on the differential decay rates of the RNAs: rapid decay of Sok-RNA results in an uninhibited pool of toxin-encoding *hok* mRNA freely accessible to be translated. This explains, in a simple way, the killing of plasmid-free segregants (Gerdes *et al.*, 1988, 1990). Basically, this induction model is valid. However, the events that precede activation of *hok* mRNA translation in plasmid-free cells are considerably more complicated and involve a unique type of RNA metabolism as described in the following sections.

Full-length *hok* mRNA is translationally inactive

The PSK model proposed above is based on a toxin–antidote principle by which the selective killing of plasmid-free cells is explained by the instability of the antidote (in this case, the Sok-RNA). However, the interaction between Sok-RNA and *hok* mRNA leads to duplex formation between the RNAs. The duplexes are rapidly cleaved by

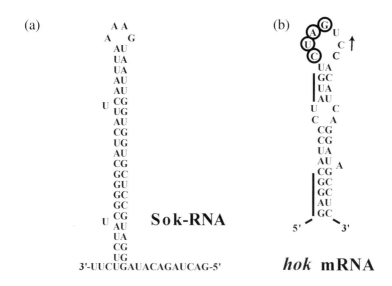

Fig. 2. (a) Secondary structure of the 64 nt Sok antisense RNA (Thisted *et al.*, 1994b; Franch *et al.*, 1997). (b) Secondary structure of the antisense RNA target stem–loop in *hok* mRNA (*sokT*) (Franch *et al.*, 1999b). Bases that conform to the YUNR U-turn motif are encircled. Bars indicate the Shine–Dalgarno region and start codon of the *mok* gene. The arrow indicates the last 3′-nucleotide that is complementary to the Sok-RNA 5′-end. (c) Step-wise pairing mechanism between Sok-RNA and its target in *hok* mRNA forming a complete duplex. Solvent-exposed bases configuring the U-turn are indicated by slashes. A proposed loop complex (Lc) comprising the nucleus for antisense/target helix formation may involve up to six loop/tail base-pairings. A stable complex intermediate constitutes one helical turn of pairing sufficient for *hok* mRNA inactivation (Franch *et al.*, 1999b).

RNase III, both *in vivo* and *in vitro* (Gerdes *et al.*, 1992). Thus, as described for numerous other antisense RNA gene systems, the proposed scheme of inhibition leads to irreversible mRNA inactivation and decay (Wagner & Simons, 1994). Therefore, the above-described model does not explain how *hok* translation can be activated in plasmid-free cells. In other words, how is irreversible inactivation of *hok* mRNA avoided? The answer to this difficult question came from the observation that *hok* mRNA exists in two forms (Gerdes *et al.*, 1990; Thisted *et al.*, 1994a). One form, denoted full-length *hok* mRNA, is translationally inactive and binds the antisense RNA very inefficiently (Thisted *et al.*, 1994a). The secondary structure of the inert full-length *hok* mRNA is shown in Fig. 3(b) and was obtained by structure probing, mutational analyses, phylogenetic comparisons and computer predictions (Thisted *et al.*, 1995; Franch & Gerdes, 1996; Franch *et al.*, 1997; Gultyaev *et al.*, 1997). The 3′-end of *hok* mRNA specifies a so-called 'fold-back-inhibition' element (*fbi*) that pairs with the very 5′-end of the mRNA. The 5′ to 3′ pairing locks the RNA in an inert configuration in which the Shine–Dalgarno element of *mok* (*mok*-SD) is sequestered by an upstream sequence. The Shine–Dalgarno element of *hok* (*hok*-SD) is also sequestered by a nearly perfect repetition of the sequence that sequesters *mok*-SD. The fold-back inhibitory element sequestrates the antisense RNA recognition element in *hok* mRNA (*sokT*), thus preventing antisense RNA binding (Fig. 3b). Therefore, full-length *hok* mRNA is inert both with respect to translation and to antisense RNA binding. Since the mRNA is inert, it can accumulate without killing of the host cells and simultaneously avoid inactivation due to antisense RNA binding. As described below, the accumulation of a reservoir of an activatable full-length *hok* mRNA is prerequisite for the PSK mechanism.

Translation of *hok* is activated by mRNA 3′ processing

A second version of the *hok* mRNA is generated by slow 3′ processing of the full-length mRNA (Gerdes *et al.*, 1990). The 3′ processing removes the 40 terminal nucleotides of the full-length mRNA (compare Fig. 3b and 3c). The truncated mRNA is stable, translationally active and binds Sok-RNA avidly (Thisted *et al.*, 1994a; Franch *et al.*, 1997). *E. coli* contains two 3′ exoribonucleases involved in mRNA decay: polynucleotide phosphorylase (PNPase) and ribonuclease II (Guarneros & Portier, 1991). The 3′ processing of *hok* mRNA was investigated in *pnp* (encoding PNPase) and *rnb* (encoding ribonuclease II) mutant cells (N. Mikkelsen & K. Gerdes, unpublished). Inactivation of either one of the exonucleases did not prevent the 3′ processing, whereas the simultaneous inactivation of both enzymes abolished processing. Thus PNPase and ribonuclease II can both accomplish the 3′ processing of *hok* mRNA. Usually, 3′ processing by PNPase and ribonuclease II leads to mRNA inactivation and decay (Cohen, 1995). Therefore, *hok* mRNA constitutes an unusual case in which the opposite is true: 3′ exonucleolytic processing activates translation.

The 3′ processing triggers structural rearrangements at the *hok* mRNA 5′-end

Single point mutations in the very 5′-end of *hok* mRNA prevented translation of *mok* and therefore also translation of *hok* (Franch & Gerdes, 1996). This mutational analysis defined a translational activation element (*tac*) at the 5′-end of the mRNA (Fig. 3a, b, c). The *tac* element is located approximately 100 nt upstream of the *mok*-SD element

(a) **nascent *hok* transcript**

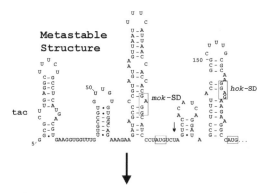

(b) **full-length *hok* mRNA**

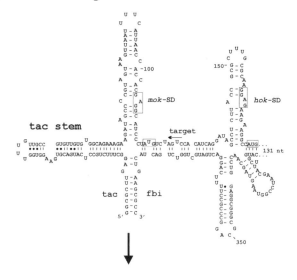

(c) **truncated, refolded *hok* mRNA**

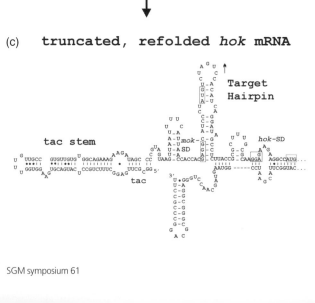

which constitutes its target of regulation. Besides being complementary to the 3' *fbi* element, *tac* is also partly complementary to the sequence that sequestrates *mok*-SD in full-length *hok* mRNA (Fig. 3b). This suggested that the 3' processing, which disrupts the *fbi–tac* pairing, might trigger refolding of the mRNA 5'-end such that *tac* pairs with the upstream inhibitory sequence. The refolding, in turn, could mediate rearrangements further downstream in the RNA with the formation of structures that would allow translation and antisense RNA binding. This proposal was supported by computer simulations of the *hok* mRNA folding pathway (Gultyaev *et al.*, 1997), and shown directly with structure probings (Franch *et al.*, 1997).

The refolded, truncated *hok* mRNA is shown in Fig. 3(c). The partial *tac* stem already present at the 5'-end of the full-length mRNA was extended, thus giving rise to the entire *tac* stem in which the very 5'-end pairs with the upstream inhibitory sequence. Thus the refolding disrupted the base-pairing known to prevent translation.

STRUCTURE AND FUNCTION OF THE Sok ANTISENSE RNA TARGET

Structure of the antisense RNA target

The refolding of the 5'-end of *hok* mRNA was accompanied by significant structural rearrangements in the translation initiation regions located downstream of the *tac* stem. A stem–loop structure, denoted the antisense RNA target hairpin, appeared in the folding simulations (Fig. 3c). Structural analyses confirmed the existence of the target hairpin in truncated *hok* (Franch *et al.*, 1997).

In the target hairpin, *mok*-SD is located at the bottom of the 5' part of the antisense RNA target stem (Figs 2b and 3c). Normally, such base-pairing of an SD sequence confers inhibition of translation. However, mutations which disrupted the pairing abolished translation of truncated *hok* mRNA (Franch *et al.*, 1997). Structural analyses

Fig. 3. Folding pathway of *hok* mRNA. (a) Folding of the 5'-end of the nascent transcript showing the metastable hairpin and the sequestration of the *mok* and *hok* Shine–Dalgarno (SD) regions. (b) Folding of the full-length *hok* mRNA showing the *fbi–tac* interaction, the top of the *tac* stem, the sequestrated *hok* and *mok*-SD regions, and the shielded antisense RNA target. (c) Truncated, refolded *hok* mRNA. The *mok*-SD interaction in the full-length molecule is disrupted by the formation of the stable *tac* stem. In turn, this favours formation of the antisense RNA target stem–loop in which *mok*-SD can interact with ribosomes (shown by toe-printing). Arrow denoted 'target' indicates the nucleotide that is complementary to the very first nucleotide of the 5'-end of Sok-RNA and thus marks the 3' border of the Sok target region (*sokT*) in *hok* mRNA. The secondary and tertiary structures in the *hok* mRNAs were based on mutational analyses, structural probings, phylogenetic comparisons, computer-assisted predictions and molecular modelling (Thisted *et al.*, 1995; Franch & Gerdes, 1996; Franch *et al.*, 1997, 1999b; Gultyaev *et al.*, 1997; Møller-Jensen *et al.*, 2001).

indicated that such mutations changed the overall secondary structure of the truncated RNA in favour of the inhibited structure present in full-length *hok* mRNA. This result indicates that the base-pairing in the target stem is required to maintain truncated *hok* mRNA in a translatable configuration. This conclusion was corroborated by analyses in which the *tac* stem was forced by mutation (Franch *et al.*, 1997).

The recognition reaction

Sok-RNA consists of a single-stranded 5'-tail and a stem–loop (Fig. 2a). Mutational analyses revealed that the single-stranded tail of Sok-RNA recognizes the nucleotides in the target-loop (Fig. 2b). Thus the initial recognition reaction between Sok and its target occurs via single-stranded regions (Thisted *et al.*, 1994b; Franch *et al.*, 1997). The initial recognition event is followed by more extensive duplex formation (Franch *et al.*, 1999a). The complete or partial RNA duplex is cleaved by RNase III, thereby leading to irreversible inactivation of *hok* mRNA. The antisense RNA binding reaction is outlined in Fig. 2(c).

The U-turn: a general antisense RNA recognition element

Many plasmids and chromosomes of enteric bacteria contain *hok/sok* homologous loci, most of which contain all the regulatory elements described above (Gerdes *et al.*, 1997; Pedersen & Gerdes, 1999). Sequence alignment of the *hok*-homologous mRNAs revealed a phylogenetically conserved YUNR (Y = pyrimidine; U = uracil; N = any base; R = purine) motif in the target hairpin (Franch *et al.*, 1999b). The secondary structure of the target hairpin is shown in Fig. 2(b) and nucleobases in the loop that conform to the consensus YUNR motif are encircled. Strikingly, the YUNR motifs are located at the same relative positions within the recognition loops of the *hok*-homologous mRNAs. The motif was proposed to configure a U-turn loop structure similar to that present in tRNA anticodon loops (Franch *et al.*, 1999b). The U-turn is characterized by a sharp bend in the phosphate backbone at the UpN phosphate stabilized by three non-Watson–Crick interactions involving the invariant uracil ribonucleoside. The nucleobases on the 3' side of the turning phosphate are presented in an A-form structure, creating an unpaired Watson–Crick surface predisposed for binding to a complementary set of nucleotides (indicated by slashes in the pin diagram in Fig. 2c). Furthermore, the U-turn architecture retracts the phosphodiester backbone within the loop thereby decreasing the local electronegative potential surrounding the bases presented. This, in turn, promotes the pairing to the complementary bases by reducing backbone repulsion in the initial recognition step. Mutational analyses showed that mutations that disrupt the U-turn configuration reduce the rate of antisense RNA binding to the *hok* target approximately 10-fold (Franch *et al.*, 1999b). Thus the U-turn in *hok/sok* and similar systems has evolved to facilitate rapid antisense RNA binding, perhaps to minimize detrimental expression of the lethal Hok toxin.

Inspection of known antisense RNAs and their targets revealed that almost all such systems contain YUNR motifs either in the loops of the antisense RNAs or in target loops (Franch *et al.*, 1999b; Franch & Gerdes, 2000). For example, RNA I contains U-turn motifs at the correct position in two of its three loops involved in the initial 'kissing reaction' between RNA I and II. This suggests that almost all known antisense RNAs have evolved binding pathways that involve initial recognition between the RNAs via U-turn loop structures.

PHYLOGENETICALLY CONSERVED FOLDING PATHWAY IN *hok* mRNA

Metastable hairpins in the nascent transcript

The *tac* stem, which is required for activation of *hok* translation, is located at the 5′-end of the mRNA (Fig. 3c). If formed in the nascent transcript, the *tac* stem predictably would lead to formation of the antisense RNA target stem–loop further downstream, thereby leading to premature antisense RNA binding or to translation of *hok*. Obviously, both these scenarios would be detrimental to the PSK mechanism. How, then, is premature *tac*-stem formation prevented?

It appears that *hok* mRNA has the possibility of forming a local 5′ hairpin as an alternative to the *tac* stem (Fig. 3a) (Gultyaev *et al.*, 1997). The local hairpin was called metastable since it was predicted to exist in the nascent transcript only. Similar hairpins were found in all the *hok*-homologous mRNAs and their existence was strongly supported by nucleotide covariations. Structural analyses of native RNAs (i.e. non-denatured RNAs synthesized *in vitro*) showed that indeed the metastable structure exists in *hok* mRNA during transcription (Møller-Jensen *et al.*, 2001). Functional analyses of mutations in the metastable hairpin showed that it prevents *hok* translation and antisense RNA binding during transcription (Møller-Jensen *et al.*, 2001). Furthermore, the metastable hairpin was stable until the *hok* mRNA 3′-end, including the *fbi* sequence, was synthesized. In other words, disruption of the metastable hairpin required the *fbi* sequence. Thus synthesis of the *fbi* sequence triggers refolding from the nascent configuration (Fig. 3a) to the stable full-length configuration (Fig. 3b). How this switch is accomplished at the molecular level is not yet known.

The *hok* mRNA folding pathway

Using phylogenetic comparisons and computer simulations, a 'consensus' folding pathway for the *hok* family of mRNAs was predicted (described in detail by Gultyaev *et al.*, 1997) and confirmed experimentally (Franch *et al.*, 1997; Møller-Jensen *et al.*, 2001). The main conserved intermediate structures of the folding pathway are visualized in Fig. 3. In the 5′-untranslated region, the folding of the metastable hairpin at the very 5'-end prevents *tac*-stem formation and favours the formation of the inhibited configurations of *mok*-SD and *hok*-SD (Fig. 3a). Therefore, during mRNA synthesis, the antisense RNA target hairpin is not formed (Fig. 3a versus 3c). This has two effects: the rate of antisense RNA binding is reduced and ribosome loading to the nascent transcript is prevented (Møller-Jensen *et al.*, 2001). Premature binding of Sok-RNA to *hok* mRNA would prevent accumulation of a pool of mRNA that can be activated for translation in plasmid-free cells, and premature translation of *hok* would be lethal to host cells.

A model that explains PSK

The studies on *hok* mRNA and Sok antisense RNA have yielded a profound understanding of a unique and complex post-transcriptional control system. The basis of activation of *hok* translation in plasmid-free cells is the differential stabilities of the two RNAs: *hok* mRNA is stable and Sok-RNA is unstable. However, in order for *hok* mRNA to avoid being irreversibly inactivated by RNase III cleavage, it has evolved two unique features: (i) it contains a metastable RNA hairpin at its 5'-end that acts as a safeguard that prevents premature activation of *hok*; and (ii) the existence in *hok* mRNA of the conditionally inhibitory *fbi* sequence explains how it can be reactivated for translation in plasmid-free cells. The model that explains PSK by the *hok/sok* locus is outlined in Fig. 4.

AN ANTISENSE RNA REGULATED PSK LOCUS FROM GRAM-POSITIVE PLASMIDS

Plasmids from enterococci encode small *par* regions of approximately 400 bp (Weaver *et al.*, 1993). *par* encodes two RNAs designated RNA I and RNA II (Weaver & Tritle, 1994) and stabilizes plasmids by a PSK mechanism (Weaver *et al.*, 1996, 1998). RNA I and II are convergently transcribed from promoters located at each end of *par*, and terminate at a common bidirectional terminator (Fig. 5a). The RNA I and II genes each contain two direct repeats, DRa and DRb (Fig. 5a, b). The direct repeats lead to the presence of complementary regions that are potential sites of interaction between the two RNAs. Likewise, the bidirectional terminator encodes complementary RNA structures in the two RNAs. Computer predictions and structural analyses suggested that these regions are predominantly single-stranded (see Fig. 5b) (Greenfield *et al.*, 2000). RNA I encodes a small toxin of 33 aa called Fst (*faecalis* plasmid-<u>st</u>abilizing protein). Expression of Fst is inhibited by RNA II. RNA II interacts directly with RNA I and prevents ribosome loading at the *fst* gene (Greenfield *et al.*, 2000; Greenfield & Weaver, 2000). RNA I is more stable than RNA II. Thus plasmid-free cells experience decay of RNA II leaving RNA I open to be translated. This leads to production of Fst and killing of plasmid-free cells. In many respects, the *par* system from *Enterococcus faecalis* is similar in function to *hok/sok* of plasmid R1. However, there is no indication that RNA I exists in two forms and the mechanism by which RNA I translation is activated in

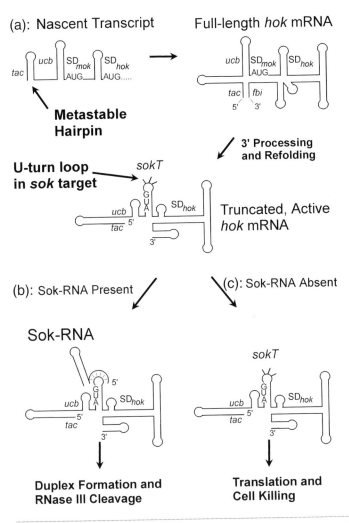

Fig. 4. Molecular model that explains activation of *hok* translation in plasmid-free cells. (a) Folding pathway of *hok* mRNA. During transcription, a metastable hairpin at the mRNA 5′-end prevents formation of the *tac* stem (translational activation). Formation of the *tac* stem in the nascent transcript leads to premature activation of translation or antisense RNA binding (Møller-Jensen et al., 2001). In the full-length transcript, the *fbi* element (fold-back inhibition) pairs with *tac* thereby locking the mRNA into an inert configuration: the *mok*-SD and *hok*-SD elements are base-paired and the Sok-RNA target region (*sokT*) is shielded by the fold-back structure (*fbi*). During steady state, a pool of inert full-length mRNA accumulates in plasmid-containing cells. The full-length mRNA is activated by slow, constitutive 3′ exonucleolytic processing, which removes the *fbi* element. The removal of *fbi* triggers refolding of the 5′-end of the mRNA with the formation of the *tac* and antisense target hairpins, the latter containing the U-turn structure. (b) In plasmid-carrying cells, the truncated mRNA is rapidly scavenged by Sok-RNA, which prevents its translation. Subsequently, the RNAs form a duplex that is cleaved by RNase III. (c) In plasmid-free cells, in which the unstable antisense RNA has decayed, translation of the truncated *hok* mRNA is allowed, thus leading to cell killing.

(a)

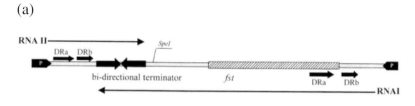

Fst: MKDLMSLVAIPIFVGLVLEMISRVLDEDDSRK

(b)

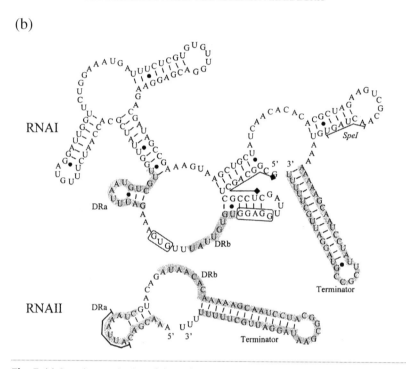

Fig. 5. (a) Genetic organization of the *E. faecalis* plasmid pAD1 *par* locus. Promoters for RNA I and RNA II are indicated by arrows on each end of the region. The open reading frame *fst* encodes a 33 aa toxic peptide (shown below the region). The two RNAs read in opposite directions across direct repeats labelled DRa and DRb (arrows) to a bidirectional terminator (converging arrows). (b) Secondary structures of RNA I and RNA II. The Shine–Dalgarno sequence and start codon of *fst* in RNA I are boxed. The complementary direct repeats (DRa and DRb) and terminator sequences are shaded (from Greenfield *et al.*, 2000, with permission from Blackwell Science).

plasmid-free cells is not yet known. RNA I and II interact via short stretches of complementarity (Fig. 5b). Since the RNAs are not fully complementary they cannot form a complete RNA duplex and this, in turn, protects the RNA complex from being cleaved by RNase III. A major question relates to the mechanism of activation of *fst* translation. One possibility is that the off-rate of the two RNAs is sufficiently high as to allow formation of free RNA I molecules in plasmid-free cells (after decay of RNA II). However, it cannot be excluded that enzymes such as an RNA helicase or an RNase participate in the activation mechanism.

CONCLUDING REMARKS

It appears that plasmids from both Gram-negative and Gram-positive bacteria encode PSK systems that increase the maintenance of their replicons. In the cases of *hok/sok* of plasmids from enterics and *par* from *E. faecalis* plasmids the loci seem evolutionarily unrelated, suggesting that this type of control mechanism has evolved independently at least twice. A major remaining question relates to the function of the multiple *hok*-homologous loci encoded by the chromosomes of enteric bacteria.

REFERENCES

Altuvia, S., Weinstein-Fischer, D., Zhang, A., Postow, L. & Storz, G. (1997). A small, stable RNA induced by oxidative stress: role as a pleiotropic regulator and antimutator. *Cell* **90**, 43–53.

Boe, L., Gerdes, K. & Molin, S. (1987). Effects of genes exerting growth inhibition and plasmid stability on plasmid maintenance. *J Bacteriol* **169**, 4646–4650.

Cohen, S. N. (1995). Surprises at the 3′-end of mRNA. *Cell* **80**, 829–832.

Coyer, J., Andersen, J., Forst, S. A., Inouye, M. & Delihas, N. (1990). *micF* RNA in *ompB* mutants of *Escherichia coli*: different pathways regulate *micF* RNA levels in response to osmolarity and temperature change. *J Bacteriol* **172**, 4143–4150.

Delihas, N. (1995). Regulation of gene expression by trans-encoded antisense RNAs. *Mol Microbiol* **15**, 411–414.

Eguchi, Y., Itoh, T. & Tomizawa, J. (1991). Antisense RNA. *Annu Rev Biochem* **60**, 631–652.

Engelberg-Kulka, H. & Glaser, G. (1999). Addiction modules and programmed cell death and antideath in bacterial cultures. *Annu Rev Microbiol* **53**, 43–70.

Franch, T. & Gerdes, K. (1996). Programmed cell death in bacteria: translational repression by mRNA end-pairing. *Mol Microbiol* **21**, 1049–1060.

Franch, T. & Gerdes, K. (2000). U-turns and regulatory RNAs. *Curr Opin Microbiol* **3**, 159–164.

Franch, T., Gultyaev, A. P. & Gerdes, K. (1997). Programmed cell death in bacteria: processing at the *hok* mRNA 3′-end triggers structural rearrangements that allow translation and antisense RNA binding. *J Mol Biol* **273**, 38–51.

Franch, T., Thisted, T. & Gerdes, K. (1999a). Ribonuclease III processing of coaxially stacked RNA helices. *J Biol Chem* **274**, 26572–26578.

Franch, T., Petersen, M., Wagner, E. G. H., Jacobsen, J. P. & Gerdes, K. (1999b). Antisense RNA regulation in prokaryotes: rapid RNA/RNA interaction facilitated by a general U-turn loop structure. *J Mol Biol* **294**, 1115–1125.

Gerdes, K. (1988). The *parB* (*hok/sok*) system of plasmid R1: a general purpose plasmid stabilization system. *Bio/Technology* **6**, 1402–1405.

Gerdes, K. (2000). Toxin-antitoxin modules may regulate synthesis of macromolecules during nutritional stress. *J Bacteriol* **182**, 561–572.

Gerdes, K., Larsen, J. E. L. & Molin, S. (1985). Stable inheritance of plasmid R1 requires two different loci. *J Bacteriol* **161**, 292–298.

Gerdes, K., Rasmussen, P. B. & Molin, S. (1986a). Unique type of plasmid maintenance function: postsegregational killing of plasmid free cells. *Proc Natl Acad Sci U S A* **83**, 3116–3120.

Gerdes, K., Bech, F. W., Jørgensen, S. T., Løbner-Olesen, A., Atlung, T., Boe, L., Karlstrøm, O., Molin, S. & von Meyenburg, K. (1986b). Mechanism of postsegregational killing by the *hok* gene product of the *parB* system of plasmid R1 and its homology with the *relF* gene product of the *E. coli relB* operon. *EMBO J* **5**, 2023–2029.

Gerdes, K., Helin, K., Christensen, O. W. & Løbner-Olesen, A. (1988). Translational control and differential RNA decay are key elements regulating postsegregational expression of the killer protein encoded by the *parB* locus of plasmid R1. *J Mol Biol* **203**, 119–129.

Gerdes, K., Thisted, T. & Martinussen, J. (1990). Mechanism of post-segregational killing by the *hok/sok* system of plasmid R1: the *sok* antisense RNA regulates the formation of a *hok* mRNA species correlated with killing of plasmid free segregants. *Mol Microbiol* **4**, 1807–1818.

Gerdes, K., Nielsen, A., Thorsted, P. & Wagner, E. G. H. (1992). Mechanism of killer gene activation: antisense RNA-dependent RNase III cleavage ensures rapid turn-over of the stable Hok, SrnB and Pnd effector mRNAs. *J Mol Biol* **226**, 637–649.

Gerdes, K., Gultyaev, A. P., Franch, T., Pedersen, K. & Mikkelsen, N. D. (1997). Antisense RNA regulated programmed cell death. *Annu Rev Genet* **31**, 1–31.

Greenfield, T. J. & Weaver, K. E. (2000). Antisense RNA regulation of the pAD1 *par* post-segregational killing system requires interaction at the 5′ and 3′ ends of the RNAs. *Mol Microbiol* **37**, 661–670.

Greenfield, T., Ehli, E., Kirshenmann, T., Franch, T., Gerdes, K. & Weaver, K. E. (2000). The antisense RNA of the par locus of pAD1 regulates the expression of a 33-amino-acid toxic peptide by an unusual mechanism. *Mol Microbiol* **37**, 652–660.

Guarneros, G. & Portier, C. (1991). Different specificities of ribonuclease II and polynucleotide phosphorylase in 3′ mRNA decay. *Biochimie* **73**, 543–549.

Gultyaev, A. P., Franch, T. & Gerdes, K. (1997). Programmed cell death in bacteria: coupled nucleotide covariations reveal a phylogenetically conserved folding pathway in the *hok* family of mRNAs. *J Mol Biol* **273**, 26–37.

Lease, R. A. & Belfort, M. (2000). Riboregulation by DsrA RNA: trans-actions for global economy. *Mol Microbiol* **38**, 667–672.

Lease, R. A., Cusick, M. E. & Belfort, M. (1998). Riboregulation in *Escherichia coli*: DsrA RNA acts by RNA : RNA interactions at multiple loci. *Proc Natl Acad Sci U S A* **95**, 12456–12461.

Majdalani, N., Cunning, C., Sledjeski, D., Elliott, T. & Gottesman, S. (1998). DsrA RNA regulates translation of RpoS message by an anti-antisense mechanism, independent of its action as an antisilencer of transcription. *Proc Natl Acad Sci U S A* **95**, 12462–12467.

Mikkelsen, N. D. & Gerdes, K. (1997). Sok antisense RNA from plasmid R1 is functionally inactivated by RNase E cleavage and polyadenylated by poly(A)polymerase I. *Mol Microbiol* **26**, 311–320.

Møller-Jensen, J., Franch, T. & Gerdes, K. (2001). Temporal translational control by a metastable RNA structure. *J Biol Chem* **276**, 35707–35713.

Pedersen, K. & Gerdes, K. (1999). Multiple *hok* loci in *Escherichia coli. Mol Microbiol* **32**, 1090–1102.

Takayanagi, K., Maeda, S. & Mizuno, T. (1991). Expression of *micF* involved in porin synthesis in *Escherichia coli*: two distinct cis-acting elements respectively regulate *micF* expression positively and negatively. *FEMS Microbiol Lett* **67**, 39–44.

Thisted, T. & Gerdes, K. (1992). Mechanism of post-segregational killing by the *hok/sok* system of plasmid R1: Sok antisense RNA regulates *hok* gene expression indirectly through the overlapping *mok* gene. *J Mol Biol* **223**, 41–54.

Thisted, T., Nielsen, A. & Gerdes, K. (1994a). Mechanism of post-segregational killing: translation of Hok, SrnB, and PndA mRNAs of plasmids R1, F and R483 is activated by 3′-end processing. *EMBO J* **13**, 1950–1959.

Thisted, T., Sørensen, N., Wagner, E. G. H. & Gerdes, K. (1994b). Mechanism of postsegregational killing: Sok antisense RNA interacts with Hok mRNA *via* its 5′-end single-stranded leader and competes with the 3′-end of Hok mRNA for binding to the *mok* translational initiation region. *EMBO J* **13**, 1960–1968.

Thisted, T., Sørensen, N. & Gerdes, K. (1995). Mechanism of post-segregational killing: secondary structure analysis of the entire Hok mRNA from plasmid R1 suggests a fold-back inhibitory structure that prevents translation and antisense RNA binding. *J Mol Biol* **247**, 859–873.

Wagner, E. G. & Simons, R. W. (1994). Antisense RNA control in bacteria, phages and plasmids. *Annu Rev Microbiol* **48**, 713–742.

Wassarman, K. M., Zhang, A. & Storz, G. (1999). Small RNAs in *Escherichia coli. Trends Microbiol* **7**, 37–45.

Weaver, K. E. & Tritle, D. J. (1994). Identification and characterization of an *Enterococcus faecalis* plasmid pAD1-encoded stability determinant which produces two small RNA molecules necessary for its function. *Plasmid* **32**, 168–181.

Weaver, K. E., Clewell, D. B. & An, F. (1993). Identification, characterization, and nucleotide sequence of a region of *Enterococcus faecalis* pheromone-responsive plasmid pAD1 capable of autonomous replication. *J Bacteriol* **175**, 1900–1909.

Weaver, K. E., Jensen, K. D., Colwell, A. & Sriam, S. L. (1996). Functional analysis of the *Enterococcus faecalis* plasmid pAD1-encoded stability determinant *par. Mol Microbiol* **20**, 53–63.

Weaver, K. E., Walz, K. D. & Hiene, M. S. (1998). Isolation of a derivative of *Escherichia coli/Enterococcus faecalis* shuttle vector pAM401 temperature sensitive for maintenance in *E. faecalis* and its use in evaluating the mechanism of pAD1 *par*-dependent plasmid stabilization. *Plasmid* **40**, 225–232.

Yarmolinsky, M. B. (1995). Programmed cell death in bacterial populations. *Science* **267**, 836–837.

Control of signal transduction in the sporulation phosphorelay

James A. Hoch

Division of Cellular Biology, Department of Molecular and Experimental Medicine, The Scripps Research Institute, 10550 North Torrey Pines Road, La Jolla, CA 92037, USA

SIGNAL RECOGNITION IS THE ESSENCE OF ADAPTABILITY

While it is now clear from genomic sequencing studies that bacteria are replete with genes for a large variety of enzymic pathways to adapt to countless environmental situations, it is also evident that bacteria are miserly in the expression of non-essential genes, spending a great deal of effort to ensure no energy is lost on frivolous protein synthesis. The connections between the environment and gene expression that regulate this parsimonious behaviour are the signal transduction systems. A major type of one signal transduction system commonly used for this purpose is the two-component system (Hoch & Silhavy, 1995).

Two-component systems consist of a sensor kinase and a response regulator. The N-terminal domain of a sensor kinase is a signal-sensing domain whose amino acid sequence and structure differ from sensor kinase to sensor kinase, depending on the stimulus to which the sensor kinase is dedicated (Fig. 1). C-terminal to this domain is the autokinase domain, which may be subdivided into a phosphotransferase subdomain and an ATP-binding–hydrolysing subdomain. Stimulus recognition by the sensor kinase induces ATP hydrolysis and phosphoryl transfer to a histidine residue on the phosphotransferase domain. The majority of the sensor kinases have signal-sensing domains embedded in the cytoplasmic membrane, which may indicate that most sense extracellular signals.

The second component of the system is the response regulator. Most of these are transcription activators or repressors or both. Response regulators consist of two

SGM symposium 61: Signals, switches, regulons and cascades: control of bacterial gene expression.
Editors D. A. Hodgson, C. M. Thomas. Cambridge University Press. ISBN 0 521 81388 3 ©SGM 2002.

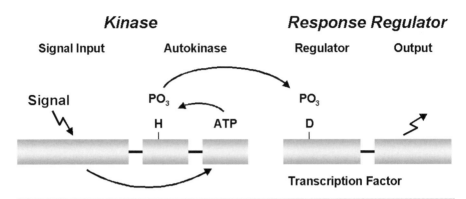

Fig. 1. Domain organization and signal transduction in a two-component system. Major domains of the sensor kinase and response regulator are shown. The autokinase domain consists of two subdomains: a histidine-containing phosphotransferase subdomain and an ATP-binding subdomain. Ligand binding by the signal input domain activity activates the series of steps shown that lead to binding of the transcription factor through its output domain to a DNA target.

domains: an N-terminal regulatory domain and a C-terminal DNA-binding domain. Response regulators exist to translocate the signal sensed by the membrane-bound sensor kinase to the chromosome, where it can regulate gene activity. Signal propagation to the response regulator is by transfer of the phosphoryl group from the sensor kinase to an aspartate of the regulatory domain of the response regulator, the consequence of which is the activation of the DNA-binding and transcriptional functions of the C-terminal domain of the response regulator.

A phosphorelay is a more complicated version of a two-component system with two more phosphorylatable domains (Fig. 2). In these systems, the phosphoryl group is transferred in the order His-Asp-His-Asp. Phosphorelays have the same type of sensor kinase and response regulator-transcription factor as two-component systems but a regulatory domain and a phosphotransferase domain are sandwiched between them. These domains may be on separate proteins as in the sporulation phosphorelay of *Bacillus subtilis* (Burbulys *et al.*, 1991) or combined with the sensor kinase as in the BglS phosphorelay of *Bordetella pertussis* (Uhl & Miller, 1996). Other domain combinations are found in eukaryotes where phosphorelays and not two-component systems are the rule (Thomason & Kay, 2000). It was originally proposed that the additional domains of a phosphorelay provide more points of control on the phosphorylation of the transcription factor (Burbulys *et al.*, 1991). This is certainly true for the sporulation phosphorelay.

Two-component systems have been adapted to regulate an enormous number of genes in virtually all bacteria. Basic requirements of cellular life such as nitrogen or phosphate

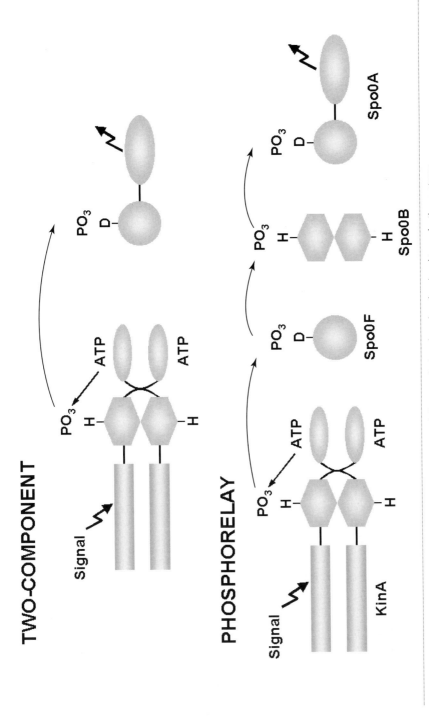

Fig. 2. Phosphoryl group transfer in generic two-component and the sporulation phosphorelay signal transduction systems.

Two-component Regulation

Phosphate	Respiration	Haemagglutinins
Nitrogen	Transport	Toxins
Osmotic Response	Autolysins	Capsules
Chemotaxis	Haemolysins	Surface Proteins

Phosphorelay Regulation

– Cell Cycle and Development in *Bacillus* and *Caulobacter*
– Virulence in *Bordetella pertussis*
– Anaerobiosis in *Escherichia coli*
– Osmotic, Oxidative and Developmental Regulation
in *Dictyostelium*, Yeast and Fungi
– Ethylene Response (Ripening) in Plants

Fig. 3. Some general cellular properties regulated by two-component systems and phosphorelays.

availability are regulated by such systems (Fig. 3). The direction of movement in chemotaxis toward or away from sources of food or toxic components is two-component system regulated. In pathogens, these systems control the expression of virulence factors that allow successful infection.

Phosphorelays are used for cellular lifestyle changes such as sporulation in *Bacillus* species or cell cycle control in *Caulobacter* species (Domian *et al.*, 1997). The switch from oxidative to anaerobic growth in *Escherichia coli* (Tsuzuki *et al.*, 1995) or complex activation of virulence in *Bordetella pertussis* (Uhl & Miller, 1996) is also subject to phosphorelay regulation. In a similar fashion, eukaryotes adapted phosphorelays for control of developmental events as exotic as fruit ripening in plants (Chang *et al.*, 1993). While regulation of processes within the cell or between cells is an exceedingly complex signalling situation, two-component and phosphorelay systems are woven in the overall fabric of regulation and comprise a major means by which opportunistic bacteria adapt to a variety of environmental challenges.

MECHANISMS IN TWO-COMPONENT SYSTEMS

A brief overview of the biochemical mechanisms by which these systems operate should be helpful in understanding their regulation. Sensor kinases in many two-component systems have the capacity to both phosphorylate and dephosphorylate the response regulator to which they are paired. Thus the sensor kinase acts as a switch to both turn on and turn off the response regulator. This may be important to fine-tune the activity of the response regulator to the signal level and also to prevent its activity if phosphorylated by other sensor kinases or small molecule acyl phosphates such as acetyl phosphate. The mechanism of activation of sensor kinases by signal ligand binding remains a mystery since a complete crystal structure of a sensor kinase has not been accomplished. Given the extreme variety of signal-sensing domains of sensor kinases, it might be surprising if any two of them used the same mechanism for activation.

Despite the gap in understanding how signals activate sensor kinases, some appreciation is being gained of the structure of the ATP-binding and phosphotransfer domains of sensor kinases. The ATP-binding domain is similar in structure to ATP-binding domains of other proteins such as DNA gyrase or Hsp70 (Bilwes *et al.*, 2001). The phosphotransfer domain is generated by its dimerization surface to form a four-helix bundle similar to that of the Spo0B phosphotransferase (Varughese *et al.*, 1998; Bilwes *et al.*, 1999; Tomomori *et al.*, 1999). The active site histidine protrudes from one helix of each protomer. These two domains must be connected by a flexible linker, but how they interact to bring the ATP in juxtaposition with the histidine to effect phosphorylation has not been determined.

Once phosphorylated, the phosphotransferase domain of the kinase and the response regulator interact to transfer the phosphoryl group. The interaction of the phosphotransferase domain of Spo0B with the Spo0F regulatory domain of the sporulation phosphorelay has been visualized in a co-crystal and has provided valuable insight into the nature of the interacting surfaces of the two proteins (Zapf *et al.*, 2000). How such surfaces evolve molecular specificity while two-component systems were amplified during evolution to form the complement now seen in present day bacteria has also been answered (Hoch & Varughese, 2001). Briefly, interaction surfaces in two-component systems retain precise orientation of the donor histidine with the recipient aspartate, by coming together in the same way in all two-component systems, and generate individuality by changing a defined subset of the residues making up the interaction surface. Such surfaces are mosaics of conserved and variable residues.

The phosphorylated regulatory domain of response regulators no longer is capable of inhibiting the activity of the DNA-binding domain, which frees this domain to carry out its functions. It is a subject of active investigation to determine how phosphorylation disinhibits the output domain. Crystallographic studies of the phosphorylated regulatory domain of the sporulation transcription factor Spo0A have revealed the active movement of the side chain of a conserved catalytic threonine residue, which allows a surface residue to rotate inward (Lewis *et al.*, 1999). Moreover, this movement is thought to promote dimerization between two phosphorylated regulatory domains, which may be a prerequisite for RNA polymerase interaction and activation of transcription. Recent crystal studies of the C-terminal domain of Spo0A bound to its DNA target show that two molecules interact in a head-to-tail fashion on tandem repeats of the '0A box' 5'-TGNCGAA-3' (H. Zhao, T. Msadek, J. Zapf, Madhusudan, J. A. Hoch & K. I. Varughese, unpublished). Since most Spo0A-controlled promoters have more than one box of this type, dimerization may be important for its activity.

The basic outline of how two-component systems work is now clear but many of the details are obscure. Some of these details include how sensor kinases bind ligands and activate autophosphorylation, how response regulators respond to phosphorylation, and how response regulators interact with the transcription machinery to turn on gene expression.

SIGNAL INPUT IN THE SPORULATION PHOSPHORELAY

The sporulation phosphorelay serves to illustrate how complex regulation of a signal transduction pathway can be, even in a 'simple' bacterium. Sporulation is the last thing a bacterium wants to do. It is a desperation measure undertaken when it is clear that the nutrients required to sustain growth and division have been exhausted. When thinking of developmental regulation it is important to remember this is a cellular process that results in substantial commitment of resources to whole-cell morphogenesis. Therefore, it should not be surprising that the signal transduction pathway leading to the induction of sporulation is under exquisite control by a variety of positive and negative forces.

The general concept of a link between initiation of sporulation and the cell cycle evolved from the early work of Mandelstam and co-workers (Mandelstam & Higgs, 1974), who showed that sporulation could only be initiated at a certain point in the cell cycle (Fig. 4). At this point, the cell assesses its nutrient situation and other, as yet undefined, parameters, and makes a decision to either grow and divide or initiate sporulation and stationary phase. A decision to grow and divide presumably cannot be reversed whereas early sporulation events may be. What signals are being sensed at this point remain to be discovered. However, the sensor kinases that feed phosphate to the phosphorelay are undoubtedly responsible for sensing these signals. In *B. subtilis*, there are five sensor kinases, KinA–KinE, with the potential to phosphorylate the Spo0F response regulator of the phosphorelay (Jiang *et al.*, 2000). Each of these has a unique signal-sensing domain, which has been interpreted to mean that each sensor kinase

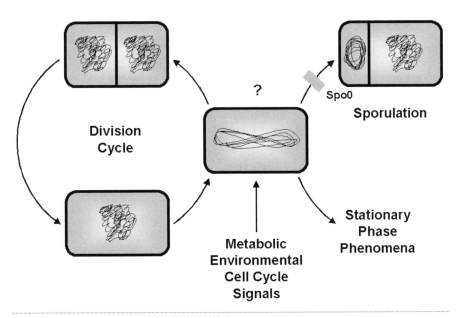

Fig. 4. Cell cycle decision point between growth and sporulation. At some point in the division cycle, the decision to grow or enter stationary phase and sporulation is made. This decision is made based on signals from several sources.

senses a different signal (Fig. 5). This suggests that a variety of environmental conditions may initiate sporulation and one or more of the sensor kinases will recognize the signal(s) emanating from that condition. There is a predominance of PAS subdomains within the signal-sensing domains of KinA, KinC and KinE. PAS domains serve a number of recognition roles in signalling proteins, including sensing oxygen concentrations, light, and oxidation/reduction potentials of the cell, as well as being involved in dimerization of proteins (Taylor & Zhulin, 1999). In KinA, the major sensor kinase used in laboratory media, at least one PAS domain may bind ATP and, perhaps, sense energy charge (Stephenson & Hoch, 2001). KinB, KinC and KinD are membrane proteins and may receive stimuli from extracellular conditions or ligands associated with the cell wall or membrane. However, at this time no definitive signals have been identified for any of the sporulation sensor kinases.

Early analysis of the genomic data generated for *Bacillus* species related to *B. subtilis*, such as *Bacillus anthracis* and *Bacillus halodurans*, has revealed that the complement of sensor kinases with potential to phosphorylate Spo0F in these species are very much different in their signal-sensing domains from those sensor kinases of *B. subtilis* (J. A. Hoch, unpublished). These three species occupy much different ecological niches in nature and therefore must be exposed to and sense different signals for sporulation.

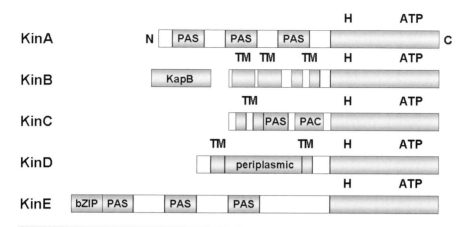

Fig. 5. Structures of the five sensor kinases recognizing signals for the initiation of sporulation. The autokinase domains of all five sensor kinases are highly conserved in amino acid sequence. The signal-sensing domain of each sensor kinase is unique. Recognized regions in these domains include transmembrane segments (TM), putative redox-sensing regions (PAS and PAC), coiled-coiled regions (bZIP) and a periplasmic region.

The five sensor kinases of *B. subtilis*, when activated by signals, pass their phosphoryl groups to a common intermediate response regulator, SpoOF (Fig. 6). This protein lacks an output domain and, as far as can be determined, serves only to pass the phosphoryl group to the SpoOB phosphotransferase. However, SpoOF is the target of negative regulatory forces that will be described later. SpoOB passes its phosphoryl group to SpoOA, resulting in activation of its DNA-binding and transcriptional properties. SpoOA is both a repressor and an activator of gene transcription and binds to DNA specifically through a 0A box (Strauch *et al.*, 1990).

Phosphorylated SpoOA represses the transcription of the *abrB* gene, which is highly expressed in cells growing in nutrient-rich conditions. AbrB is a repressor of the expression of a large array of genes encoding stationary phase functions (Strauch & Hoch, 1993). The major consequence of SpoOA~P-mediated repression of *abrB* is to release these genes from AbrB-mediated arrest to their expression. SpoOA~P is an activator of transcription of the stage II genes of sporulation and other genes whose functions are yet to be determined (Spiegelman *et al.*, 1995). The stage II genes encode proteins of the sigma factor cascade and their regulators that determine compartment-specific gene expression and timing of gene expression during development (Stragier & Losick, 1996). While details of the sigma cascade and compartmentalized gene expression (Shapiro & Losick, 2000) make fascinating reading and are a tour de force of science, space considerations limit any discussion of sporulation past stage 0.

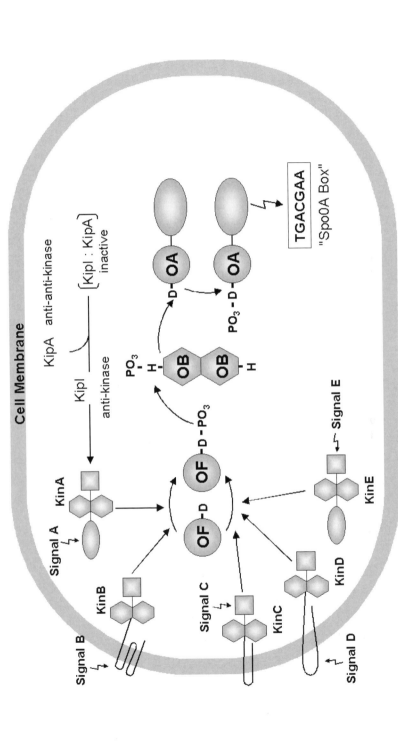

Fig. 6. Schematic of the sporulation phosphorelay, the sensor kinases and the KipI-KipA regulation within a cell. All of the sensor kinases are capable of phosphate input into the phosphorelay. KipI is an antikinase known to inhibit KinA. The Spo0A transcription factor binds to the Spo0A box located in promoters it regulates.

REGULATION OF PHOSPHATE FLOW IN THE PHOSPHORELAY

Regulation of the cellular concentration of Spo0A~P is crucial in the decision between growth and sporulation. The cell has evolved several means to ensure that the Spo0A~P level reflects a consensus of cellular signals either for or against growth or sporulation. At the sensor kinase level, signal concentration must be the major factor promoting sporulation, but the presence of PAS domain on three of those sensor kinases suggests that redox sensing or ATP levels might modulate sensor kinase activity irrespective of signal concentration. KinA activity is also subject to control by a protein inhibitor of its activity whose concentration is regulated by nitrogen availability (Fig. 6). The inhibitor, KipI, binds to the ATP-binding subdomain of KinA and prevents ATP hydrolysis and KinA activation (Wang *et al.*, 1997). This anti-kinase activity of KipI does not affect the transfer of phosphoryl groups to Spo0F from phosphorylated KinA. KipI activity is in turn regulated by an anti-anti-kinase, KipA, expressed from the same operon as KipI. A KipI:KipA complex is inactive as an inhibitor. The details of how KipI chooses between these two proteins have not been worked out.

The major negative regulation of the phosphorelay occurs at the response regulators Spo0F and Spo0A. The aspartyl-phosphate of activated response regulators is a meta-stable mixed anhydride that hydrolyses rather easily to return the response regulator to the inactive form. Some response regulators such as CheY involved in chemotaxis turn over the phosphoryl group in a matter of seconds but both Spo0F~P and Spo0A~P are longer lived. The cell has evolved a complex array of proteins that promote the turn-over of the phosphoryl group at both Spo0F and Spo0A.

Spo0A~P is specifically hydrolysed by a related family of small proteins including Spo0E, YisI and YnzD (Perego, 2001) (Fig. 7). Spo0E is expressed as an 85-amino-acid protein and can be mutationally activated by deletion of the C-terminal 25 residues (Ohlsen *et al.*, 1994). It is believed that these C-terminal residues modulate the activity of Spo0E on Spo0A~P and it has been suggested that this region of Spo0E may be the site of regulation by other effector proteins (Perego & Hoch, 1991). The YisI and YnzD proteins were identified by homology to Spo0E and were proven to be Spo0A~P phosphatases by *in vitro* studies (Perego, 2001). Both of these proteins lack the putative C-terminal regulatory region. The genes for these proteins are transcribed differently: *ynzD* is induced by glucose; *yisI* is constitutive; and *spo0E* is induced by nutrient deprivation. While transcription regulation may explain the activity of YnzD, there must be other factors controlling the activity of these proteins on Spo0A~P.

Spo0F~P is dephosphorylated by several proteins of the Rap family of phosphatases (Perego, 1998). The best studied of these is RapA. RapA activity on Spo0F~P is regulated by transcription of the *rapA* gene and by an export–import control circuit

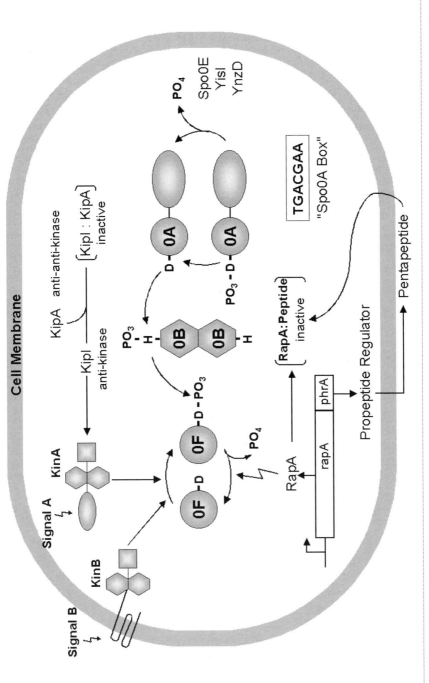

Fig. 7. Negative regulators of the sporulation phosphorelay. The Spo0E, YisI and YnzD proteins are phosphatases of Spo0A~P. RapA and some other members of the Rap family are phosphatases of Spo0F~P. RapA activity is regulated by a pentapeptide arising from the *phrA* gene product and processed as shown.

involving an inhibitor peptide (Perego, 1997) (Fig. 7). The *rapA* gene is co-transcribed with the *phrA* gene, which encodes a protein of 44 amino acids. This protein has the properties of an exported protein, including an N-terminal signal sequence and a consensus signal peptidase site. The inhibitor part of this protein is the C-terminal five amino acids. Secretion and proteolysis of PhrA gives rise to the C-terminal pentapeptide located external to the cytoplasmic membrane (Perego, 1997). What its role is in this location remains a mystery, but it may have something to do with cell–cell communication (Perego & Hoch, 1996). The pentapeptide is re-imported by the oligopeptide permease to the cytoplasm, where it is a specific inhibitor of the RapA phosphatase. Thus it appears that transcription of RapA and PhrA produces both proteins simultaneously and RapA is active on SpoOF~P as long as the PhrA pentapeptide remains external to the cytoplasm. One regulatory possibility for this scenario is that this forms a timing circuit for regulating RapA activity and other factors control the timing interval of this circuit. Regardless of the reason behind this control mechanism, the pentapeptide is very important for inhibiting RapA activity since a deletion of the *phrA* gene or a mutation blocking the oligopeptide permease severely decreases sporulation because of uncontrolled RapA activity. It is worth considering this activity in the context of the broad description of peptide signal/sensor molecules in the chapter by Morrison in this book.

There seems little doubt that the physiological role of the RapA phosphatase is to prevent sporulation. Because transcription of *rapA* is controlled by ComA~P (Mueller *et al.*, 1992), a positive regulator of the competence system, it is likely that RapA is produced to ensure sporulation does not initiate in that portion of the population in the competent state since both sporulation and competence occur in stationary phase in *B. subtilis*. Another Rap phosphatase, RapB, is produced in early exponential phase in the presence of glucose (M. Perego, unpublished). RapB acts as an additional control on sporulation during growth.

Thus Rap phosphatases serve to help guarantee that sporulation does not initiate in conditions favourable to growth or other competing states. There is a delicate balance between growth and sporulation. If any of these restraints on sporulation is deleted allowing sporulation to initiate at inappropriate times, there is strong selective pressure to block sporulation by acquisition of secondary mutations in sporulation phosphorelay genes (Spiegelman *et al.*, 1990).

The sporulation phosphorelay illustrates how signal transduction systems that regulate whole-cell events are subject to complex interacting regulation. The basis for this regulation is sensor kinase–phosphatase competition, a theme found throughout life at all levels. In signal transduction, it is easier to identify the participants than it is to divine how they are regulated at the protein level. Transcription regulation takes on unwarranted importance simply because it is easy to measure, while assaying protein activities is much more difficult in an *in vivo* situation. The challenge is understanding the flow of information and how regulators and their effector molecules influence this flow.

ACKNOWLEDGEMENTS

This research was supported, in part, by Grant GM19416 from the National Institute of General Medical Sciences, National Institutes of Health, USPHS. This is publication 14288-MEM from the Department of Molecular and Experimental Medicine at The Scripps Research Institute.

REFERENCES

Bilwes, A. M., Alex, L. A., Crane, B. R. & Simon, M. I. (1999). Structure of CheA, a signal-transducing histidine kinase. *Cell* **96**, 131–141.

Bilwes, A. M., Quezada, C. M., Croal, L. R., Crane, B. R. & Simon, M. I. (2001). Nucleotide binding by the histidine kinase CheA. *Nat Struct Biol* **8**, 353–360.

Burbulys, D., Trach, K. A. & Hoch, J. A. (1991). The initiation of sporulation in *Bacillus subtilis* is controlled by a multicomponent phosphorelay. *Cell* **64**, 545–552.

Chang, C., Kwok, S. F., Bleecker, A. B. & Meyerowitz, E. M. (1993). Arabidopsis ethylene-response gene *ETR1*: similarity of product to two-component regulators. *Science* **262**, 539–544.

Domian, I. J., Quon, K. C. & Shapiro, L. (1997). Cell type-specific phosphorylation and proteolysis of a transcriptional regulator controls the G1-to-S transition in a bacterial cell cycle. *Cell* **90**, 415–424.

Hoch, J. A. & Silhavy, T. J. (1995). *Two-Component Signal Transduction*. Washington, DC: American Society for Microbiology.

Hoch, J. A. & Varughese, K. I. (2001). Keeping signals straight in phosphorelay signal transduction. *J Bacteriol* **183**, 4941–4949.

Jiang, M., Shao, W., Perego, M. & Hoch, J. A. (2000). Multiple histidine kinases regulate entry into stationary phase and sporulation in *Bacillus subtilis*. *Mol Microbiol* **38**, 535–542.

Lewis, R. J., Brannigan, J. A., Muchova, K., Barak, I. & Wilkinson, A. J. (1999). Phosphorylated aspartate in the structure of a response regulator protein. *J Mol Biol* **294**, 9–15.

Mandelstam, J. & Higgs, S. A. (1974). Induction of sporulation during synchronized chromosome replication in *Bacillus subtilis*. *J Bacteriol* **120**, 38–42.

Mueller, J. P., Bukusoglu, G. & Sonenshein, A. L. (1992). Transcriptional regulation of *Bacillus subtilis* glucose starvation-inducible genes: control of *gsiA* by the ComP-ComA signal transduction system. *J Bacteriol* **174**, 4361–4373.

Ohlsen, K. L., Grimsley, J. K. & Hoch, J. A. (1994). Deactivation of the sporulation transcription factor SpoOA by the SpoOE protein phosphatase. *Proc Natl Acad Sci U S A* **91**, 1756–1760.

Perego, M. (1997). A peptide export-import control circuit modulating bacterial development regulates protein phosphatases of the phosphorelay. *Proc Natl Acad Sci U S A* **94**, 8612–8617.

Perego, M. (1998). Kinase-phosphatase competition regulates *Bacillus subtilis* development. *Trends Microbiol* **6**, 366–370.

Perego, M. (2001). A new family of aspartyl-phosphate phosphatases targeting the sporulation transcription factor SpoOA of *Bacillus subtilis*. *Mol Microbiol* **42**, 133–144.

Perego, M. & Hoch, J. A. (1991). Negative regulation of *Bacillus subtilis* sporulation by the *spoOE* gene product. *J Bacteriol* **173**, 2514–2520.

Perego, M. & Hoch, J. A. (1996). Cell-cell communication regulates the effects of protein aspartate phosphatases on the phosphorelay controlling development in *Bacillus subtilis*. *Proc Natl Acad Sci U S A* **93**, 1549–1553.

Shapiro, L. & Losick, R. (2000). Dynamic spatial regulation in the bacterial cell. *Cell* **100**, 89–98.

Spiegelman, G., Van Hoy, B., Perego, M., Day, J., Trach, K. & Hoch, J. A. (1990). Structural alterations in the *Bacillus subtilis* SpoOA regulatory protein which suppress mutations at several spo0 loci. *J Bacteriol* **172**, 5011–5019.

Spiegelman, G. B., Bird, T. H. & Voon, V. (1995). Transcription regulation by the *Bacillus subtilis* response regulator SpoOA. In *Two-Component Signal Transduction*, pp. 159–179. Edited by J. A. Hoch & T. J. Silhavy. Washington, DC: American Society for Microbiology.

Stephenson, K. & Hoch, J. A. (2001). PAS-A domain of phosphorelay sensor kinase A: a catalytic ATP-binding domain involved in the initiation of development in *Bacillus subtilis*. *Proc Natl Acad Sci U S A* **98**, 15251–15256.

Stragier, P. & Losick, R. (1996). Molecular genetics of sporulation in *Bacillus subtilis*. *Annu Rev Genet* **30**, 297–341.

Strauch, M. A. & Hoch, J. A. (1993). Transition state regulators: sentinels of *Bacillus subtilis* post-exponential gene expression. *Mol Microbiol* **7**, 337–342.

Strauch, M. A., Webb, V., Spiegelman, G. & Hoch, J. A. (1990). The SpoOA protein of *Bacillus subtilis* is a repressor of the *abrB* gene. *Proc Natl Acad Sci U S A* **87**, 1801–1805.

Taylor, B. L. & Zhulin, I. B. (1999). PAS domains: internal sensors of oxygen, redox potential, and light. *Microbiol Mol Biol Rev* **63**, 479–506.

Thomason, P. & Kay, R. (2000). Eukaryotic signal transduction via histidine-aspartate phosphorelay. *J Cell Sci* **113**, 3141–3150.

Tomomori, C., Tanaka, T., Dutta, R. & 10 other authors (1999). Solution structure of the homodimeric core domain of *Escherichia coli* histidine kinase EnvZ. *Nat Struct Biol* **6**, 729–734.

Tsuzuki, M., Ishige, K. & Mizuno, T. (1995). Phosphotransfer circuitry of the putative multi-signal transducer, ArcB, of *Escherichia coli*: *in vitro* studies with mutants. *Mol Microbiol* **18**, 953–962.

Uhl, M. A. & Miller, J. F. (1996). Integration of multiple domains in a two-component sensor protein: the *Bordetella pertussis* BvgAS phosphorelay. *EMBO J* **15**, 1028–1036.

Varughese, K. I., Madhusudan, Zhou, X. Z., Whiteley, J. M. & Hoch, J. A. (1998). Formation of a novel four-helix bundle and molecular recognition sites by dimerization of a response regulator phosphotransferase. *Mol Cell* **2**, 485–493.

Wang, L., Grau, R., Perego, M. & Hoch, J. A. (1997). A novel histidine kinase inhibitor regulating development in *Bacillus subtilis*. *Genes Dev* **11**, 2569–2579.

Zapf, J. W., Sen, U., Madhusudan, Hoch, J. A. & Varughese, K. I. (2000). A transient interaction between two phosphorelay proteins trapped in a crystal lattice reveals the mechanism of molecular recognition and phosphotransfer in signal transduction. *Structure Fold Des* **8**, 851–862.

A stranglehold on a transcriptional activator by its partner regulatory protein – the case of the NifL–NifA two-component regulatory system

Ray Dixon

Department of Molecular Microbiology, John Innes Centre, Colney Lane, Norwich NR4 7UH, UK

INTRODUCTION

The discovery of two-component regulatory systems has provided a landmark in the history of microbial gene regulation. The ubiquitous nature of these systems and their abundance in individual bacterial genomes, coupled with the elucidation that two-component systems utilize a common phosphotransfer chemistry for sensory communication, has led to the realization that these components play a major role in bacterial signal transduction in response to diverse environmental signals (Perraud et al., 1999; Stock et al., 2000). In the classical two-component systems, the input signal is sensed by a histidine protein kinase (HPK) and transmitted to a response regulator (RR) to elicit the response. The signal is perceived by the sensor domain of the HPK and is transmitted to the transmitter domain of the kinase to control autophosphorylation on a conserved histidine residue (Fig. 1). The phosphoryl group is subsequently transferred to a conserved aspartate residue on the receiver domain of the RR, invoking a conformational change which controls the activity of its output domain. The output response of these systems is therefore controlled by the level of phosphorylation of the RR.

The core domains in two-component systems directly involved in phosphotransfer, namely the transmitter domain of the HPK and the receiver domain of the RR, are conserved and have a modular architecture (Robinson et al., 2000). Nevertheless, in most two-component signalling pathways there is a one-to-one relationship between a given HPK and its RR such that most HPKs do not phosphorylate multiple RR targets. Although 'cross-talk' between systems has been observed, specificity is generally maintained in interactions between an HPK and its cognate RR. Biochemical and structural

SGM symposium 61: Signals, switches, regulons and cascades: control of bacterial gene expression.
Editors D. A. Hodgson, C. M. Thomas. Cambridge University Press. ISBN 0 521 81388 3 ©SGM 2002.

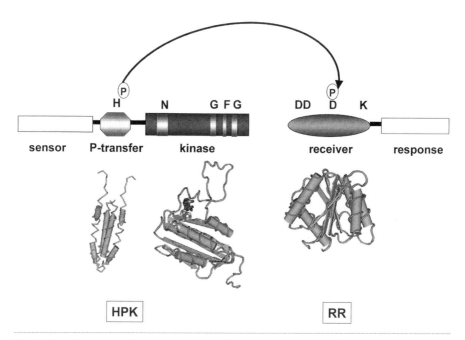

Fig. 1. Domain structure of two-component regulatory systems. Conserved domains are shown in dark fill and conserved amino acid residues are shown as single letter code. Three-dimensional structures of the 'H' domain of EnvZ (Tomomori *et al.*, 1999), the 'kinase' domain of EnvZ (Tanaka *et al.*, 1998) and the receiver domain represented by CheY (Stock *et al.*, 1989) are shown beneath the conserved domains.

studies have provided a detailed molecular description of the functions of these conserved core domains (Fig. 1). The transmitter domain of the HPKs consists of at least two subdomains. The 'H' domain contains the autophosphorylation site and is a dimer made up of two antiparallel helices forming a four-helix bundle in which the conserved histidine residues are positioned midway along opposing solvent-exposed faces of the bundle (Bilwes *et al.*, 1999; Tomomori *et al.*, 1999). The C-terminal region of the transmitter, known as the 'kinase' domain, contains four conserved regions designated N, G1, F and G2 which are implicated in nucleotide binding. The fold of the kinase domain is distinct from that of other known kinases but has an α/β sandwich fold common to a family of ATPases which includes DNA gyrase B, Hsp90 and MutL (Tanaka *et al.*, 1998; Bilwes *et al.*, 2001). The receiver domain of RRs is characterized by a conserved set of residues, including an aspartic acid residue at the site of phosphorylation, a pair of acidic residues close to the amino-terminus involved in divalent metal cation binding, and a lysine residue near the C-terminus of the domain which contacts the phosphoryl group. The active site in the receiver domain catalyses phosphotransfer since small molecules such as acetyl phosphate and carbamoyl phosphate act as *in vitro* phosphodonors in the absence of the HPK. The fold of this domain is doubly wound α/β with five α helices surrounding a five-stranded parallel β sheet (Stock *et al.*, 1989). Structural studies on phospho-activated forms of this domain indicate that reorientation of specific side chains is pivotal in driving the phosphorylation-induced conformational change (Stock & Da Re, 2000).

Despite the highly conserved architecture of the core domains of two-component regulatory systems there are some cases where transmitter and receiver domains may have diverged in terms of biochemical function. For example, unusual RRs have been identified in *Streptomyces coelicolor*. Although the proteins RedZ and WhiI are predicted to have a fold corresponding to the receiver domain, highly conserved residues in the active site are missing (Ainsa *et al.*, 1999; Guthrie *et al.*, 1998). Conversely, although the response regulator BldM has the conserved active site residues, mutational analysis suggests that the putative phosphorylation site is not required for protein function and BldM could not be phosphorylated *in vitro* by small molecule phosphodonors (Molle & Buttner, 2000). The NifL–NifA system from *Azotobacter vinelandii* provides another example of an unusual two-component regulatory system in which the sensor protein NifL has a C-terminal domain similar to the transmitter domain of HPKs, with a potential 'H' domain and a 'kinase' domain containing conserved residues corresponding to those found in the N, G1, F and G2 regions of well-characterized HPKs (Blanco *et al.*, 1993; Woodley & Drummond, 1994). In contrast, its partner protein, the transcriptional activator NifA, is not a member of the RR family and does not contain a receiver domain. Nevertheless, the sensor protein NifL stringently controls the activity of the NifA protein in response to environmental signals by a mechanism which apparently does not involve phosphotransfer but requires stoichiometric interactions between the two proteins (Austin *et al.*, 1994; Lee *et al.*, 1993). The properties of this unusual two-component system will be considered further in this article.

OVERVIEW OF THE NifL–NifA SYSTEM

Nitrogen fixation catalysed by the enzyme nitrogenase is a highly energy-dependent, oxygen-sensitive process which is only required when cells are subjected to conditions of nitrogen limitation. Consequently, the biosynthesis of this enzyme is tightly regulated at the transcriptional level. In diazotrophic representatives of the proteobacteria, transcription of the nitrogen fixation (*nif*) genes is activated by NifA, a potent transcriptional activator. The activity of NifA is regulated in response to environmental conditions (oxygen and fixed nitrogen status), and in some nitrogen-fixing organisms this control is exerted via a second regulatory protein, NifL, which is co-expressed with NifA via translational coupling of the *nifLA* operon (Dixon, 1998). The activity of NifA is thus kept in check by NifL, which acts as an anti-activator to prevent transcriptional activation under adverse environmental conditions. Unlike many two-component systems which are responsive to only a single signal input, the NifL–NifA system senses multiple environmental cues, namely the redox, carbon and nitrogen status. This regulatory protein complex must therefore integrate three antagonistic signals to control transcription of nitrogen fixation (*nif*) genes. Although this system can be described as two-component it differs from classical systems in a number of ways. Firstly, although *A. vinelandii* NifL has an HPK-like domain structure, neither autophosphorylation of NifL nor phosphotransfer from NifL to NifA has been demonstrated. Secondly, the ability of NifL to control NifA activity requires stoichiometric levels of the two proteins and formation of a NifL–NifA protein complex has been demonstrated, implying that protein–protein interaction forms the basis of signal communication (Henderson *et al.*, 1989; Money *et al.*, 1999). Thirdly, NifA is not a member of the RR family and does not have a recognizable receiver domain or phosphorylation site.

DOMAIN STRUCTURE OF NifL AND NifA

Sensor proteins of the HPK family perceive signals via an amino-terminal sensory domain and convey this signal to the C-terminal domain to activate phosphotransfer. Initial sequence analysis revealed that *A. vinelandii* NifL conforms to this overall structure, having an amino-terminal sensory domain and a C-terminal transmitter-like domain, which as mentioned above has conserved regions found in characterized HPKs. Each domain is probably composed of discrete subdomains. Purification of NifL revealed that it is a flavoprotein with FAD as prosthetic group (Hill *et al.*, 1996) and the non-covalently bound flavin was shown to be located in the amino-terminal domain (Söderbäck *et al.*, 1998). The first 150 residues of NifL constitute a PAS domain, a conserved protein motif observed in proteins which sense oxygen, redox or light represented amongst all kingdoms of life (Ponting & Aravind, 1997; Taylor & Zhulin, 1999; Zhulin *et al.*, 1997). Structural analysis of PAS domains has revealed a common profilin-like protein fold consisting of a five-stranded anti-parallel beta barrel decorated with α helices to form a hand-like structure enclosing a hydrophobic pocket (Gong *et al.*, 1998; Morais *et al.*, 1998; Pellequer *et al.*, 1998). This pocket can accommodate different cofactors required for signal transduction. For example, in Photoactive Yellow Protein (PYP), conformational changes in the PAS domain are mediated by a 4-hydroxycinnamoyl chromophore upon photoactivation (Pellequer *et al.*, 1998). In the oxygen-responsive HPK FixL, which is a haemoglobin-like protein, the haem moiety is located in the PAS domain and upon binding of oxygen, the spin state of the haem iron is conveyed to the PAS fold to mediate signal transduction (Gong *et al.*, 1998). In the case of NifL, there is strong evidence that the FAD cofactor is enclosed within the PAS domain and mediates the redox response (Söderbäck *et al.*, 1998).

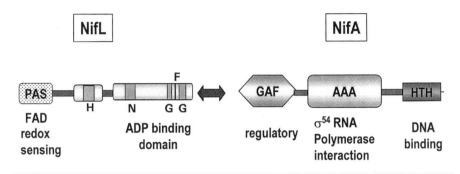

Fig. 2. Domain structure of NifL and NifA.

The remainder of the amino-terminal domain of NifL, from residues 150 to 300, is of unknown function and although conserved in NifL proteins, it does not have a counterpart in the databases. As mentioned above, the C-terminal domain of *A. vinelandii* NifL is similar to the transmitter domain of the HPK family and can be subdivided into an 'H'-like domain and a 'kinase'-like domain containing the N, G1, F and G2 motifs found in other HPKs (Fig. 2). The potential 'H'-like domain in *A. vinelandii* NifL is not as highly conserved as in other members of the family and most substitutions in the conserved histidine residue do not impair NifL function (Woodley & Drummond, 1994), commensurate with *in vitro* evidence that sensory transduction by NifL does not involve phosphorylation (Austin *et al.*, 1994). In contrast to *A. vinelandii* NifL, the C-terminal domain of *Klebsiella pneumoniae* NifL shows very limited homology to the HPK family (Blanco *et al.*, 1993; Drummond & Wootton, 1987).

NifA proteins are nitrogen-fixation-specific transcriptional activators belonging to the $\sigma^{54}(\sigma^N)$-dependent family of transcriptional activators. Members of this family bind to enhancer-like DNA target sequences, catalyse ATP hydrolysis and activate transcription by interacting with σ^{54}-RNA polymerase holoenzyme to promote the formation of open promoter complexes in which the DNA strands close to the transcription start-site are locally melted (Buck *et al.*, 2000). NifA proteins have a modular domain architecture consisting of three domains. The amino-terminal domain has a regulatory function and comprises a GAF domain, a ubiquitous motif found in signalling proteins from all kingdoms of life. This domain is observed in proteins which bind cyclic GMP, in plant phytochromes and in the formate-responsive σ^{54}-dependent activator FhlA (Aravind & Ponting, 1997). Recent structural determination of *Saccharomyces cerevisiae* YKG9, a member of the GAF family, reveals that the fold of this motif resembles that of the PAS domain, and similarities in the binding pockets of the two motifs have led to the suggestion that the GAF fold may also bind a variety of cofactors (Ho *et al.*, 2000). The central domain of NifA is common to all members of the σ^{54}-dependent family of enhancer binding proteins and contains sites for binding and hydrolysis of nucleoside triphosphates as well as determinants for the interaction with σ^{54}-RNA polymerase holoenzyme (Buck *et al.*, 2000; Morett & Segovia, 1993). This domain belongs to the AAA+ family of chaperone-like ATPases which function to remodel protein complexes (Neuwald *et al.*, 1999). The C-terminal domain of NifA has a DNA-binding function and includes a helix–turn–helix motif required for recognition of specific upstream activator sequences in *nif* promoters (Missaillidis *et al.*, 1999) (Fig. 2).

SIGNAL SENSING BY NifL

Redox/oxygen sensing

A. vinelandii is an obligate aerobe and has to reconcile aerobic respiratory energy generation with the requirement to prevent nitrogenase from oxygen damage. Consequently, NifL is required to prevent expression of nitrogenase when the organism is unable to consume excess oxygen. The occurrence of non-covalently bound flavin in NifL suggested that the FAD moiety could be involved in the redox response of this protein. NifL is a tetrameric protein containing 1 mol FAD per monomer and exhibits a classical flavin spectrum with an absorption maximum at 446 nm which is bleached upon reduction with sodium dithionite (Hill *et al.*, 1996). In the oxidized form, NifL inhibits NifA activity, but when reduced by sodium dithionite or xanthine oxidase, NifL is not competent to prevent NifA from activating transcription (Hill *et al.*, 1996; Macheroux *et al.*, 1998). The redox potential of the FAD_{red}/FAD_{ox} couple in *A. vinelandii* NifL is ~ -225 mV at pH 8 and the protein can be reduced enzymically by a variety of electron donors. Enzymic reduction of NifL with the NAD(P)H-dependent oxidoreductase Hmp occurs with stoichiometric consumption of NADH, indicative of a two-electron reduction without the appearance of the anionic flavin semiquinone as an intermediate (Macheroux *et al.*, 1998). It is possible that NifL is a general redox sensor, responding to the availability of reducing equivalents in the cell, or, alternatively, a specific electron donor may be required. The reduced form of NifL is rapidly oxidized upon exposure to air, suggesting that oxygen may act as a physiological electron acceptor. Hydrogen peroxide is generated upon oxidation, again suggesting that the redox switch involves the transfer of two electrons (Little *et al.*, 1999).

As mentioned above, the FAD cofactor is located within the PAS domain of NifL. When this domain is removed, producing a polypeptide lacking the first 150 residues of the protein, the redox response is lost since the truncated protein lacks FAD. Nevertheless, the truncated protein is still able to respond to the presence of fixed nitrogen, indicating that the PAS domain has an exclusive role in oxygen sensing and that NifL must respond to carbon and fixed nitrogen by other mechanisms (Söderbäck *et al.*, 1998).

Whereas native NifL is a tetramer, the PAS-deleted form of the protein purifies as a dimer, suggesting a role for the PAS domain in oligomerization. However, the association state of the PAS domain is not influenced by the redox state of the FAD (R. Little, A. Leech, M. Hefti & R. Dixon, unpublished results). The redox switch is thus most likely mediated via conformational changes in the PAS fold which are communicated to the other domains of NifL via intramolecular signal transduction.

NifL from *K. pneumoniae*, a facultative anaerobe, also contains FAD as prosthetic group and has a redox potential of ~ -275 mV at pH 8 (Klopprogge & Schmitz, 1999; Schmitz, 1997). Iron is necessary in the medium to prevent *K. pneumoniae* NifL from inactivating NifA under anaerobic nitrogen-limiting conditions *in vivo*. Since NifL contains neither iron nor acid-labile sulphur, this may reflect a requirement for iron in a component that is involved in signalling to NifL (Schmitz *et al.*, 1996). Recently it has been demonstrated that the presence of the global redox regulator FNR is also required to prevent *K. pneumoniae* NifL from inhibiting NifA under anaerobic conditions. Since FNR contains an oxygen-sensitive [4Fe–4S] cluster, it is possible that the iron requirement for NifL inactivation reflects the involvement of FNR in the redox signalling pathway (Grabbe *et al.*, 2001). FNR is unlikely to be competent to donate electrons directly to NifL so the involvement of this global oxygen sensor is likely to reflect its role in regulating the expression of physiological electron donors, which in *K. pneumoniae* are potentially components of the anaerobic electron chain.

Nucleotide binding and carbon sensing

The activity of *A. vinelandii* NifL, as measured by its ability to inhibit transcriptional activation by NifA, is influenced by the presence of adenosine nucleotides which increase the inhibitory activity of NifL (Eydmann *et al.*, 1995). ADP appears to be the most potent effector but the increase in inhibition is also observed with ATP, and non-hydrolysable analogues of ATP such as ATPγS and AMPPNP. The presence of adenosine nucleotides also overrides the redox response of *A. vinelandii* NifL since ADP stimulates the inhibitory activity of NifL even when the flavin is in the reduced form (Hill *et al.*, 1996). Limited proteolysis experiments indicate that the C-terminal domain of NifL is rapidly degraded in the absence of nucleotides, but in the presence of ATP a polypeptide corresponding to the 'kinase'-like subdomain of NifL is protected from cleavage. More enhanced protection is observed in the presence of ADP, whereby the complete C-terminal domain, including the 'H' and the 'kinase'-like subdomains, is more resistant to proteolysis. These observations suggest that adenosine nucleotides may increase the inhibitory activity of NifL by influencing the conformation of the C-terminal domain. Similar protection from proteolysis was also observed with the isolated 'kinase' subdomain, suggesting that the observed conformational changes are a consequence of the binding of adenosine nucleotides to this region (Söderbäck *et al.*,

1998). This is not unexpected, bearing in mind the presence of conserved sequences corresponding to the N, G1, F and G2 motifs which have been shown to be required for ATP binding in other HPKs. Mutations in conserved residues in the N, G1 and G2 regions, which significantly decrease nucleotide binding affinity, eliminate both redox and nitrogen sensing by NifL, implicating nucleotide binding to this domain as a major determinant of NifL activity (S. Perry & R. Dixon, unpublished). Moreover, the presence of adenosine nucleotides also significantly enhances the stability of protein complexes formed between NifL and NifA (Money *et al.*, 1999).

Potentially *A. vinelandii* NifL may have evolved from a classical HPK to utilize nucleotide binding as a sensing function, rather than to catalyse ATP hydrolysis and promote communication via phosphotransfer. NifL has ~10-fold higher affinity for ADP (apparent K_d 13 µM) compared with ATP (apparent K_d 130 µM), prompting the suggestion that NifL might sense the ADP/ATP ratio *in vivo* (Eydmann *et al.*, 1995; Söderbäck *et al.*, 1998). However, since the *in vivo* level of adenosine nucleotides is far in excess of these values it would be anticipated that the nucleotide-binding site(s) on NifL would be saturated under all environmental conditions. However, it has recently been demonstrated that the NifL–NifA system is directly responsive to an additional ligand, 2-oxoglutarate, which antagonizes the influence of adenosine nucleotides on NifL activity, thus relieving inhibition by NifL in the presence of ADP. This effect is specific to 2-oxoglutarate and is not observed with other components of the TCA cycle or related compounds such as 3-oxoglutarate or 2-oxobutyrate (Little *et al.*, 2000). Since 2-oxoglutarate is a key intracellular signal of the carbon status, the response of the NifL–NifA system to this metabolite may reflect an additional physiological switch which deactivates NifL in response to carbon availability. The response of the system to 2-oxoglutarate is within the physiological range, which in *Escherichia coli* varies between ~100 µM in carbon-starved cells to ~1 mM in nitrogen-starved cells. The *A. vinelandii* NifL–NifA system is apparently responsive to the intracellular level of 2-oxoglutarate *in vivo* when introduced into *E. coli* (Reyes-Ramirez *et al.*, 2001). Our recent data suggest that 2-oxoglutarate binds to NifA and that this interaction requires the amino-terminal GAF domain, commensurate with the proposed role of this domain in binding small signal molecules (R. Little & R. Dixon, unpublished data).

Nitrogen sensing

The ubiquitous signal transduction protein PII, which is found in bacteria, archaea and plants, has a key role in communicating the nitrogen status to diverse receptors to regulate nitrogen assimilation and the transcription of nitrogen regulated genes (Arcondeguy *et al.*, 2001; Ninfa & Atkinson, 2000). However, only in recent years has a link between this protein and the NifL–NifA system been established. In enteric bacteria, the nitrogen status is conveyed to the PII protein via uridylylation and this is likely to be the case in most proteobacteria. Covalent modification of PII is catalysed by the enzyme uridylyltransferase (UTase), which carries out both the forward reaction (uridylylation) and the reverse reaction (de-uridylylation) according to the level of intracellular metabolites. Glutamine, a key signal of the nitrogen status, allosterically modulates the activity of UTase, such that under nitrogen-sufficient conditions, when the level of this metabolite is relatively high, the enzyme removes uridylyl groups from PII. Conversely, under nitrogen-limiting conditions, when the level of glutamine is relatively low (~200 μM), the transferase reaction of the enzyme is favoured leading to uridylylation of PII (Ninfa *et al.*, 2000). In enteric bacteria, PII influences the activity of the enzyme adenylyltransferase, thus controlling the adenylylation state of glutamine synthetase and hence controlling nitrogen assimilation (Jiang *et al.*, 1998b). PII also interacts with the HPK NtrB to modulate the phosphorylation state of the RR NtrC, thus controlling the transcription of nitrogen regulated genes (Jiang & Ninfa, 1999; Jiang *et al.*, 1998a). However, there are likely to be hitherto undiscovered receptors which also interact with PII. The situation is further complicated by the recent finding that many bacteria contain more than one gene encoding a PII-like protein. For example, in enteric bacteria, the well-studied 'classical' PII protein is encoded by the *glnB* gene and a second PII-like protein, encoded by *glnK*, is transcribed under the control of NtrC and hence is itself nitrogen-regulated (van Heeswijk *et al.*, 1996). Although these proteins have similar structures and functions, their roles are not identical. In *E. coli*, GlnK appears only to be important for the regulation of those genes which are expressed under nitrogen starvation and require a high level of NtrC-P for transcriptional activation (Atkinson & Ninfa, 1998, 1999).

Transcription of the *K. pneumoniae nifLA* operon is activated by NtrC-P and hence is subject to nitrogen control, but the activity of the NifL–NifA system is itself also responsive to the nitrogen status and this response is mediated via NifL. It was initially thought, however, that nitrogen signalling to NifL did not involve the PII protein since mutations in *glnB* (encoding PII) and in *glnD* (encoding uridylyltransferase) did not influence the ability of NifL to perceive the nitrogen status (Edwards & Merrick, 1995; Holtel & Merrick, 1989). The more recent discovery of the second PII-like protein encoded by *glnK* prompted a re-evaluation of the mechanism of N-signalling to *K. pneumoniae* NifL. The first indication of the involvement of GlnK came from studies in *E. coli* where it was observed that *ntrC* mutations influenced NifL activity, even though transcription of the *nifLA* operon was engineered to be independent of nitrogen control (He *et al.*, 1997). This implicated a role for *glnK* in nitrogen control of NifL activity, since, as mentioned above, *glnK* transcription is NtrC-dependent. Subsequent work revealed that NifL inhibited NifA activity even under conditions of nitrogen starvation in *glnK* mutants, suggesting that GlnK had a role, either direct or indirect, in preventing NifL from inhibiting NifA under nitrogen-limiting conditions (He *et al.*, 1998; Jack *et al.*, 1999). This role for GlnK in relieving inhibition by NifL was shown to be independent of its uridylylation state, thus providing an explanation for the earlier observation that NifL remains responsive to fixed nitrogen in a *glnD* mutant (Edwards & Merrick, 1995). Although this nitrogen response can to a certain extent be explained by the increased level of expression of GlnK under nitrogen-limiting conditions, it does not explain the relatively rapid response of NifL when cells are subjected to high concentrations of fixed nitrogen (He *et al.*, 1998). Although under physiological conditions NifL appears to be specifically responsive to GlnK, some response is also observed when GlnB is overexpressed (Arcondéguy *et al.*, 1999). The situation is further complicated by the observation that GlnK and GlnB form heterotrimers *in vivo* (Forchhammer *et al.*, 1999; van Heeswijk *et al.*, 2000), so it is possible that the influence of GlnK on inhibition by NifL could be modulated via the uridylylation state of GlnB. When *K. pneumoniae* NifL is purified from cells grown under nitrogen excess, a specific absorbance at 420 nm is retained after denaturation and refolding of the protein, perhaps indicative of covalent modification of NifL in response to the nitrogen source (Klopprogge & Schmitz, 1999).

The *A. vinelandii* nitrogen fixation regulatory proteins respond to fixed nitrogen using a mechanism which is clearly different to that in *K. pneumoniae*. This organism contains a UTase/UR enzyme, encoded by a homologue of the *E. coli glnD* gene (Contreras *et al.*, 1991), but only a single PII-like protein has been identified, encoded by the *glnK* gene (Meletzus *et al.*, 1998). Mutations in both of these genes are apparently lethal, although some *glnD* mutations can be suppressed by the Y407F mutation in glutamine synthetase which prevents adenylylation of this enzyme. This is presumably because GS is essential for nitrogen assimilation in *A. vinelandii* under N-limiting conditions and if GS is adenylylated and consequently inactivated by the interaction of non-uridylylated PII with adenylyltransferase, the organism is unable to grow (Colnaghi *et al.*, 2001). *glnD* mutations also prevent nitrogen fixation in *A. vinelandii*, but this effect can be suppressed by secondary mutations which inactivate NifL, suggesting that the presence of a functional UTase/UR is necessary to prevent NifL from inhibiting NifA (Contreras *et al.*, 1991). Thus, unlike in enteric bacteria, *A. vinelandii* UTase/UR apparently has a role in the regulation of nitrogen fixation. However, the role of the PII-like protein encoded by the *glnK* gene has not been easy to ascertain since it has not been possible to obtain 'knockout' mutations in this gene (Meletzus *et al.*, 1998). When introduced into *E. coli*, the *A. vinelandii* NifL–NifA system is responsive to both oxygen and fixed nitrogen, suggesting that the *A. vinelandii* regulatory proteins interact appropriately with the enteric signal transduction components (Söderbäck *et al.*, 1998). This has provided an opportunity to compare the responses of the *A. vinelandii* and *K. pneumoniae* NifL–NifA systems in an enteric background. Surprisingly, under nitrogen-limiting conditions, the activity of NifL and NifA was not influenced in a *glnK* background or even in a *glnB glnK* double mutant background. The activity of *A. vinelandii* NifL was also not affected in an *ntrC* mutant background (Reyes-Ramirez *et al.*, 2001). Thus, unlike the case of *K. pneumoniae*, the enteric PII-like proteins are apparently not required to prevent *A. vinelandii* NifL from inhibiting NifA. Moreover, in a *glnB ntrC* double mutant strain, NifL did not respond to nitrogen regulation such that NifA was active in both the absence and presence of excess fixed nitrogen, suggesting that the absence of the PII-like proteins influences the ability of *A. vinelandii* NifL to respond to the nitrogen status. The activity of NifL in relation to the nitrogen status also correlated with the level of 2-oxoglutarate. In excess fixed nitrogen, the 2-oxoglutarate level was less than 50 μM but increased to ~1 mM under nitrogen-limiting conditions. In the *glnB ntrC* mutant, however, a high level of 2-oxoglutarate was observed in both nitrogen-limiting and nitrogen excess conditions, mirroring the inability of NifL to inhibit NifA in response to fixed N in this strain. This provides circumstantial evidence that 2-oxoglutarate may control *A. vinelandii* NifL–NifA activity *in vivo*, in accord with the *in vitro* observations (Reyes-Ramirez *et al.*, 2001). As mentioned above, 2-oxoglutarate acts as an allosteric effector, preventing inhibition of NifA activity by NifL in the presence of adenosine nucleotides. When purified *E. coli* PII (Ec PII) was added to a reaction mixture containing NifL, NifA, 2-oxoglutarate and ATP, the inhibitory activity of NifL was restored, demonstrating that PII acts negatively to activate NifL and inhibit NifA (Little *et al.*, 2000). This response was not seen when GTP replaced ATP, reflecting the requirement for adenosine nucleotides for NifL inhibition. It was also not observed with fully uridylylated *E. coli* PII (Ec PII-UMP), suggesting that modification of PII might modulate the interaction. *E. coli* GlnK was also ineffective in stimulating inhibition by NifL. The PII-like protein encoded by *A. vinelandii glnK* (Av GlnK) also activated the inhibitory function of NifL and this activation was influenced by the uridylylation state of Av GlnK, one modification per trimer being sufficient to cause a significant decrease in the level of inhibition (Little *et al.*, 2000). Current results indicate that Av GlnK interacts with NifL rather than NifA (R. Little & R. Dixon, unpublished observations).

These observations suggest a model for the response of the *A. vinelandii* NifL–NifA system to the nitrogen status in which an increase in the 2-oxoglutarate concentration under nitrogen-limiting conditions serves to activate NifA, provided that Av GlnK is uridylylated. Under conditions of nitrogen excess, the concentration of 2-oxoglutarate decreases and Av GlnK is de-uridylylated, providing conditions for activation of the inhibitory form of NifL by this PII-like protein. Although no equivalent biochemical experiments have been performed with *K. pneumoniae* NifL–NifA, the mechanism of the nitrogen response is clearly different in each system since in the latter case the *in vivo* experiments suggest that GlnK is required to inactivate NifL under nitrogen-limiting conditions whereas in the *A. vinelandii* system non-modified Av GlnK is required to activate NifL under conditions of nitrogen excess (Fig. 3).

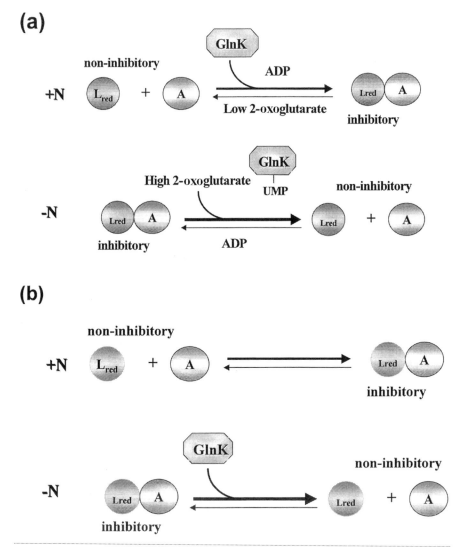

Fig. 3. Model for control of the NifL–NifA regulatory system in response to nitrogen status by the PII-like signal transduction proteins. (a) Regulation in *A. vinelandii*. Under conditions of nitrogen excess (+N), Av GlnK acts to increase the inhibitory function of NifL, thus forming a complex in which the transcriptional activator NifA is inactivated. Under nitrogen-limiting conditions (−N), Av GlnK is uridylylated and the concentration of 2-oxoglutarate increases, leading to dissociation of the complex, leaving NifA free to activate transcription. (b) Regulation in *K. pneumoniae*. Under conditions of nitrogen excess (+N), phosphorylated NtrC (which is required to activate *glnK* transcription) is limiting and there is insufficient GlnK available to prevent complex formation between NifL and NifA. Under conditions of nitrogen limitation (−N), GlnK prevents complex formation between NifL and NifA, thus allowing NifA to activate transcription. It should be noted that the role of GlnK in preventing formation of the NifL–NifA complex might be indirect in this case since no detailed biochemical experiments have been performed with the *K. pneumoniae* system.

INTERACTION OF NifL WITH NifA

As mentioned above, the binding of adenosine nucleotides to *A. vinelandii* NifL promotes complex formation with NifA (Money *et al.*, 1999). This interaction requires magnesium ions and the C-terminal domain of NifL but is not detectable with a truncated polypeptide comprising the isolated 'kinase' subdomain of NifL, congruent with the observation that this subdomain is not competent to inhibit NifA activity (Söderbäck *et al.*, 1998). The isolated amino-terminal domain of NifL also does not form complexes with NifA but an interaction is detectable with the complete C-terminal domain of NifL (residues 257–519) in a yeast two-hybrid assay (Lei *et al.*, 1999). Protein footprinting experiments suggest that the interaction with NifA protects NifL from proteolytic cleavage in the central region adjacent to the potential 'H' subdomain, implying that complex formation either protects this region or generates conformational changes within it (Money *et al.*, 2001). Taken together, these results suggest that the central region of *A. vinelandii* NifL (residues 260–360) may be involved in the interaction with NifA. Experiments with *K. pneumoniae* NifL apparently reveal a different mode of interaction since NifA activity is inhibited by a fusion protein in which the maltose-binding protein is linked to the C-terminal subdomain (residues 369–495) of NifL (Narberhaus *et al.*, 1995). These differences may arise as a consequence of the sequence divergence in the C-terminal region of these proteins since *K. pneumoniae* NifL lacks significant similarity to the HPK family and has not been shown to bind nucleotides.

The amino-terminal GAF domain of *K. pneumoniae* NifA appears to be involved in the response to NifL, since in its absence, NifA activity is more susceptible to inhibition by NifL *in vivo* (Drummond *et al.*, 1990). The GAF domain of *A. vinelandii* NifA also plays an important role in the interaction with NifL, since a truncated version of NifA lacking this domain does not form stable complexes *in vitro* (Money *et al.*, 2001). However, although the truncated protein is not responsive to the oxidized form of NifL, it is inhibited by NifL when ADP is present, indicating that the truncated NifA protein is still able to interact with NifL. The GAF domain may also regulate the nucleoside triphosphatase activity of the central AAA+ domain of NifA, in response to NifL, since the ATPase activity of NifA is no longer inhibited by NifL when the GAF domain is absent (Barrett *et al.*, 2001). Protein footprinting experiments indicate that the C-terminal region of the GAF domain is less susceptible to proteolytic cleavage in the presence of NifL, suggesting that the conformation of the domain may be altered in response to the interaction (Money *et al.*, 2001). Thus NifL may alter the conformation of the GAF domain to bring about intramolecular repression of the ATPase activity of the central domain of NifA (Barrett *et al.*, 2001). However, since the isolated central domains of both *K. pneumoniae* and *A. vinelandii* NifA are still responsive to NifL *in vitro* (Barrett *et al.*, 2001; Berger *et al.*, 1994) and the interaction with the central domain has been detected in the yeast two-hybrid system (Lei *et al.*, 1999), major contact sites for the interaction probably reside within the central domain of NifA. The interaction with NifL also inhibits the DNA-binding function of *A. vinelandii* NifA, but this is likely to be an indirect effect consequent upon alterations in the oligomerization state or the conformation of NifA (Barrett *et al.*, 2001).

CONCLUDING REMARKS

Adaptive responses in bacteria often require rapid signalling events which can be promoted by the phosphotransfer chemistry associated with most two-component regulatory systems. However, in some cases, regulatory complexes need to be relatively stable in order to maintain a given regulatory status over a relatively long time period. In the case of the NifL–NifA system, the transcriptional activator is sequestered in a complex by the sensory protein until physiological conditions are appropriate for nitrogen fixation. The ability of NifL to perceive multiple signals provides an integrated response to ensure that the activator is only free to productively interact with σ^{54}-RNA polymerase when all appropriate conditions have been met. Under oxidizing conditions, the switch is dominated by the 'ON' state of the flavin in NifL which ensures that NifA is inactivated in the binary NifL–NifA complex irrespective of the nitrogen status. Under reducing conditions, the switch is dominated by the PII-like regulatory proteins, thus only allowing dissociation of the complex under conditions of extreme nitrogen limitation.

ACKNOWLEDGEMENTS

I thank the BBSRC for financial support and Mike Merrick for his comments on the manuscript.

REFERENCES

Ainsa, J. A., Parry, H. D. & Chater, K. F. (1999). A response regulator-like protein that functions at an intermediate stage of sporulation in *Streptomyces coelicolor* A3(2). *Mol Microbiol* **34**, 607–619.

Aravind, L. & Ponting, C. P. (1997). The GAF domain: an evolutionary link between diverse phototransducing proteins. *Trends Biochem Sci* **22**, 458–459.

Arcondéguy, T., van Heeswijk, W. C. & Merrick, M. (1999). Studies on the roles of GlnK and GlnB in regulating *Klebsiella pneumoniae* NifL-dependent nitrogen control. *FEMS Microbiol Lett* **180**, 263–270.

Arcondeguy, T., Jack, R. & Merrick, M. (2001). PII signal transduction proteins, pivotal players in microbial nitrogen control. *Microbiol Mol Biol Rev* **65**, 80–105.

Atkinson, M. R. & Ninfa, A. J. (1998). Role of the GlnK signal transduction protein in the regulation of nitrogen assimilation in *Escherichia coli*. *Mol Microbiol* **29**, 431–447.

Atkinson, M. R. & Ninfa, A. J. (1999). Characterization of the GlnK protein of *Escherichia coli*. *Mol Microbiol* **32**, 301–313.

Austin, S., Buck, M., Cannon, W., Eydmann, T. & Dixon, R. (1994). Purification and *in vitro* activities of the native nitrogen fixation control proteins NifA and NifL. *J Bacteriol* **176**, 3460–3465.

Barrett, J., Ray, P., Sobczyk, A., Little, R. & Dixon, R. (2001). Concerted inhibition of the transcriptional activation functions of the enhancer-binding protein NifA by the anti-activator NifL. *Mol Microbiol* **39**, 480–494.

Berger, D. K., Narberhaus, F. & Kustu, S. (1994). The isolated catalytic domain of NifA, a bacterial enhancer-binding protein, activates transcription *in vitro*: activation is inhibited by NifL. *Proc Natl Acad Sci U S A* **91**, 103–107.

Bilwes, A. M., Alex, L. A., Crane, B. R. & Simon, M. I. (1999). Structure of CheA, a signal-transducing histidine kinase. *Cell* **96**, 131–141.

Bilwes, A. M., Quezada, C. M., Croal, L. R., Crane, B. R. & Simon, M. I. (2001). Nucleotide binding by the histidine kinase CheA. *Nat Struct Biol* **8**, 353–360.

Blanco, G., Drummond, M., Woodley, P. & Kennedy, C. (1993). Sequence and molecular analysis of the *nifL* gene of *Azotobacter vinelandii*. *Mol Microbiol* **9**, 869–880.

Buck, M., Gallegos, M. T., Studholme, D. J., Guo, Y. & Gralla, J. D. (2000). The bacterial enhancer-dependent sigma^{54} (sigma^{N}) transcription factor. *J Bacteriol* **182**, 4129–4136.

Colnaghi, R., Rudnick, P., He, L., Green, A., Yan, D., Larson, E. & Kennedy, C. (2001). Lethality of *glnD* null mutations in *Azotobacter vinelandii* is suppressible by prevention of glutamine synthetase adenylylation. *Microbiology* **147**, 1267–1276.

Contreras, A., Drummond, M., Bali, A., Blanco, G., Garcia, E., Bush, G., Kennedy, C. & Merrick, M. (1991). The product of the nitrogen fixation regulatory gene *nfrX* of *Azotobacter vinelandii* is functionally and structurally homologous to the uridylyl-transferase encoded by *glnD* in enteric bacteria. *J Bacteriol* **173**, 7741–7749.

Dixon, R. (1998). The oxygen-responsive NifL-NifA complex: a novel two-component regulatory system controlling nitrogenase synthesis in γ-proteobacteria. *Arch Microbiol* **169**, 371–380.

Drummond, M. H. & Wootton, J. C. (1987). Sequence of *nifL* from *Klebsiella pneumoniae*: mode of action and relationship to two families of regulatory proteins. *Mol Microbiol* **1**, 37–44.

Drummond, M. H., Contreras, A. & Mitchenall, L. A. (1990). The function of isolated domains and chimaeric proteins constructed from the transcriptional activators NifA and NtrC of *Klebsiella pneumoniae*. *Mol Microbiol* **4**, 29–37.

Edwards, R. & Merrick, M. (1995). The role of uridylyltransferase in the control of *Klebsiella pneumoniae nif* gene regulation. *Mol Gen Genet* **247**, 189–198.

Eydmann, T., Söderbäck, E., Jones, T., Hill, S., Austin, S. & Dixon, R. (1995). Transcriptional activation of the nitrogenase promoter in vitro: adenosine nucleosides are required for inhibition of NifA activity by NifL. *J Bacteriol* **177**, 1186–1195.

Forchhammer, K., Hedler, A., Strobel, H. & Weiss, V. (1999). Heterotrimerization of PII-like signalling proteins: implications for PII-mediated signal transduction systems. *Mol Microbiol* **33**, 338–349.

Gong, W., Hao, B., Mansy, S. S., Gonzalez, G., Gilles-Gonzalez, M. A. & Chan, M. K. (1998). Structure of a biological oxygen sensor: a new mechanism for heme-driven signal transduction. *Proc Natl Acad Sci U S A* **95**, 15177–15182.

Grabbe, R., Klopprogge, K. & Schmitz, R. A. (2001). Fnr is required for NifL-dependent oxygen control of nif gene expression in *Klebsiella pneumoniae*. *J Bacteriol* **183**, 1385–1393.

Guthrie, E. P., Flaxman, C. S., White, J., Hodgson, D. A., Bibb, M. J. & Chater, K. F. (1998). A response-regulator-like activator of antibiotic synthesis from *Streptomyces coelicolor* A3(2) with an amino-terminal domain that lacks a phosphorylation pocket. *Microbiology* **144**, 727–738.

He, L. H., Soupene, E. & Kustu, S. (1997). NtrC is required for control of *Klebsiella pneumoniae* NifL activity. *J Bacteriol* **179**, 7446–7455.

He, L., Soupene, E., Ninfa, A. & Kustu, S. (1998). Physiological role for the GlnK protein of enteric bacteria: relief of NifL inhibition under nitrogen-limiting conditions. *J Bacteriol* **180**, 6661–6667.

van Heeswijk, W., Hoving, S., Molenaar, D., Stegeman, B., Kahn, D. & Westerhoff, H. (1996). An alternative P_{II} protein in the regulation of glutamine synthetase in *Escherichia coli*. *Mol Microbiol* **21**, 133–146.

van Heeswijk, W. C., Wen, D., Clancy, P., Jaggi, R., Ollis, D. L., Westerhoff, H. V. & Vasudevan, S. G. (2000). The *Escherichia coli* signal transducers PII (GlnB) and GlnK form heterotrimers in vivo: fine tuning the nitrogen signal cascade. *Proc Natl Acad Sci U S A* **97**, 3942–3947.

Henderson, N., Austin, S. A. & Dixon, R. A. (1989). Role of metal ions in negative regulation of nitrogen fixation by the *nifL* gene product from *Klebsiella pneumoniae*. *Mol Gen Genet* **216**, 484–491.

Hill, S., Austin, S., Eydmann, T., Jones, T. & Dixon, R. (1996). *Azotobacter vinelandii* NifL is a flavoprotein that modulates transcriptional activation of nitrogen-fixation genes via a redox-sensitive switch. *Proc Natl Acad Sci U S A* **93**, 2143–2148.

Ho, Y. S., Burden, L. M. & Hurley, J. H. (2000). Structure of the GAF domain, a ubiquitous signaling motif and a new class of cyclic GMP receptor. *EMBO J* **19**, 5288–5299.

Holtel, A. & Merrick, M. J. (1989). The *Klebsiella pneumoniae* P_{II} protein (*glnB* gene product) is not absolutely required for nitrogen regulation and is not involved in NifL-mediated *nif* gene regulation. *Mol Gen Genet* **217**, 474–480.

Jack, R., De Zamaroczy, M. & Merrick, M. (1999). The signal transduction protein GlnK is required for NifL-dependent nitrogen control of *nif* gene expression in *Klebsiella pneumoniae*. *J Bacteriol* **181**, 1156–1162.

Jiang, P. & Ninfa, A. J. (1999). Regulation of autophosphorylation of *Escherichia coli* nitrogen regulator II by the PII signal transduction protein. *J Bacteriol* **181**, 1906–1911.

Jiang, P., Peliska, J. A. & Ninfa, A. J. (1998a). Reconstitution of the signal-transduction bicyclic cascade responsible for the regulation of Ntr gene transcription in *Escherichia coli*. *Biochemistry* **37**, 12795–12801.

Jiang, P., Peliska, J. A. & Ninfa, A. J. (1998b). The regulation of *Escherichia coli* glutamine synthetase revisited: role of 2-ketoglutarate in the regulation of glutamine synthetase adenylylation state. *Biochemistry* **37**, 12802–12810.

Klopprogge, K. & Schmitz, R. A. (1999). NifL of *Klebsiella pneumoniae*: redox characterization in relation to the nitrogen source. *Biochim Biophys Acta* **1431**, 462–470.

Lee, H.-S., Narberhaus, F. & Kustu, S. (1993). In vitro activity of NifL, a signal transduction protein for biological nitrogen fixation. *J Bacteriol* **175**, 7683–7688.

Lei, S., Pulakat, L. & Gavini, N. (1999). Genetic analysis of *nif* regulatory genes by utilizing the yeast two-hybrid system detected formation of a NifL-NifA complex that is implicated in regulated expression of *nif* genes. *J Bacteriol* **181**, 6535–6539.

Little, R., Hill, S., Perry, S., Austin, S., Reyes-Ramirez, F., Dixon, R. & Macheroux, P. (1999). Properties of NifL, a regulatory flavoprotein containing a PAS domain. In *Flavins and Flavoproteins 1999*, pp. 737–740. Edited by S. Ghisla, P. Kroneck, P. Macheroux & H. Sund. Berlin: Rudolf Weber.

Little, R., Reyes-Ramirez, F., Zhang, Y., van Heeswijk, W. C. & Dixon, R. (2000). Signal transduction to the *Azotobacter vinelandii* NifL-NifA regulatory system is influenced directly by interaction with 2-oxoglutarate and the PII regulatory protein. *EMBO J* **19**, 6041–6050.

Macheroux, P., Hill, S., Austin, S., Eydmann, T., Jones, T., Kim, S.-O., Poole, R. & Dixon, R. (1998). Electron donation to the flavoprotein NifL, a redox-sensing transcriptional regulator. *Biochem J* 332, 413–419.

Meletzus, D., Rudnick, P., Doetsch, N., Green, A. & Kennedy, C. (1998). Characterization of the *glnK-amtB* operon of *Azotobacter vinelandii*. *J Bacteriol* 180, 3260–3264.

Missaillidis, S., Jaseja, M., Ray, P., Chittock, R., Wharton, C. W., Drake, A. F., Buck, M. & Hyde, E. I. (1999). Secondary structure of the C-terminal DNA-binding domain of the transcriptional activator NifA from *Klebsiella pneumoniae*: spectroscopic analyses. *Arch Biochem Biophys* 361, 173–182.

Molle, V. & Buttner, M. J. (2000). Different alleles of the response regulator gene *bldM* arrest *Streptomyces coelicolor* development at distinct stages. *Mol Microbiol* 36, 1265–1278.

Money, T., Jones, T., Dixon, R. & Austin, S. (1999). Isolation and properties of the complex between the enhancer binding protein NifA and the sensor NifL. *J Bacteriol* 181, 4461–4468.

Money, T., Barrett, J., Dixon, R. & Austin, S. (2001). Protein-protein interactions in the complex between the enhancer binding protein NifA and the sensor NifL from *Azotobacter vinelandii*. *J Bacteriol* 183, 1359–1368.

Morais, J. H., Lee, A., Cohen, S. L., Chait, B. T., Li, M. & Mackinnon, R. (1998). Crystal structure and functional analysis of the HERG potassium channel N terminus: a eukaryotic PAS domain. *Cell* 95, 649–655.

Morett, E. & Segovia, L. (1993). The sigma54 bacterial enhancer-binding protein family: mechanism of action and phylogenetic relationship of their functional domains. *J Bacteriol* 175, 6067–6074.

Narberhaus, F., Lee, H.-S., Schmitz, R. A., He, L. & Kustu, S. (1995). The C-terminal domain of NifL is sufficient to inhibit NifA activity. *J Bacteriol* 177, 5078–5087.

Neuwald, A. F., Aravind, L., Spouge, J. L. & Koonin, E. V. (1999). AAA+: a class of chaperone-like ATPases associated with the assembly, operation, and disassembly of protein complexes. *Genome Res* 9, 27–43.

Ninfa, A. & Atkinson, M. (2000). PII signal transduction proteins. *Trends Microbiol* 8, 172–179.

Ninfa, A. J., Jiang, P., Atkinson, M. R. & Peliska, J. A. (2000). Integration of antagonistic signals in the regulation of nitrogen assimilation in *Escherichia coli*. *Curr Top Cell Regul* 36, 31–75.

Pellequer, J. L., Wager-Smith, K. A., Kay, S. A. & Getzoff, E. D. (1998). Photoactive yellow protein: a structural prototype for the three-dimensional fold of the PAS domain superfamily. *Proc Natl Acad Sci U S A* 95, 5884–5890.

Perraud, A. L., Weiss, V. & Gross, R. (1999). Signalling pathways in two-component phosphorelay systems. *Trends Microbiol* 7, 115–120.

Ponting, C. P. & Aravind, L. (1997). PAS: a multifunctional domain family comes to light. *Curr Biol* 7, R674–677.

Reyes-Ramirez, F., Little, R. & Dixon, R. (2001). Role of *Escherichia coli* nitrogen regulatory genes in the nitrogen response of the *Azotobacter vinelandii* NifL-NifA complex. *J Bacteriol* 183, 3076–3082.

Robinson, V. L., Buckler, D. R. & Stock, A. M. (2000). A tale of two components: a novel kinase and a regulatory switch. *Nat Struct Biol* 7, 626–633.

Schmitz, R. A. (1997). NifL of *Klebsiella pneumoniae* carries an N-terminally bound FAD cofactor, which is not directly required for the inhibitory function of NifL. *FEMS Microbiol Lett* 157, 313–318.

Schmitz, R., He, L. & Kustu, S. (1996). Iron is required to relieve inhibitory effects of NifL on transcriptional activation by NifA in *Klebsiella pneumoniae*. *J Bacteriol* **178**, 4679–4687.

Söderbäck, E., Reyes-Ramirez, F., Eydmann, T., Austin, S., Hill, S. & Dixon, R. (1998). The redox-and fixed nitrogen-responsive regulatory protein NifL from *Azotobacter vinelandii* comprises discrete flavin and nucleotide-binding domains. *Mol Microbiol* **28**, 179–192.

Stock, A. M., Mottonen, J. M., Stock, J. B. & Schutt, C. E. (1989). Three-dimensional structure of CheY, the response regulator of bacterial chemotaxis. *Nature* **337**, 745–749.

Stock, A. M., Robinson, V. L. & Goudreau, P. N. (2000). Two-component signal transduction. *Annu Rev Biochem* **69**, 183–215.

Stock, J. & Da Re, S. (2000). Signal transduction: response regulators on and off. *Curr Biol* **10**, R420–424.

Tanaka, T., Saha, S. K., Tomomori, C. & 12 other authors (1998). NMR structure of the histidine kinase domain of the *E. coli* osmosensor EnvZ. *Nature* **396**, 88–92.

Taylor, B. L. & Zhulin, I. B. (1999). PAS domains: internal sensors of oxygen, redox potential, and light. *Microbiol Mol Biol Rev* **63**, 479–506.

Tomomori, C., Tanaka, T., Dutta, R. & 10 other authors (1999). Solution structure of the homodimeric core domain of *Escherichia coli* histidine kinase EnvZ. *Nat Struct Biol* **6**, 729–734.

Woodley, P. & Drummond, M. (1994). Redundancy of the conserved His residue in *Azotobacter vinelandii* NifL, a histidine protein kinase homologue which regulates transcription of nitrogen fixation genes. *Mol Microbiol* **13**, 619–626.

Zhulin, I. B., Taylor, B. L. & Dixon, R. (1997). PAS domain S-boxes in Archaea, Bacteria and sensors for oxygen and redox. *Trends Biochem Sci* **22**, 331–333.

Is anybody here? Cooperative bacterial gene regulation via peptide signals between Gram-positive bacteria

Donald A. Morrison

Laboratory for Molecular Biology, Department of Biological Sciences, University of Illinois at Chicago, Chicago, IL 60607, USA

INTRODUCTION

Bacterial metabolism reflects constant adjustment of gene activity. Adjustments respond not only to the chemical composition of the cell interior but also to the chemical composition of its surrounding environment. Many regulators respond to changes in extracellular conditions indirectly, via levels of intracellular molecules derived from external substrates by transport and metabolism; a smaller number of important molecules are sensed more directly, through specific membrane-mounted sensors that have direct access to the milieu, sensors that are often members of the histidine protein kinase/response regulator families of two-component systems (Parkinson, 1993). A few special classes of extracellular molecules sensed by bacteria are important for reflecting not simply the chemical composition of the environment, but its biological composition. These include the quorum-sensing signals, so named after it was realized that their production, release and sensing allowed regulation of gene activity in response to a culture's population density and, thus, development of traits requiring cooperative behaviour (Fuqua *et al.*, 1994). Most known quorum-sensing systems control secreted products, turning them on (or off) when population density reaches a certain level – i.e. when a 'quorum' is sensed as present. In contrast, one of the first cell–cell signals found to monitor population density acts to control competence for genetic transformation (Tomasz & Hotchkiss, 1964). The survival value of this mode of regulation is readily apparent in the case of the many bacterial traits that are dependent on the action of soluble secreted products and others where mass action is important.

The molecular mechanisms of quorum-sensing systems uncovered to date fall into three contrasting groups. One, found in the Gram-negative bacteria, employs a lipophilic

SGM symposium 61: Signals, switches, regulons and cascades: control of bacterial gene expression.
Editors D. A. Hodgson, C. M. Thomas. Cambridge University Press. ISBN 0 521 81388 3 ©SGM 2002.

signal that diffuses across the cell membrane and is coupled with a transcriptional activator that is directly responsive to that signal. A well-known example is the control of bioluminescence in *Vibrio fischeri* by 3-oxo-hexanoyl-L-homoserine lactone, termed AI or autoinducer (Fuqua *et al.*, 1994). Another signalling strategy, found only in the Gram-positive bacteria, operates quite differently, coupling an actively secreted membrane-impermeant peptide with a two-component sensor–regulator pair to produce and sense an external signal reflecting the population level of the producer cells. Five classes of such signals were discovered nearly simultaneously in 1994 and 1995, aided by emerging homologies between their most strongly conserved elements, the secretion and sensor proteins. A third strategy combines features of the first two, using small peptide signals that are sampled from the environment by oligopeptide permeases and act via intracellular targets and effectors (Lazazzera & Grossman, 1998).

There has been much recent progress in identifying quorum-sensing regulatory components for disparate bacterial traits, and some steps have been taken toward linking them to a broader context. This chapter describes elements of five families of quorum-sensing systems mediated by extracellular peptides and surface receptors, chosen to illustrate the variety of gene sets linked to population density in this way, and to highlight a common core structure of the quorum-sensing regulatory circuits. Each peptide quorum-sensing system in these families comprises at least the following components: (1) a peptide signal; (2) a receptor for the signal; (3) a regulator dependent on that receptor; and (4) a molecular target for the regulator. The target is somehow linked to a set of regulated genes, which typically include some elements of the quorum-sensing circuit and at least one other gene representing the trait subject to regulation by quorum sensing.

PEPTIDE SIGNALS IN QUORUM SENSING

The peptide signals of Gram-positive quorum-sensing systems known to depend on surface receptors range in size from 7 to 34 amino acid residues, and are encoded within genes. Several of these have been purified on the basis of their biological activity, allowing their structure to be determined. Once the conserved patterns in peptide sequences were recognized, more were identified from gene sequences. The sequences of representative peptide signals in five classes are shown in Table 1. In each case, the primary translation product is processed to produce the mature peptide signal; in many, its release is known to depend on a specialized ABC transporter, but there are multiple routes of processing and secretion.

The nisin family

An elegant quorum-sensing circuit is employed to regulate synthesis of the bacteriocin nisin. Nisin is the archetype of a large class of bacteriocins, the lantibiotics, which undergo extensive post-translational modifications. Ring structures are formed by dehydration of cysteine in reaction with threonine or serine residues to form thioether bridges and dehydrated serine or threonine residues (de Vos *et al.*, 1995). Genetic analysis revealed that the cell-to-cell signal controlling nisin synthesis is in fact nisin itself. The antibacterial activity and the signalling activity of the peptide are distinct, depending on different critical residues of the peptide, although ring formation is required for both activities (Kuipers *et al.*, 1995). While there is an apparent economy in using a secreted molecule as its own cell-to-cell signal, the principle is not so readily applied in cases of a more complex secreted product, and is simply not applicable where the cooperative trait involves no secreted products. Indeed, for each of the remaining cases of quorum sensing discussed here (except plantaricin A, the exception that illustrates the rule), the signal is encoded in a separate gene, distinct from the genes for regulated products.

The Agr-associated system of *Staphylococcus aureus*

A small peptide signal with less modification regulates accessory gene expression in *Staphylococcus aureus*. In this species, many virulence genes are regulated through a density-dependent circuit utilizing genes of the Agr locus (Ji *et al.*, 1995). The signal in this case is a 7-aa peptide, the active form of which contains a single thiolactone ring formed by a thioester linkage of the terminal carboxyl group and the fifth subterminal amino acid, cysteine (Mayville *et al.*, 1999). At least four alleles of this signal exist in natural isolates, each encoded as an internal segment of a prepeptide precursor (Jarraud *et al.*, 2000). It is proposed that the thiolactone ring participates in activating the receptor by establishing a covalent attachment. Interestingly, despite wide sequence variety, each allelic form of this signal is inhibitory to the receptors of cells carrying other alleles (Ji *et al.*, 1997). A general route for synthesis of this class of modified peptides (Mayville *et al.*, 1999) has opened many avenues for characterizing staphylococcal quorum sensing.

Competence regulation in *Bacillus subtilis*

Another low-molecular-mass peptide signal with a single modified residue contributes to density-dependent regulation of competence in *Bacillus subtilis*. A small secreted peptide, ComX, comprises 10 aa with no indication of internal rings, but does carry an uncharacterized lipid modification of its single tryptophan. It is processed by release of the COOH-terminal portion of a 55-aa precursor (Magnuson *et al.*, 1994). Four alleles of *comX* have been found in a small number of strains of *B. subtilis* surveyed, each encoding a divergent signal sequence, but with a single conserved tryptophan (Tortosa *et al.*, 2001). Until the structure of this modification is determined, ComX remains available only from culture supernatants. No other trait has yet been found to be regulated via this class of quorum-sensing signal, but it would not be surprising if the system is found to have been adapted for control of other functions.

Table 1. Peptide pheromones and precursors

Species	Gene and pherotype	Sequence of peptide signal precursor*	Citation
Lactococcus lactis	*nisA* MG1614	MSTKDFNLDLVSVSKKDSGASPR-**ITSISLCTPGCKTGALMGCNMKTATCHCSIHVSK**	Kuipers *et al.* (1995)
Staphylococcus aureus	*agrD* 1	MNTLFNLFFDFITGILKNIGNIAA-**YSTCDFIM**-DEVEVPKELTQLHE	Ji *et al.* (1995)
	agrD 2	MNTLVNMFFDFIIKLAKAIGIVG-**GVNACSSLF**-DEPKVPAELTNLYDK	Ji *et al.* (1995)
	agrD 3	MKKLLNKVIELLVDFFNSIGYRAA-**YINCDFLL**-DEAEVPKELTQLHE	Ji *et al.* (1995)
	agrD 4	MNTLLNIFFDFITGVLKNIGNVAS-**YSTCYFIM**-DEVEIPKELTQLHE	Jarraud *et al.* (2000)
Bacillus subtilis	*comX* 168	MQDLINVFLNYPEALKKLKNKEACLIGFDVQETETIIKAYNDYYL-**ADPITRQWGD**	Magnuson *et al.* (1994)
	comX ROFF1	MQELISYLLKYPEVLKKLKKLKSNEASLIGFSSDETQLIIEGFEGIEVKRGNAGKWGPE	Tortosa *et al.* (2001)
	comX RSB1	MQEMVGYLIKYPNVLREVMEGNACLLGVDKDQSECIINGFKGLEIYSMMDWHY	Tortosa *et al.* (2001)
	comX ROOO2	MQEIVGYLVKNPEVLDEVMEGRASLLNIDKDQLKSIVDAFRGMQIYTNGNWVPS	Tortosa *et al.* (2001)
Lactobacillus sake	*spplP* LTH673	MMIFKKLSEKELQKINGG-**MAGNSSNFIHKIKQIFTHR**	Brurberg *et al.* (1997)
Lactobacillus plantarum	*plnA* C11	MKIQIGKGMKQLSNKEMQKIVGG-**KSSAYSLQMGATAIKQVKKLFKKWGW**	Diep *et al.* (1995)
Lactobacillus sakei	*orf4* Lb706	MKLNYIEKKQLTNKQLKLIIGG-**TNRNYYGKPNKDIGTCIWSGFRHC**	Diep *et al.* (2000)
Carnobacterium piscicola	*cbaX* LV17A	MKIKTITRKQLIQIKGG-**SINSQIGKATSSISKCVFSFFKKC**	Franz *et al.* (2000)
	cbnS LV17B	MKIKTITKKQLIQIKGG-**SKNSQIGKSTSSISKCVFSFFKKC**	Quadri *et al.* (1997)
Enterococcus faecium	*entF* CTC492	MLNNVQIKSLKKLKGG-**AGTKPQGKPASNLVECVFSLFKKCN**	Nilsen *et al.* (1998)

Streptococcus pneumoniae	spiP knr7/87	MDKKQNLTSFQELTTTELNQITGG-**GLWEDLLYNINRYAHYIT**	Reichmann & Hakenbeck (2000)
	comC 1	MKNTVKLEQFVALKEKDLQKIKGG-**EMRLSKFFRDFILQRKK**	Pozzi et al. (1996)
	comC 2	MKNTVKLEQFVALKEKDLQKIKGG-**EMRISRIILDFLFLRKK**	Pozzi et al. (1996)
Streptococcus gordonii	comC 7865	MKKKNKQNLLPKELQQFEILTERKLEQVTGG-**DIRHRINNSIWRDIFLKRK**	Håvarstein et al. (1997)
	comC challis	MKKKNKQNLLPKELQQFEILTERKLEQVTGG-**DVRSNKIRLWWENIFFNKK**	Håvarstein et al. (1997)
Streptococcus oralis	comC 11427	MKNTEKLEQFKEVTEAELQEIRGG-**DKRLPYFFKHLFSNRTK**	Håvarstein et al. (1997)
	comC 20066	MKNTVKLEQFKEVTEAELQEIRGG-**DWRISETIRNLIFPRRK**	Håvarstein et al. (1997)
Streptococcus mitis	comC B6	MKNTVKLEQFVALKEKELQKIKGG-**EMRKPDGALFNLFRRR**	Håvarstein et al. (1997)
	comC HU8	MKNTVKLEQFVALKEKDLQKIQGG-**EMRKSNNNFFHFLRRI**	Håvarstein et al. (1997)
Streptococcus mutans	comC NG8	MKKTLSLKNDFKEIKTDELEIIIGG-**SGSLSTFFRLFNRSFTQALGK**	Li et al. (2001)

*The mature peptide signal is in bold face. Underlined conserved features: modified residues in nisin, modified cysteine in Agr signals, modified tryptophan in Bacillus competence signals, conserved position 3 in streptococcal competence peptides.

Non-modified peptide signals associated with bacteriocin activity

In addition to the modified peptide pheromones, many peptide signals have been discovered that are apparently not modified at all, beyond post-translational proteolysis to release the mature peptide; corresponding synthetic peptides complement signal-deficient mutants directly. A number of these peptides regulate production of small non-lantibiotic peptide bacteriocins. Production of many small peptide bacteriocins proceeds via a chromosomally encoded precursor with an N-terminal extension that is removed by proteolytic scission after the GG positions in a recognition motif, M--LS--EL--I-GG (Håvarstein et al., 1994). The responsible protease occupies a special domain at the N-terminus of an ABC transporter required for secretion of mature bacteriocin (Håvarstein et al., 1995b). In most known cases, production of the bacteriocin has proven to be regulated by a quorum-sensing peptide pheromone that is distinct from the bacteriocin but is processed in the same way at a typical GG motif. The sakacin A pheromone is 23 aa long. A different bacteriocin, sakacin P, is also regulated by a GG-motif peptide pheromone (IP-673) of 19 aa. The inducing peptide (IP) for carnobacteriocin A, made by *Carnobacterium piscicola*, is a 24-aa peptide released from a 41-aa precursor. In *Lactobacillus plantarum* strain C11, three bacteriocins are regulated by a somewhat longer peptide pheromone, the 26-aa PlnA, also processed at a GG site from a longer precursor. Interestingly, PlnA exhibits both activity as a pheromone signal and moderate activity as a bacteriocin, which could represent an intermediate stage in evolutionary development of separate bacteriocin and quorum-sensing signal molecules (Hauge et al., 1998). Its pheromone activity was uncovered by analysis of culture supernatants with a capacity to maintain bacteriocin production in laboratory culture (Diep et al., 1995). The genes encoding these bacteriocin IPs are closely linked to other genes needed for regulation and production of the cognate bacteriocin. The GG peptide signal genes are typically linked to two genes for an ABC transporter and protease responsible for maturation and export, while in the cases of modified or non-GG processed signals, at least one gene required for the maturation of peptide signal is tightly linked to the signal gene (Kleerebezem et al., 1997).

Regulation of competence in streptococci

Another large group of unmodified quorum-sensing peptide signals, also synthesized by the GG leader/ABC transporter/protease mechanism, was discovered in the streptococci (Håvarstein et al., 1995a, 1997). They regulate competence for genetic transformation, but are not known to have any role in bacteriocin production. These unmodified peptide signals range in size approximately from 15 to 25 aa and have 20–30-aa leaders. Two examples from each of four streptococcal species are shown in Table 1; more than a dozen additional peptides in this class have been described (e.g. Whatmore et al., 1999). For one of them, a role in mediating cell–cell communication in exponential-phase laboratory cultures of *Streptococcus pneumoniae* to bring about sudden coordinated expression of competence was discovered in the early sixties (Tomasz & Hotchkiss, 1964), but its structure remained unknown until 1995 (Håvarstein *et al.*, 1995a). As for the *Bacillus* competence pheromones, these peptides regulate a trait that depends on no other known secreted proteins.

It is interesting to speculate on how a separate peptide pheromone signal arose, if a peptide product can be used directly as a population-sensing signal. One possible scenario would be that a primitive nisin-like system evolved first, dependent on the regulated peptide product itself as the population signal. A gene duplication event then could have allowed one copy of the peptide gene to evolve as a regulated product while the other copy evolved separately for optimal signalling properties. PlnA, which exhibits both activities (Diep *et al.*, 1995), may have arisen recently in just this way.

RECEPTOR/REGULATOR ELEMENTS AND THEIR ORGANIZATION

A two-component regulatory system is typically associated with each of the peptide pheromone loci described above. Its two proteins complete the quorum-sensing circuit: one acts as the receptor for the peptide signal and as a histidine protein kinase, while the other, the response regulator, regulates target genes in response to changes in the level of the peptide signal, communicated from the receptor. The patterns of organization of these key genes for signals listed in Table 1 are compared in Fig. 1. In those cases where it has been directly established, receptor identification has usually been accomplished by receptor specificity switching experiments. Since the receptors belong to a subfamily of histidine protein kinases associated with two-component regulatory systems, and they are linked to members of a family of response regulators found in the same two-component systems, it is thought that regulation follows a signal transduction chain depending on phosphotransfer from the former to a conserved aspartate in the latter proteins, but such transfer has been demonstrated biochemically in only a few cases. Other classes of evidence implicating these genes in signal transmission include mutations leading to constitutive expression of target genes and mapping of response regulator protein binding sites adjacent to promoters of pheromone responsive genes.

In three of these systems, identical sets of downstream genes are regulated in different strains by pheromone peptides of different sequence (Pozzi *et al.*, 1996; Tortosa *et al.*, 2001; Ji *et al.*, 1997). In an arrangement that seems likely to reduce activation of cells of one pherotype by signal from another, a small hypervariable region of DNA encompasses the peptide signal gene and part of the receptor gene.

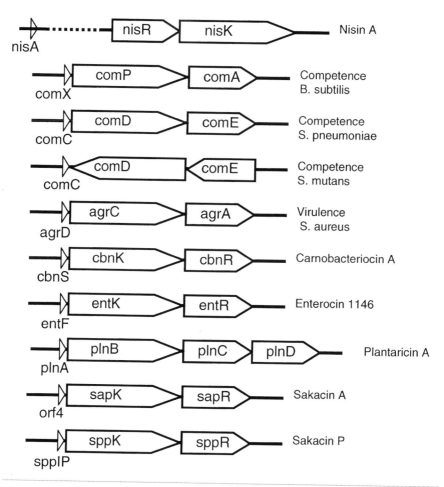

Fig. 1. Organization of genes for signals, sensors and response regulators in peptide quorum-sensing circuits listed in Table 1. Arrow, inducing peptide gene; large pentagon, histidine protein kinase gene; small pentagon(s), response regulator gene.

REGULATORY TARGETS OF THE RESPONSE REGULATORS

For the three-component (pheromone/receptor/response regulator) regulatory loci described above, experimental or circumstantial evidence implicates short (9–10 bp) direct repeats, with centres 20 bp apart, as the site of action for the response regulators (Table 2). These sites are located upstream of known or probable promoters (Diep *et al.*, 1996). Evidence for the common view that the response regulator acts as a transcriptional activator is stronger for some loci than others, while direct evidence for a role of phosphotransfer is available in only a few cases. However, the response regulator has been directly shown to bind to DNA direct repeats in a number of well-studied

Table 2. Direct repeat sequences at binding sites for response regulators

Sequences of direct repeats are in bold. The consensus −10 site at each promoter is underlined. Genes regulated by the promoters are indicated to the left.

Response regulator	Target gene	Sequence of promoter and direct repeats	Citation
SppR	spplP	AGAGTTCT **TAACGTTAAT** CCGAAAAAAAC **TAACGTTAATA** TTAAAAAATAAGATCCGCTTGTGAATTATG <u>TATAAT</u> TTGATTAGA	Risoen et al. (2000)
	sppT	TGCAGCAT **TAACGTTAAT** TTTGATAAACG **TAACGTTAATG** GATAATCATCCTGTTTACAAATAGTGTATGA <u>CATAAT</u> TAAGTAATT	
	sppA	GCGCATAT **TAACGTTTAA** CCGATAAAGTT **GAACGTTAATA** TTTTTTTTGCGCAGAAATGGTAAATTGAAG <u>CATAAT</u> AGTCTTGTA	
CbnR	cbnS	GTTTATTCA **TTCAAGACCG** TATTCGATGTAG **TTCAGGATGT** TTTTTCATATATAATAAAATTTTATGC <u>CATACT</u> TTAATTAAACAAG	Quadri et al. (1997)
	cbnB2	TTTATTTCA **TTTATGAATT** CAAATACCCTGG **TTCAAGATGT** ATTTTCCAAAAAAATGTTCAGATATGA <u>TATAGT</u> TTTTTTGAAATAC	
PlnDC	plnA	ATGGTGATT **CACGTTTA** AATTTAAAAAATG **TACGTTAAT** AGAAATAATTCCTCCGTACTTCAAAAACACAT <u>TATCCT</u> AAAAGCGAGG	Risoen et al. (1998)
	plnE	TTGGTATTT **GACGTTAA** GAGAACGTTTTTT **TACTTTTAT** AATTTTTCAACAATCTGGTAAAAAATAAAT <u>TAAACT</u> AAATTTGTTC	
	plnJ	CTTTCAAGT **TACGTTAA** ATCGATTAAATAG **TACGATAAC** AAATTTAAAATAATTTTTTTAAATTGTAGCG <u>TATATT</u> AATAAGTGCA	
ComE	comC	TTAAAAAAG **TACACTTTGG** GAGAAAAAATG**ACAGTTGAG** AGAATTTATCTCAAAACGAAATTCCATTTTG <u>TATAAT</u> GGTTTTTGTAA	Ween et al. (1999)
	comA	CCAAAAAGT **GCAGTTGGGA** GGGAGATAGGC**TCATTTGGGA** AGGAAGTCCAGTTTTTGTTTAGTGATTGGGG <u>TAAGAT</u> AGTTGTTATCA	

cases, including plantaricin A regulation (Risoen *et al.*, 1998), streptococcal competence control (Ween *et al.*, 1999) and sakacin P synthesis (Risoen *et al.*, 2000). The binding site for the response regulator ComA in *Bacillus* also contains direct repeats, but the function of this regulator may be distinct from the majority of cases discussed here, as the repeated sequences are longer (cggcatcccgc – 32 – cggcatcccgc) and more widely spaced, with two more turns of the DNA helix between repeats (Roggiani & Dubnau, 1993).

Where are these direct-repeat-associated promoters found? That is, what do the response regulators directly control? As Table 3 shows, each response regulator appears to regulate two or three types of genes. It may regulate the pheromone signal gene, the receptor and response regulator gene pair, genes for signal processing and transport, and, finally, at least one promoter outside the quorum-sensing circuit. In bacteriocin systems, these last directly control the genes providing the ultimate output – bacteriocins, as well as maturases and immunity proteins that protect the producer from its own toxic products. In the Agr system, the regulator controls the gene for a small RNA molecule which, in turn, serves as a global regulator, coordinating a multigenic response that activates some genes and represses others, although the interaction of AgrA with the regulatory sites may be indirect (Novick *et al.*, 1995; Cheung *et al.*, 1997). In *Bacillus* competence control, the regulator ComA activates a gene (*comS, srfA*) encoding an intermediate link that triggers a shift to autostimulatory production of and action by a transcriptional activator which targets sites upstream of promoters for late competence genes (Roggiani & Dubnau, 1993; van Sinderen *et al.*, 1995). Finally, in the streptococcal competence system, the response regulators also target the promoter of an intermediary global regulator, in this case a putative alternative sigma factor which coordinates a global response by dozens of genes dependent on non-canonical promoters.

AUTOCATALYTIC SIGNALS

A common formulation in the literature on quorum sensing speaks of AI 'reaching a critical level', whereupon target genes are activated. The idea of a critical level and an all-or-none response to AI signal is probably an oversimplification of the real process, as no biochemical element of the quorum-sensing circuit has been described which exhibits an all-or-none response; indeed observed responses by isolated components are approximately proportional to signal levels (e.g. de Ruyter *et al.*, 1996). Nonetheless, population-wide responses to quorum-sensing signals can be very sudden; in the case of streptococcal competence, the level of competence throughout a mid-exponential culture can rise 10 000-fold in just 5 min while the basal level of competence pheromone is presumably doubling only once per 40 min doubling time.

Table 3. Peptide pheromone quorum-sensing circuits

Product or regulation	Genes controlled by promoters targeted by response regulator						Citation
	Inducing peptide	HK/RR	Export and modification	Secreted product	Global regulator	Other	
Nisin A	nisA	–	nisBTC	nisA	–	nisI, nisFEG	de Ruyter et al. (1996)
Accessory gene regulation, Staphylococcus aureus	agrD	agrCA	agrB	None	RNAIII	–	Ji et al. (1995)
Competence, B. subtilis	–	–	–	None	(comS) comK		Roggiani & Dubnau (1993)
Carnobacteriocin B2	cbnS	cbnKR	cbnTP	cbnB2, cbnBM1	–	cbiB2, cbiBM1, cbnXY	Franz et al. (2000)
Plantaricin A	plnA	plnBCID	plnGH	plnJK, plnEF, plnA	–	plnLR, plnMOP, plnI	Diep et al. (1996); Anderssen et al. (1998)
Sakacin A	orf4	sapKR	sapTE	sapA	–	saiA	Diep et al. (2000)
Sakacin P	sppIP	sppKR	sppTE	sppA	–	orfX, sppiA	Brurberg et al. (1997)
Competence, Streptococcus pneumoniae	comC	comD/E	comAB	None	comX	–	Peterson et al. (2000); Pestova et al. (1996)
Competence, Streptococcus mutans	comC	comDE	cslAB	None	comX	–	Li et al. (2001); Petersen & Scheie (2000)

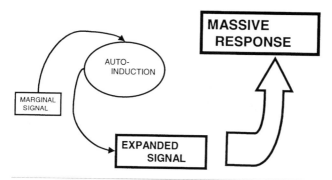

Fig. 2. Autoinduction of a quorum-sensing signal to ensure maximal participation in cooperative activity.

In the earliest studies of autoinduction, it was noticed that part of the response to AI is production of more AI. In the quorum-sensing systems described here (except for the *Bacillus* competence control system, where only a small fraction of the population responds), analysis of gene transcription has shown that the autoinducer signal is self-regulated, such that its rate of production is increased as part of the response to increasing levels of autoinducer (Table 3). Why should this be so? Is there a general reason that activation of the sensor and effector is insufficient for an effective quorum-sensing response, but requires an accompanying hyperproduction of the initiating signal? The answer may perhaps be found in observing that autostimulation could convert a response to a (by definition) marginal signal into a simultaneous response by many or all members of a population, even if many of its members were initially not equally receptive to the signal. If it is worthwhile for a cell to make a product only when in the company of other potential producers, then it would seem likely that it would generally also be worthwhile for it to ensure that its neighbours actually make the product too. For a trait which is expressed at high population density, the expression of the trait will be more effective when all members of the population are recruited to contribute despite possible heterogeneity in responsiveness or in signal distribution. Thus once the signal has been sensed by some cells and expression of the coordinated trait is begun in those cells, the population will benefit and, most importantly for evolution of the regulatory circuit, those cells will benefit if all cells nearby are drafted for the effort. This logic suggests that most quorum-sensing peptides are serving two roles, both as a census-taker and as a fire-alarm bell (Fig. 2).

What, then, would be the roles of the commonly observed AI-induced overexpression of AI secretion machinery and AI receptor proteins? They could support two independent strategies for coordinating the response: more signal or more sensitivity to signal.

Activated expression of secretion machinery obviously aids accelerated signal production, and avoids the expense of maintaining that machinery in quantities not needed for basal signal elaboration. In contrast, hyperproduction of signal receptor could have the effect of rendering a responding cell rapidly more sensitive to signal (although this has not been demonstrated), and sooner capable of saturating the target promoters, again tending to the most massive response possible. While overproduction of signal recruits every available cell, overproduction of receptor may recruit every available molecule of RNA polymerase.

TARGET REGULONS, SIMPLE TO COMPLEX

The quorum-sensing circuits described above drive biological outputs that depend on gene sets ranging from simple to complex. In the simpler cases, such as the nisin and bacteriocin systems, the ultimate output is controlled by three or four promoters for about six genes for bacteriocin synthesis, maturation and secretion as well as at least one immunity gene. In the more complex cases, the added complexity is achieved not by directing the response regulator to many additional promoters, but rather by passing the signal through one or more intermediate regulators that amplify, filter and distribute the signal (Fig. 3).

In the case of Agr regulation, only two or three promoters are thought to be regulated directly by the RR protein (Novick *et al.*, 1995). One of these drives transcription of the *agrABCD* genes for components of the quorum-sensing circuit itself. Another controls an intermediate global regulator in the form of a regulatory RNA (RNAIII), which upregulates a significantly larger array of genes, mainly secreted virulence factors such as toxin-1, alpha-toxin, beta-haemolysin, proteases and lipases, and represses others, especially surface-associated factors such as protein A, and fibronectin and collagen binding proteins.

The ComPA receptor and effector proteins influence control of genetic competence in *B. subtilis*, through a series of linked proteins. The ultimate targets of its activity are a set of operons encoding the apparatus for DNA transport and genetic recombination. The only direct target of ComA for competence regulation is the gene *comS*. Its product acts by diverting ComK from a proteolytic fate, switching on a self-sustaining expression of ComK, a transcription factor that activates its own promoter and that of the 'late' competence genes (Hahn *et al.*, 1996; Turgay *et al.*, 1998).

Streptococcal competence, depending on late genes largely homologous to the *Bacillus* competence genes, is linked to the pheromone quorum sensing in a different, apparently more direct, way. The quorum-sensing regulator ComE also activates expression of just one non-quorum-sensing gene, but its product, ComX, appears to act directly as an

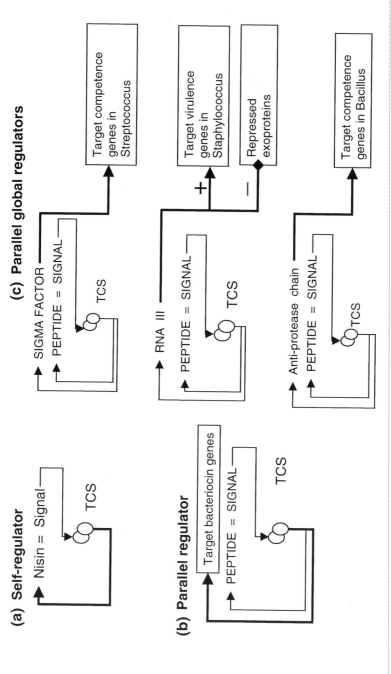

Fig. 3. Peptide pheromone gene regulation: simple to complex. Gene products and activation (→, +) or inhibition (−) pathways linking the three-component quorum-sensing regulatory system to target gene sets for five classes of quorum-sensing systems. (a) Nisin autoregulation. (b) Bacteriocin and immunity protein regulation through a separate peptide signal. (c) Peptide signal control of global regulators: sigma factor in streptococcal competence; regulatory RNA in accessory gene regulation in *Staphylococcus*; anti-protease regulation of competence in *B. subtilis*. Heavy arrow, link to biological output; TCS, two-component sensor/effector system.

alternative sigma factor that allows recognition of promoters for more than 16 operons, among which eight are known to affect competence (Lee & Morrison, 1999).

PERSPECTIVES

In the parliamentary sense, a quorum is, in itself, neither sufficient nor necessary for a body to conduct business (Robert, 2000). It is not sufficient unless a meeting has been properly called, for example, while it is not necessary for certain actions, such as adjournment, or (in practice) when a count of the quorum has not been explicitly requested by a member. Rather, a quorum is required under certain conditions and for the conduct of certain kinds of activity. Similarly, bacterial quorum-sensing systems are likely to be part of a larger regulatory network that calls them into play, modulates their activity, and influences the nature of the response they coordinate. Thus a clear understanding of the biology of any quorum-sensing system is likely to entail a determination of factors affecting signal production, signal sensing and regulatory output, as well as a detailed analysis of the mechanisms of these events themselves. Multiple factors influencing basal expression of the quorum-sensing circuits can already be seen for Agr, *Bacillus* competence control, and the streptococcal competence system (Martin *et al.*, 2000).

Many questions remain unanswered about the design of peptide quorum-sensing systems. Why are all Gram-positive pheromone signals peptides? Why are all the signals processed? Are the peptide structures 'tuned' to the niche where they function, so as to integrate specific information about the extracellular milieu? Is this why so few peptide products serve as their own quorum-sensing signals?

The phenomenon of pherotype differentiation remains an intriguing mystery. In the Agr virulence control system, the species of *S. aureus* is divided into four or more pherotypes. Among these, one pheromone inhibits response to the other pheromone types; the role this may play in the ecology or epidemiology of the organism is unknown. Both peptide-controlled competence systems also display several pherotypes per species. Their role in reproductive isolation is also unknown. Tortosa *et al.* (2001) have suggested that in *B. subtilis*, pherotype elaboration achieves a degree of isolation, but the small number of pherotypes per species in the streptococcal competence systems (Whatmore *et al.*, 1999) means that any isolation achieved this way would be very weak, and perhaps suggests a different significance to signal peptide sequence changes, or that pherotype differentiation may quite simply reflect a coordinated genetic drift.

As the components and modes of operation of quorum-sensing peptide circuits are discovered, they offer a rich source of material for designing genetic systems for regulated gene expression. In one well-developed example, several plasmid-borne systems have been developed that allow any desired gene to be placed under control of a *nisA* promoter, so that its expression depends entirely on the level of the signal, nisin, added to culture medium (Bryan *et al.*, 2000). It remains to be seen how important cross-talk with other regulators will be in its applications.

The peptide-mediated cell-to-cell signal systems discussed here have come to be thought of as quorum-sensing systems largely because of the behaviour they generate under conditions of pure laboratory cultures, and it is indeed clear in most cases that in such cultures the peptide signal and regulatory circuit can link gene expression to population density. Nonetheless, it is important to keep in mind that their function in the context of the natural ecology of each species generally remains to be determined. While a quorum-sensing role certainly makes intuitive sense for many of the known sets of regulated genes, an emitted signal necessarily passes through and becomes part of the cell's environment; as such their sensing has the potential to reflect features of that environment, including confinement, competitors, proteases and other sinks. Indirect indications of such non-quorum-sensing roles may emerge in cases where strict regulation of the basal expression level of peptide signals come to light. While it is clear that the signals discussed here act for quorum sensing in laboratory culture, and it seems likely that this is part of their function in nature, we have no clear idea of how their structure is adapted to their function. A 20-aa peptide is so complex, one must wonder if they are also adapted to serve as probes for other aspects of the cell's environment besides counting cells. Confounding this evaluation are the differences between conventional laboratory cultures and the natural niches of bacterial species. In nature, a peptide signal may face undocumented proteases, avid binding surfaces, competing peptides or fluid flow, any of which could decrease the signal. It may also face special factors strengthening the signal, such as confinement, receptor cofactor or conditions stimulating peptide gene transcription.

Why peptides? It is intriguing to attempt to understand what special features may be associated with the use of peptides as cell-to-cell signals. Two aspects of the differences between peptide signals and acyl homoserine lactone signals come to mind. First, a distinct range of physico-chemical characteristics is available for the two classes of compound. If the signal interacts with the cell milieu, the increased flexibility of peptide structures could be valuable. Second, and perhaps of broader significance, peptides of length 15–22 aa have a wide range of possible structures. This enormous structure-space is accessible, furthermore, without a need for a great diversity of synthetic pathways, as would be needed to approach a comparable diversity of acyl lactones. It is achieved rather by the economical path of codon evolution in a small gene sequence. This diversity means that there are many options for using a signal to 'sense' aspects of the environment and that highly selective specificity can easily be achieved. It also means that many 'private' signal circuits can evolve. It does seem that bacteria are taking advantage of this possibility, from the wide variety of signal gene alleles found already in the Agr, ComC and ComX regulatory systems, although it is not yet clear what biologically valuable 'purpose' these variations are serving.

ACKNOWLEDGEMENTS

Parts of the work on competence regulation in Streptococcus pneumoniae discussed in this chapter were supported by the National Science Foundation (MCB-9506785 and MCB-9722821). It is a pleasure to acknowledge the many contributions to the pneumococcal competence story by research students at the University of Illinois at Chicago and by collaborators at many other laboratories.

REFERENCES

Anderssen, E. L., Diep, D. B., Nes, I. F., Eijsink, V. G. & Nissen-Meyer, J. (1998). Antagonistic activity of *Lactobacillus plantarum* C11: two new two-peptide bacteriocins, plantaricins EF and JK, and the induction factor plantaricin A. *Appl Environ Microbiol* **64**, 2269–2272.

Brurberg, M. B., Nes. I. F. & Eijsink, V. G. (1997). Pheromone-induced production of antimicrobial peptides in *Lactobacillus*. *Mol Microbiol* **26**, 347–360.

Bryan, E. M., Bae, T., Kleerebezem, M. & Dunny, G. M. (2000). Improved vectors for nisin-controlled expression in Gram-positive bacteria. *Plasmid* **44**, 183–190.

Cheung, A. L., Bayer, M. G. & Heinrichs, J. H. (1997). Sar genetic determinants necessary for transcription of RNAII and RNAIII in the *agr* locus of *Staphylococcus aureus*. *J Bacteriol* **179**, 3963–3971.

Diep, D. B., Håvarstein, L. S. & Nes, I. F. (1995). A bacteriocin-like peptide induces bacteriocin synthesis in *Lactobacillus plantarum* C11. *Mol Microbiol* **18**, 631–639.

Diep, D. B., Håvarstein, L. S. & Nes, I. F. (1996). Characterization of the locus responsible for the bacteriocin production in *Lactobacillus plantarum* C11. *J Bacteriol* **178**, 4472–4483.

Diep, D. B., Axelsson, L., Grefsli, C. & Nes, I. F. (2000). The synthesis of the bacteriocin sakacin A is a temperature-sensitive process regulated by a pheromone peptide through a three-component regulatory system. *Microbiology* **146**, 2155–2160.

Franz, C. M. A. P., van Belkum, M. J., Worobo, R. W., Vederas, J. C. & Stiles, M. E. (2000). Characterization of the genetic locus responsible for production and immunity of carnobacteriocin A: the immunity gene confers cross-protection to enterocin B. *Microbiology* **146**, 621–631.

Fuqua, W. C., Winans, S. C. & Greenberg, E. P. (1994). Quorum-sensing in bacteria: the LuxR-LuxI family of cell density-responsive transcriptional regulators. *J Bacteriol* **176**, 269–275.

Hahn, J., Luttinger, A. & Dubnau, D. (1996). Regulatory inputs for the synthesis of ComK, the competence transcription factor of *Bacillus subtilis*. *Mol Microbiol* **21**, 763–775.

Hauge, H. H., Mantzilas, D., Moll, G. N., Konings, W. N., Driessen, A. J. M., Eijsink, V. G. H. & Nissem-Meyer, J. (1998). Plantaricin A is an amphiphilic alpha-helical bacteriocin-like pheromone which exerts antimicrobial and pheromone activities through different mechanisms. *Biochemistry* **37**, 16026–16032.

Håvarstein, L. S., Holo, H. & Nes, I. F. (1994). The leader peptide of colicin V shares consensus sequences with leader peptides that are common among peptide bacteriocins produced by gram-positive bacteria. *Microbiology* **140**, 2383–2389.

Håvarstein, L. S., Coomaraswamy, G. & Morrison, D. A. (1995a). An unmodified heptadecapeptide pheromone induces competence for genetic transformation in *Streptococcus pneumoniae. Proc Natl Acad Sci U S A* **92**, 11140–11144.

Håvarstein, L. S., Diep, D. B. & Nes, I. F. (1995b). A family of bacteriocin ABC transporters carry out proteolytic processing of their substrates concomitant with export. *Mol Microbiol* **16**, 229–240.

Håvarstein, L. S., Hakenbeck, R. & Gaustad, P. (1997). Natural competence in the genus streptococcus: evidence that streptococci can change pherotype by interspecies recombinational exchanges. *J Bacteriol* **179**, 6589–6594.

Jarraud, S., Lyon, G. J., Figueiredo, A. M., Gerard, L., Vandenesch, F., Etienne, J., Muir, T. W. & Novick, R. P. (2000). Exfoliatin-producing strains define a fourth *agr* specificity group in *Staphylococcus aureus. J Bacteriol* **182**, 6517–6522.

Ji, G., Beavis, R. C. & Novick, R. P. (1995). Cell density control of staphylococcal virulence mediated by an octapeptide pheromone. *Proc Natl Acad Sci U S A* **92**, 12055–12059.

Ji, G., Beavis, R. & Novick, R. P. (1997). Bacterial interference caused by autoinducing peptide variants. *Science* **276**, 2027–2030.

Kleerebezem, M., Quadri, L., Kuipers, O. P. & de Vos, W. M. (1997). Quorum sensing by peptide pheromones and two-component signal transduction systems in Gram-positive bacteria. *Mol Microbiol* **24**, 895–904.

Kuipers, O. P., Beerthuyzen, M., de Ruyter, P., Luesink, E. & de Vos, W. M. (1995). Autoregulation of nisin biosynthesis in *Lactococcus lactis* by signal transduction. *J Biol Chem* **270**, 27299–27304.

Lazazzera, B. A. & Grossman, A. D. (1998). The ins and outs of peptide signaling. *Trends Microbiol* **6**, 288–294.

Lee, M. S. & Morrison, D. A. (1999). Identification of a new regulator in *Streptococcus pneumoniae* linking quorum sensing to competence for genetic transformation. *J Bacteriol* **181**, 5004–5016.

Li, Y. H., Lau, P. C., Lee, J. H., Ellen, R. P. & Cvitkovitch, D. G. (2001). Natural genetic transformation of *Streptococcus mutans* growing in biofilms. *J Bacteriol* **183**, 897–908.

Magnuson, R., Solomon, J. & Grossman, A. D. (1994). Biochemical and genetic characterization of a competence pheromone from *B. subtilis. Cell* **77**, 207–216.

Martin, B., Prudhomme, M., Alloing, G., Granadel, C. & Claverys, J. P. (2000). Cross-regulation of competence pheromone production and export in the early control of transformation in *Streptococcus pneumoniae. Mol Microbiol* **38**, 867–878.

Mayville, P., Ji, G., Beavis, R., Yang, H., Goger, M., Novick, R. P. & Muir, T. W. (1999). Structure-activity analysis of synthetic autoinducing thiolactone peptides from *Staphylococcus aureus* responsible for virulence. *Proc Natl Acad Sci U S A* **96**, 1218–1223.

Nilsen, T., Nes, I. F. & Holo, H. (1998). An exported inducer peptide regulates bacteriocin production in *Enterococcus faecium* CTC492. *J Bacteriol* **180**, 1848–1854.

Novick, R. P., Projan, S. J., Kornblum, J., Ross, H., Ji, G., Kreisworth, B., Vandenesch, F. & Moghazeh, S. (1995). The agr P-2 operon: an autocatalytic sensory transduction system in *Staphylococcus aureus. Mol Gen Genet* **248**, 446–458.

Parkinson, J. S. (1993). Signal transduction schemes of bacteria. *Cell* **73**, 857–871.

Pestova, E. V., Håvarstein, L. S. & Morrison, D. A. (1996). Regulation of competence for genetic transformation in *Streptococcus pneumoniae* by an auto-induced peptide pheromone and a two-component regulatory system. *Mol Microbiol* **21**, 853–862.

Petersen, F. C. & Scheie, A. A. (2000). Genetic transformation in *Streptococcus mutans* requires a peptide secretion-like apparatus. *Oral Microbiol Immunol* **15**, 329–334.

Peterson, S., Cline, R. T., Tettelin, H., Sharov, V. & Morrison, D. A. (2000). Gene expression analysis of the *Streptococcus pneumoniae* competence regulons by use of DNA microarrays. *J Bacteriol* **182**, 6192–6202.

Pozzi, G., Masala, L., Iannelli, F., Manganelli, R., Håvarstein, L. S., Piccoli, L., Simon, D. & Morrison, D. A. (1996). Competence for genetic transformation in encapsulated strains of *Streptococcus pneumoniae*: two allelic variants of the peptide pheromone. *J Bacteriol* **178**, 6087–6090.

Quadri, L. E., Kleerebezem, M., Kuipers, O. P., de Vos, W. M., Roy, K. L., Vederas, J. C. & Stiles, M. E. (1997). Characterization of a locus from *Carnobacterium piscicola* LV17B involved in bacteriocin production and immunity: evidence for global inducer-mediated transcriptional regulation. *J Bacteriol* **179**, 6163–6171.

Reichmann, P. & Hakenbeck, R. (2000). Allelic variation in a peptide inducible two-component system of *Streptococcus pneumoniae*. *FEMS Microbiol Lett* **190**, 231–236.

Risoen, P. A., Håvarstein, L. S., Diep, D. B. & Nes, I. F. (1998). Identification of the DNA-binding sites for two response regulators involved in control of bacteriocin synthesis in *Lactobacillus plantarum* C11. *Mol Gen Genet* **259**, 224–232.

Risoen, P. A., Brurberg, M. B., Eijsink, V. G. & Nes, I. F. (2000). Functional analysis of promoters involved in quorum sensing-based regulation of bacteriocin production in *Lactobacillus*. *Mol Microbiol* **37**, 619–628.

Robert, H. M. (editor) (2000). *Robert's Rules of Order*, 10th edn. Boulder: Perseus.

Roggiani, M. & Dubnau, D. (1993). ComA, a phosphorylated response regulator protein of *Bacillus subtilis*, binds to the promoter region of *srfA*. *J Bacteriol* **175**, 3182–3187.

de Ruyter, P. G., Kuipers, O. P., Beerthuyzen, M. M., van Alen-Boerrigter, I. & de Vos, W. M. (1996). Functional analysis of promoters in the nisin gene cluster of *Lactococcus lactis*. *J Bacteriol* **178**, 3434–3439.

van Sinderen, D., Luttinger, A., Kong, L., Dubnau, D., Venema, G. & Hamoen, L. (1995). ComK encodes the competence transcription factor, the key regulatory protein for competence development in *Bacillus subtilis*. *Mol Microbiol* **15**, 455–462.

Tomasz, A. & Hotchkiss, R. D. (1964). Regulation of the transformability of pneumococcal cultures by macromolecular cell products. *Proc Natl Acad Sci U S A* **51**, 480–485.

Tortosa, P., Logsdon, L., Kraigher, B., Itoh, Y., Mandic-Mulec, I. & Dubnau, D. (2001). Specificity and genetic polymorphism of the Bacillus competence quorum-sensing system. *J Bacteriol* **183**, 451–460.

Turgay, K., Hahn, J., Burghoorn, J. & Dubnau, D. (1998). Competence in *Bacillus subtilis* is controlled by regulated proteolysis of a transcription factor. *EMBO J* **17**, 6730–6738.

de Vos, W. M., Kuipers, O. P., van der Meer, J. R. & Siezen, R. J. (1995). Maturation pathway of nisin and other lantibiotics: post-translationally modified antimicrobial peptides exported by Gram-positive bacteria. *Mol Microbiol* **17**, 427–437.

Ween, O., Gaustad, P. & Håvarstein, L. S. (1999). Identification of DNA binding sites for ComE, a key regulator of natural competence in *Streptococcus pneumoniae*. *Mol Microbiol* **33**, 817–827.

Whatmore, A. M., Barcus, V. A. & Dowson, C. G. (1999). Genetic diversity of the streptococcal competence (*com*) gene locus. *J Bacteriol* **181**, 3144–3154.

Quorum sensing in Gram-negative bacteria: global regulons controlled by cell-density-dependent chemical signalling

Neil A. Whitehead,[1] Abigail K. P. Harris,[1] Paul Williams[2] and George P. C. Salmond[1]

[1]Department of Biochemistry, University of Cambridge, Cambridge, UK

[2]Institute of Infections and Immunity, Queen's Medical Centre, University of Nottingham, Nottingham, UK

INTRODUCTION

It could be argued that a commonly held pre-1990s view of the microbial world was that bacteria functioned essentially as individual cells, albeit clonal. There were some very important and exciting examples of bacterial co-operativity known in which intercellular communication was either defined or hypothesized. However, such systems were mostly studied precisely because they were prokaryotic curiosities or were useful models of developmental processes in microbes. Similarly, the exotic phenomenon of bioluminescence in marine bacteria such as *Vibrio fischeri* and *Vibrio harveyi* was physiologically intriguing, if evolutionarily bizarre, but was probably considered by most scientists to represent an esoteric aspect of marine microbiology. Therefore, apart from a few researchers who had 'seen the light' (Greenberg *et al.*, 1979), it was not obvious that the molecular processes underpinning bacterial bioluminescence were likely to have profound implications as paradigms for some core aspects of prokaryotic gene regulation and bacterial physiology. Over the past 10 years it has become very clear that the fundamental molecular signalling systems regulating gene expression in bacterial bioluminescence control are widespread in Gram-negative bacteria. In that sense, our vision of eubacteria as single organisms operating with unicellular physiologies has been gradually replaced with a view of microbial populations acting in primitive multicellularity (Dunny & Winans, 1999; Whitehead & Salmond, 2000; Williams *et al.*, 2000; Schauder & Bassler, 2001; Swift *et al.*, 2001; Whitehead *et al.*, 2001a; Winzer & Williams, 2001; Withers *et al.*, 2001).

SGM symposium 61: Signals, switches, regulons and cascades: control of bacterial gene expression. Editors D. A. Hodgson, C. M. Thomas. Cambridge University Press. ISBN 0 521 81388 3 ©SGM 2002.

The ability of bacteria to sense their population densities via small molecule signalling is now commonly called 'quorum sensing'. This phenomenon occurs in Gram-positive and Gram-negative bacteria. In the former, the main signals identified to date are γ-butyrolactones and small peptides (reviewed by Kleerebezem *et al.*, 1997; Shapiro, 1998; Dunny & Winans, 1999). In Gram-negative bacteria, the main signals known currently are acylated derivatives of homoserine lactone (acyl HSLs), many of which are freely diffusible between cells and are generated and sensed by LuxRI systems.

LuxRI-TYPE QUORUM-SENSING SYSTEMS

The majority of characterized acyl-HSL-based quorum-sensing systems are reliant upon three primary constituents: the acyl HSL signal, the signal generator and the signal response regulator. The first organism in which all of these components were uncovered and for which a model was proposed was the LuxRI bioluminescence regulatory scheme of *V. fischeri*. In this system, the diffusible acyl HSL signal [N-(3-oxohexanoyl)-L-homoserine lactone; OHHL] is synthesized by the LuxI signal generator protein. The LuxR signal response protein is then believed to interact with the OHHL molecule once the signal has reached a certain concentration. The consequence of this interaction is that the LuxR protein is activated and rendered capable of binding to DNA in the promoter region of an operon containing *lux* genes, encoding bioluminescence, at a 20 bp inverted repeat sequence known as the *lux* box. From there, LuxR promotes transcription of the *lux* genes, thereby inducing subsequent heightened bioluminescence (Dunlap, 1999).

Over the past decade, screening programmes using acyl HSL sensors have revealed that many micro-organisms produce acyl-HSL-like signalling molecules. Similarly, proteins bearing functional homology to the LuxR and LuxI proteins have been discovered in many Gram-negative bacteria. The difficulty in purifying full-length LuxR has meant that studies using LuxR homologues from other systems have been necessary to shed light on the exact molecular mechanisms underlying acyl-HSL-based quorum-sensing systems. In this regard, the TraRI system of the plant pathogen *Agrobacterium tumefaciens* has revealed many aspects of the workings of LuxI and LuxR homologues and the interactions associated between the latter and acyl HSL molecules, as will be discussed later in this chapter.

LuxR- and LuxI-type proteins

The TraI enzyme catalyses the synthesis of the acyl HSL N-(3-oxooctanoyl)-L-homoserine lactone (OOHL), using S-adenosylmethionine (SAM) and the acyl carrier protein (ACP) adduct of 3-oxooctanoic acid as its substrates (Moré *et al.*, 1996). The *in vitro* synthesis of acyl HSLs by other LuxI homologues has also been analysed. Such studies also indicate that these enzymes synthesize acyl HSLs using, preferentially, SAM and

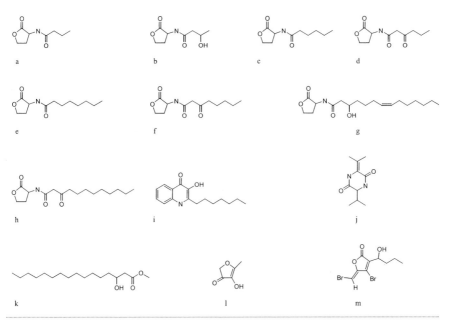

Fig. 1. Some quorum-sensing molecules. (a) BHL; (b) HBHL; (c) HHL; (d) OHHL; (e) OHL; (f) OOHL; (g) *N*-(3-hydroxy-7-*cis*-tetradecenoyl)-L-homoserine lactone (HtdeDHL); (h) OdDHL; (i) PQS; (j) the DKP cyclo(-ΔAla-L-Val); (k) 3-hydroxypalmitic acid methyl ester; (l) the putative structure of AI-2, 4-hydroxy-5-methyl-furanone; (m) the halogenated furanone 4-bromo-5-(bromomethylene)-3-(1′-hydroxybutyl)-2(5*H*)-furanone.

the appropriately charged ACP as substrates (Schaefer *et al.*, 1996; Jiang *et al.*, 1998; Hoang *et al.*, 1999; Parsek *et al.*, 1999). Depending upon the LuxI homologue, the resultant acyl HSL molecules can possess acyl chains of 4–14 carbons and they can contain double bonds and C_3 oxo- or hydroxy- groups (Fig. 1a–h). Acyl HSLs with chains containing an odd number of carbons have been identified (Lithgow *et al.*, 2000). Many LuxI homologues can direct the synthesis of more than one acyl HSL but the molecule(s) produced cannot yet be predicted via analysing the sequence of these enzymes.

LuxR homologues are around 250 amino acid residues in size. Early studies revealed that LuxR itself is modular, possessing a carboxy-terminal DNA-binding domain (Choi & Greenberg, 1991, 1992a) and an amino-terminal acyl-HSL-binding domain (Shadel *et al.*, 1990; Slock *et al.*, 1990; Hanzelka & Greenberg, 1995). Early predictions suggested that the protein multimerizes upon association with OHHL. Multimerization has been predicted to render LuxR capable of interacting with DNA from whence it can induce transcription (Choi & Greenberg, 1992b). We now know that this model (Fig. 2) holds true for some LuxR homologues. For instance, studies using TraR have

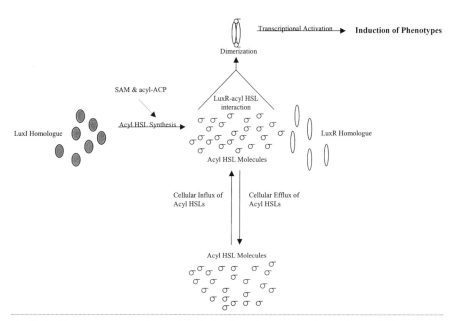

Fig. 2. LuxRI-type quorum-sensing model. The LuxI homologue synthesizes acyl HSL, which subsequently interacts with a LuxR homologue. This results in multimerization of the LuxR homologue, allowing it to bind to DNA and activate transcription of genes encoding various cellular functions.

revealed that when this protein interacts with its cognate acyl HSL, it dimerizes; removal of the acyl HSL results in dissociation of these dimers (Qin *et al.*, 2000; Zhu & Winans, 2001). A multimerization domain has been mapped to residues 49–156 of TraR (Qin *et al.*, 2000). DNA band shift assays and footprinting experiments have revealed that DNA-binding activity is dependent upon the prior OOHL-induced activation of the protein. This activity is also dependent upon the presence of *lux*-box-type sequences, known as *tra* boxes, within the DNA targets (Zhu & Winans, 1999, 2001; Luo & Farrand, 1999; Qin *et al.*, 2000).

DNase I protection studies using RNA polymerase and a truncated LuxR protein (LuxRΔN), missing its amino-terminal region, have indicated that both proteins are needed for efficient binding to the *lux* box (Stevens *et al.*, 1994; Stevens & Greenberg, 1997). Together, RNA polymerase and LuxRΔN have also been shown to activate transcription of the *lux* operon *in vitro* (Stevens & Greenberg, 1997). Based upon these results and the location of the *lux* box with respect to the transcriptional start site of the *lux* operon, LuxR has been predicted to be an ambidextrous transcriptional activator (Egland & Greenberg, 1999). Such proteins make contact with the α-subunit C-terminal domain of RNA polymerase (Rhodius & Busby, 1998), and it has been shown that this domain is necessary for full LuxR-dependent activation of the *lux* operon *in vivo* (Stevens *et al.*, 1999). Conversely, TraR alone has been shown to activate transcription from the promoters of two genes *in vitro* (Zhu & Winans, 1999). Interestingly, all mutations that affect transcriptional activity of TraR but do not affect its ability to bind DNA (positive-control mutants) have been mapped to the amino-terminal half of the protein (Luo & Farrand, 1999). This again differs from the situation with LuxR, where positive-control mutants have been mapped to the carboxy-terminal half of the protein (Egland & Greenberg, 2001). Taken together, these observations point to the possibility that, although belonging to the same family, LuxR and TraR activate transcription via two distinct mechanisms. However, it should be noted that care must be taken not to extrapolate data accumulated using LuxRΔN, as this protein may not behave in the same manner as its full-length parent. Therefore, further studies using purified full-length LuxR are necessary before the exact mechanism of transcriptional activation by this protein is to be elucidated.

Exceptions to the rules. As with many groupings of regulatory proteins, there are usually examples of elements within the family that are exceptions to the model and investigation of acyl HSL signalling systems has revealed many such cases. For instance, LuxI homologues are not the only proteins that can synthesize acyl HSL molecules. Indeed, the existence of at least two other protein types that can perform this function has been demonstrated. The first of these groupings consists of AinS of *V. fischeri*, LuxM of *V. harveyi* and VanM of *Vibrio anguillarum* (Bassler *et al.*, 1993; Gilson *et al.*, 1995; Milton *et al.*, 2001). These acyl HSL synthase proteins share sequence homology with each other but not with LuxI homologues. The substrate specificity of AinS has been investigated and it has been shown to be similar to LuxI homologues in that this enzyme utilizes SAM and the appropriately charged ACP for the synthesis of its primary acyl HSL (Hanzelka *et al.*, 1999). Recently, another type of acyl HSL synthase protein (named HdtS) that is unlike both LuxI- and AinS-type proteins has been identified in *Pseudomonas fluorescens* (Laue *et al.*, 2000). The biochemical specificity of this enzyme has yet to be investigated.

Although it was originally believed that, like OHHL, all acyl HSLs were freely diffusible between cells (Kaplan & Greenberg, 1985), it is now apparent that this may not necessarily be the case. In particular, the longer chain acyl HSLs may rely on a degree of active cellular influx and/or efflux. Such an efflux system has been shown to be required for the effective transport of a 12-carbon-chain acyl HSL out of *Pseudomonas aeruginosa* cells (Pearson *et al.*, 1999). Although examples of other similar acyl HSL translocation systems have yet to be discovered, their presence may prevent the aggregation of highly hydrophobic acyl HSLs in cellular membranes.

It is now apparent that not all acyl-HSL-based quorum-sensing bacteria have the same type of signal-sensing apparatus. At least one organism (*V. harveyi*) utilizes a two-component sensor-kinase-type protein (termed LuxN) to interact with acyl HSL signals as opposed to a LuxR homologue (Bassler *et al.*, 1993; Freeman *et al.*, 2000). Moreover, not all LuxR homologues act as transcriptional activators. Rather, some behave as transcriptional repressors. At least one of these repressors (EsaR of *Pantoea stewartii*) has the ability to form dimers in the absence of its cognate acyl HSL (Qin *et al.*, 2000). Similarly, some activatory LuxR homologues of the *Erwinia* species are also capable of forming dimers and/or binding to DNA in the absence of acyl HSLs (Nasser *et al.*, 1998; Reverchon *et al.*, 1998; Welch *et al.*, 2000). Hence, within the family of LuxR proteins, there may exist subgroups: those that do and those that do not require activation by acyl HSLs before DNA binding can ensue (Qin *et al.*, 2000).

Quorum sensing and the regulation of virulence factors

The widespread nature of the involvement of quorum-sensing systems in microbial physiology first became apparent in the early 1990s. At this time, independent research groups showed that the expression of exoenzymes by two different micro-organisms [*P. aeruginosa* and *Erwinia carotovora* subspecies *carotovora* (*Ecc*)] was reliant upon the production of acyl HSL signalling molecules (Jones *et al.*, 1993; Passador *et al.*, 1993; Pirhonen *et al.*, 1993).

Pseudomonas aeruginosa. The quorum-sensing system of *P. aeruginosa* has now been intensively investigated, probably due to its importance as a ubiquitous opportunistic pathogen of man. The micro-organism is particularly associated with infections of people with severe burns and is also prevalent in the lungs of patients afflicted with cystic fibrosis (Pollack, 1990; Govan & Deretic, 1996). The expression of multiple virulence factors is co-ordinated by the action of two complete LuxRI-type systems in *P. aeruginosa*. The first of these comprises LasR and LasI, the latter being responsible for catalysing the synthesis of the acyl HSL signal N-(3-oxododecanoyl)-L-homoserine lactone (OdDHL) (Gambello & Iglewski, 1991; Passador *et al.*, 1993; Pearson *et al.*, 1994). In the presence of inducing levels of OdDHL, LasR up-regulates the transcription of a large number of genes encoding virulence factors, and also of *rhlR* (Latifi *et al.*, 1995, 1996; Pesci *et al.*, 1997). The latter gene encodes a subordinate LuxR homologue that, in conjunction with N-butanoyl-L-homoserine lactone (BHL) synthesized by its partner LuxI homologue (RhlI), is responsible for up-regulating expression of a further gene set.

Together, the two quorum-sensing systems regulate the expression of an array of phenotypes. Indeed, recent work by Whiteley *et al.* (1999) has revealed that the expression of as many as 4 % of the genes within the genome of *P. aeruginosa* may be affected by either or both of the Las and Rhl quorum-sensing systems. Genes encoding functions involved with the production of such factors as alkaline protease, elastase, lectin, exotoxin, pyocyanin, pyoverdine, superoxide dismutase, rhamnolipid, hydrogen cyanide, haemolysin, chitinase, type IV pili and components of a type II secretion apparatus have all been shown to be regulated by the systems (Toder *et al.*, 1991; Jones *et al.*, 1993; Gambello *et al.*, 1993; Passador *et al.*, 1993; Ochsner *et al.*, 1994; Brint & Ohman, 1995; Latifi *et al.*, 1995; Ochsner & Reiser, 1995; Winson *et al.*, 1995; Chapon-Hervé *et al.*, 1997; Pearson *et al.*, 1997; Stintzi *et al.*, 1998; Glessner *et al.*, 1999; Hassett *et al.*, 1999; Pessi & Haas, 2000; Winzer *et al.*, 2000). Likewise, the formation of normal monospecies biofilms by *P. aeruginosa* depends upon the organism possessing fully functional quorum-sensing systems (Davies *et al.*, 1998). This is partly due to the quorum-sensing-dependent expression of type IV pili, required during the initial stages of biofilm formation (O'Toole & Kolter, 1998; de Kievit *et al.*, 2001). Some (but not all) of the quorum-sensing controlled genes are preceded by *lux*-box-type elements. How activated LasR and RhlR differentiate between these DNA signatures is not known.

It is perhaps unsurprising that, with so many exofactors controlled by the Las and Rhl regulators, *P. aeruginosa* mutants disrupted in their quorum-sensing systems are often less virulent than the equivalent wild-type strain when applied to model systems (Tang *et al.*, 1996; Rumbaugh *et al.*, 1999; Tan *et al.*, 1999; Jander *et al.*, 2000; Pearson *et al.*, 2000; Wu *et al.*, 2001). However, a contributing factor towards this negative effect could also be the fact that some acyl HSLs, and OdDHL in particular, have the potential to act as virulence factors *per se*. For example, it has been reported that OdDHL posssesses multifaceted immunosuppressive properties. It can also act as an inhibitor of vasoconstrictor tone in isolated blood vessels and cause marked changes in the heart rate of normal conscious rats (Telford *et al.*, 1998; Lawrence *et al.*, 1999; Williams *et al.*, 2000; Gardiner *et al.*, 2001).

The reason for *P. aeruginosa* cells controlling expression of virulence factors via quorum sensing is believed to be that the organism needs to wait for cell densities to become large enough to ensure a successful attack on a host (Salmond *et al.*, 1995). Interestingly, a recently characterized LuxR homologue, called QscR, may ensure that the Las and Rhl quorum-sensing systems are not initialized at low cell densities via repression of transcription of *lasI* (Chugani *et al.*, 2001). It is not known if QscR responds to, or is activated by, an acyl HSL signal.

Like the quorum-sensing systems of other bacteria, those of *P. aeruginosa* lie within a networked hierarchy of other regulators. For instance, the expression of *lasR* is partly dependent upon a CRP homologue, Vfr (Albus *et al.*, 1997). Both the *las* and *rhl* systems are positively regulated by GacA, a response regulator particularly associated with pseudomonads (Reimmann *et al.*, 1997). RsaL, encoded by a gene downstream of *lasR*, represses transcription of *lasI* (de Kievit *et al.*, 1999). Similarly, disruption of the *ppk* gene, encoding polyphosphate kinase, reduces synthesis of both BHL and OdDHL by *P. aeruginosa* (Rashid *et al.*, 2000). Interestingly, activation of the stationary phase sigma factor, RpoS, was initially thought to be under the regulatory control of the RhlRI system (Latifi *et al.*, 1996). However, the authors of a recent study reject this observation and provide evidence to show that RpoS is not in fact controlled by the Rhl quorum-sensing system, but instead may negatively affect transcription of *rhlI* (Whiteley *et al.*, 2000). Although the reason for these conflicting observations is not known, it is clear that RpoS is heavily involved in the regulation of phenotypes currently known also to be regulated by quorum sensing in *P. aeruginosa* (Suh *et al.*, 1999; Winzer *et al.*, 2000).

As well as acyl HSLs, *P. aeruginosa* synthesizes two further types of signalling molecule. The expression of the first such compound, 2-heptyl-3-hydroxy-4-quinolone (PQS) (Fig. 1i), is controlled by the Las and Rhl systems (Pesci *et al.*, 1999; McKnight *et al.*, 2000). The molecule has the ability to induce transcription of some genes within the quorum-sensing regulon of *P. aeruginosa*. Diketopiperazines (DKPs) (Fig. 1j) are also produced, along with other bacterial species, by *P. aeruginosa* (Holden *et al.*, 1999). The precise physiological role of these compounds is equivocal, but high concentrations of certain DKPs can antagonize or activate acyl-HSL-dependent quorum-sensing systems of other organisms. This is presumably achieved by DKPs interacting with LuxR homologues.

Erwinia spp. The plant pathogen *Ecc* secretes a barrage of plant cell wall degrading enzymes [including pectate lyases, carboxymethyl cellulases (CMCases) and polygalacturonases] and subsequently causes soft-rot of multiple plant species such as potato, carrot and turnip (Perombelon & Kelman, 1980; Py *et al.*, 1998; Thomson *et al.*, 1999). Production of these exoenzymes is, in part, dependent upon the synthesis of a threshold level of OHHL (Jones *et al.*, 1993; Pirhonen *et al.*, 1993). The protein responsible for the synthesis of this acyl HSL has been designated CarI (or HslI, ExpI and OhlI) (Jones *et al.*, 1993; Pirhonen *et al.*, 1993; Swift *et al.*, 1993; Chatterjee *et al.*, 1995). This LuxI homologue is also necessary for the up-regulation of antibiotic synthesis in certain strains of *Ecc* (Bainton *et al.*, 1992a, b). A LuxR homologue dedicated to the regulation of exoenzyme synthesis in *Ecc* has yet to be identified. To date, only two candidate LuxR homologues have been discovered in the organism. The first (CarR) is involved in the up-regulation of antibiotic synthesis alone; its disruption has no effect on exoenzyme synthesis (McGowan *et al.*, 1995). The second (ExpR) is encoded by a gene that overlaps *carI* and is convergently transcribed with it (Andersson *et al.*, 2000). The effect of disrupting *expR* on exoenzyme synthesis is negligible and, in some strains, non-existent. Interestingly, overexpression of CarR in wild-type *Ecc* results in partial reduction of exoenzyme biosynthesis (McGowan *et al.*, 1995). It is possible that this is due to the sequestration of OHHL by the excess CarR, thereby preventing activation of an as yet unidentified exoenzyme-committed LuxR homologue. It is also noteworthy that a complex network of global regulators is integrated into this quorum-sensing system and, if an exoenzyme-dedicated LuxR homologue does not exist in *Ecc*, this network may have evolved to regulate OHHL and exoenzyme levels (for detailed reviews see Whitehead & Salmond, 2000; Swift *et al.*, 2001).

Other *Erwinia* species produce acyl HSLs and express LuxRI-type proteins. Strains of *E. carotovora* subsp. *atroseptica*, *Erwinia herbicola* and *Erwinia chrysanthemi* all synthesize acyl HSLs (Cha *et al.*, 1998; Bainton *et al.*, 1992b). The quorum-sensing system of the latter has been partially characterized and is embedded within an integrated network of regulators including CRP and PecS (Nasser *et al.*, 1998; Reverchon *et al.*, 1998). Disruption of the *luxI* or *luxR* homologue (*expI* and *expR*) has a minor effect upon exoenzyme synthesis in the organism (Nasser *et al.*, 1998). However, *in vitro* studies have demonstrated that ExpR can bind to the regulatory regions of *expI*, *expR* and some exoenzyme genes (Nasser *et al.*, 1998; Reverchon *et al.*, 1998). The protein can only bind to the promoter of *expR* in the absence of acyl HSL; in the presence of the signal, such binding activity is annulled (Reverchon *et al.*, 1998). It is therefore possible that ExpR acts to repress transcription of *expR* in the absence of acyl HSL. The closely related plant pathogen *Pantoea stewartii* subsp. *stewartii* (formerly *Erwinia stewartii*) utilizes a quorum-sensing system to direct synthesis of an extracellular polysaccharide (EPS) (Beck von Bodman & Farrand, 1995; Beck von Bodman *et al.*, 1998). In this case, the LuxR homologue (EsaR) represses production of this important virulence factor in the absence of acyl HSL signal (Beck von Bodman *et al.*, 1998). Studies have shown that EsaR forms dimers in these conditions and it is thought likely that these dimers are able to bind DNA and block transcription of genes encoding EPS production (Qin *et al.*, 2000). Although repression of EPS synthesis is relieved in the presence of OHHL (Beck von Bodman *et al.*, 1998), the compound does not cause the dissociation of EsaR dimers (Qin *et al.*, 2000), and so how repression is alleviated is unknown.

As is the case with *P. aeruginosa*, it is believed that by controlling exoenzyme synthesis in a cell-density-dependent manner, *Ecc* populations enhance their chances of overwhelming plants without invoking a host defence response at early stages of infection. Interestingly, Fray *et al.* (1999) have generated tobacco plants that can express acyl HSLs via a LuxI homologue located within their chloroplasts. These workers have reasoned that such transgenic plants would be capable of inducing low-level populations of *Ecc* to 'uncover' themselves by expressing exoenzymes at early stages of infection, thereby raising an advanced plant response. Another anti-quorum-sensing strategy has been pursued by Dong *et al.* (2001). These workers have created transgenic plants that express an enzyme, AiiA, which can degrade acyl HSLs. Encouragingly, initial results show that the extent of soft-rot caused by *Ecc* infections is significantly reduced in such transgenic potato and tobacco plants compared to that in control plants (Dong *et al.*, 2001). AiiA was originally identified and isolated from a strain of *Bacillus* found in soil samples (Dong *et al.*, 2000). Although the specificity of the enzyme towards all acyl HSLs has yet to be determined, it appears to hydrolyse the ester bond of the HSL ring of those tested (Dong *et al.*, 2001).

Further surveys have led to the discovery of a strain of *Variovorax* with acyl HSL degradase activity (Leadbetter & Greenberg, 2000). The identification and application of the acyl HSL degrading apparatus expressed by this organism may also prove fruitful in the battle against bacterial infections of both plants and animals. However, it is noteworthy that some plants may already have evolved mechanisms to interact with bacterial acyl-HSL-based quorum-sensing communication schemes. Indeed, Teplitski *et al.* (2000) recently showed that plants such as pea and tomato expressed unidentified compounds that could interrelate with such systems.

Other micro-organisms. The human pathogen *Burkholderia cepacia* regulates protease and siderophore production via its CepRI quorum-sensing system (Lewenza *et al.*, 1999; Gotschlich *et al.*, 2001; Lewenza & Sokol, 2001; Lutter *et al.*, 2001). Like *P. aeruginosa*, *B. cepacia* is commonly associated with lung infections of CF patients. It is therefore likely that a degree of signalling cross-talk takes place between these two acyl-HSL-producing micro-organisms in this environment. The fish pathogens *Aeromonas salmonicida* and *Aeromonas hydrophila* express LuxRI homologues that are involved in the production of a secreted protease (Swift *et al.*, 1997, 1999a, b). Likewise, proteolytic activity of a *Serratia liquefaciens* quorum-sensing mutant is reduced compared to its wild-type progenitor (Eberl *et al.*, 1996). This is due to the expression of the Lip type I secretion system of this organism being quorum-sensing-dependent (Riedel *et al.*, 2001). *Chromobacterium violaceum* regulates the production of chitinolytic enzymes via N-hexanoyl-L-homoserine lactone (HHL) (Chernin *et al.*, 1998) and virulence factor expression by the insect pathogen *Xenorhabdus nematophilus* is also induced in the presence of an unidentified acyl-HSL-like signalling molecule (Dunphy *et al.*, 1997).

Quorum-sensing control of antibiotic synthesis

Some strains of *Ecc* produce the simple β-lactam antibiotic 1-carbapen-2-em-3-carboxylic acid (carbapenem) during late-exponential phase of growth (Parker *et al.*, 1982). In such strains, the carbapenem synthesis and resistance functions are encoded by the *car* cluster of genes (*carA–H*) (McGowan *et al.*, 1995, 1996, 1997). In *Ecc* ATCC 39048, the production of this antibiotic is regulated by a quorum-sensing system involving

OHHL (synthesized by CarI) and the transcriptional activator CarR (Swift *et al.*, 1993; McGowan *et al.*, 1995, 1999). *carR* is located approximately 150 bp upstream of the *car* cluster (McGowan *et al.*, 1995). The transcriptional start site of *carA* has been mapped to the intergenic region between *carR* and the *car* cluster (G. P. C. Salmond, unpublished observations). Unlike TraR, CarR exists as a dimer in the absence of OHHL and, *in vitro*, is capable of binding to the *carA–carR* intergenic region in this form (Welch *et al.*, 2000). Further *in vitro* studies have shown that in the presence of OHHL, each CarR dimer interacts with two molecules of OHHL, promoting the formation of multimeric CarR–OHHL complexes which are more resistant to proteolysis than their non-ligand associated counterparts (Welch *et al.*, 2000). The multimeric structures can still interact with the *carA–carR* intergenic region *in vitro*, but how (or if) they promote transcription of the *car* genes *in vivo* is not known. In this sense, it is interesting that overexpression of CarR in a *carI⁻* strain of *Ecc* does result in antibiotic synthesis (McGowan *et al.*, 1995). The addition of exogenous OHHL to *Ecc* ATCC 39048 cultures at early stages of growth also induces precocious production of carbapenem (Williams *et al.*, 1992). This indicates that OHHL is the limiting factor for carbapenem synthesis at the beginning of growth. In the environment, this could allow *Ecc* cells to rapidly respond to the presence of acyl HSLs produced by competing micro-organisms.

Other bacteria control antibiotic production via quorum-sensing systems. For instance, the expression of carbapenem in *Serratia* sp. ATCC 39006 is co-ordinated by a quorum-sensing system involving proteins designated CarR and SmaI (Cox *et al.*, 1998; Thomson *et al.*, 2000). Interestingly, *carR* and the antibiotic synthesis genes in this organism share a high level of sequence similarity with those of *Ecc*. It is therefore likely that these genes have been mobilized between the two micro-organisms by means of horizontal transfer (Gray & Garey, 2001). It is consequently noteworthy that the acyl HSLs recognized by CarR in the *Serratia* system (BHL and HHL) are different from those recognized by CarR in the *Ecc* system (OHHL). *Erwinia carotovora* subsp. *betavascolorum* and *C. violaceum* also use LuxRI homologues (EcbRI and CviRI, respectively) to up-regulate the synthesis of antimicrobial compounds (Throup *et al.*, 1995a; Costa & Loper, 1997).

The fluorescent micro-organism *Pseudomonas aureofaciens*, which is widely found in soil and within plant rhizospheres, produces three phenazine antibiotics: phenazine-1-carboxylic acid, 2-hydroxyphenazine-1-carboxylic acid and 2-hydroxyphenazine (collectively termed phenazine). The production of phenazine is regulated by a quorum-sensing system involving the LuxR homologue PhzR and the acyl HSL HHL (synthesized by PhzI) (Pierson *et al.*, 1994; Wood & Pierson, 1996; Wood *et al.*, 1997). Phenazine is a potent inhibitor of *Gaeumannomyces graminis* var. *tritici*, the fungal pathogen that causes 'take-all' disease of wheat (Thomashow & Weller, 1988). Consequently, *P. aureofaciens* has the ability to protect wheat from infection by this fungal pathogen by acting as a biocontrol agent (Weller & Cook, 1983). *Pseudomonas fluorescens* and *Pseudomonas chlororaphis* also regulate phenazine expression via PhzRI homologues (Mavrodi *et al.*, 1998; Chin-A-Woeng *et al.*, 2001). The first of these two micro-organisms produces an array of acyl HSLs in a strain-dependent manner (Shaw *et al.*, 1997; Cha *et al.*, 1998; Laue *et al.*, 2000; Elasri *et al.*, 2001). A strain of *P. fluorescens* also regulates biosynthesis of the polyketide antibiotic mupirocin via a quorum-sensing system composed of LuxRI homologues designated MupRI (El-Sayed *et al.*, 2001).

It has been postulated that antibiotic production and hence biocontrol activity by *P. aureofaciens* is induced by infection of host plants by *G. graminis* var. *tritici* (Pierson & Pierson, 1996). In this scenario, the wounding of the plant root by the fungus would cause a release of root exudates. These nutrients would then nourish the local population of *P. aureofaciens*, allowing cells to multiply and eventually reach the quorate population size necessary for antibiotic expression to be induced. This would subsequently protect the plant from further attack. In contrast, it has been speculated that the quorum-sensing regulation of antibiotic production in *Ecc* is used as part of the co-ordinated pathogenic attack on plants (Salmond *et al.*, 1995). The coupling of virulence and carbapenem expression to population size ensures that bacterial competition for nutrients from carbapenem-sensitive co-inhabitants in a newly elicited plant wound is minimized.

Quorum-sensing regulation of conjugal transfer

A. tumefaciens is a pathogen that induces crown gall tumours in plants via movement and integration of oncogenic DNA into the nucleus of its host (for reviews see Farrand, 1998; Zhu *et al.*, 2000). Genes encoding proteins for the synthesis of chemical metabolites known as opines are encoded upon this oncogenic DNA. Secreted opines from the tumours act as a food source for agrobacterial cells and also as a signal for the induction of conjugal transfer of Ti (tumour-inducing) plasmids (Dessaux *et al.*, 1998). Two different regulatory mechanisms for control of conjugal transfer exist for two different classes of Ti plasmids that encode genes for catabolism of two different classes of opines – octopine and agrocinopines. Conjugal transfer of octopine-type Ti plasmids is induced by secretion of this opine by the host plant. This induces expression of a LysR-type transcriptional activator called OccR in *A. tumefaciens* cells (Habeeb *et al.*, 1991; Wang *et al.*, 1992). OccR then activates expression of octopine catabolism genes and also *traR*, a gene located within an octopine-inducible operon, which encodes a LuxR homologue required for up-regulation of conjugal transfer (Fuqua & Winans, 1994, 1996a; Piper *et al.*, 1999). Upon interacting with OOHL, synthesized by a protein encoded by the Ti-plasmid-borne *traI* gene (Fuqua & Winans, 1994; Moré *et al.*, 1996), TraR dimerizes and is rendered capable of promoting transcription of *tra* and *trb* operons, encoding conjugal transfer functions. *traI* lies within the fore-mentioned *trb* operon and therefore, as TraI levels increase, an autoinduction ensues which subsequently leads to high-level expression of conjugal transfer. Regions showing sequence similarity to *lux* boxes (called *tra* boxes) can be found upstream of at least three operons regulated by TraR (Fuqua & Winans, 1994, 1996a, b). In the case of conjugal transfer of nopaline-type Ti plasmids, the regulator AccR serves to repress the expression of opine catabolism genes and *traR* (which lies within an AccR-regulated operon) (Beck von Bodman *et al.*, 1992; Kim & Farrand, 1997; Piper *et al.*, 1999). However, in the presence of agrocinopines, repression of these genes is relieved, leaving TraR free to activate transcription of the *tra* and *trb* genes once a critical concentration of OOHL has been synthesized by TraI (Piper *et al.*, 1993; Zhang *et al.*, 1993; Hwang *et al.*, 1994).

Conjugal transfer of both octopine- and nopaline-type Ti plasmids therefore comes under the control of two interlinked regulatory cascades involving plant-derived opines and agrobacterial-derived OOHL. Opine induction ensures that the ability to metabolize that particular opine is passed to all members of a population via transfer of the relevant Ti plasmids (Piper *et al.*, 1999). Removal of the opine control (for instance, by mutating *accR* in nopaline-type Ti plasmids) lowers the population size at which cells can perform Ti plasmid conjugal transfer (Piper & Farrand, 2000). Therefore, the maintenance of cell-density-dependent transfer is somewhat reliant upon opine-mediated control of *traR* expression. As Ti plasmids are inefficiently inherited by recipient cells (Piper & Farrand, 2000), maximizing the number of donors via quorum-sensing-dependent control of conjugation should result in increased chances of plasmid transfer.

The molecular mechanisms involved in TraR activation and consequential gene induction are well-studied and have been discussed elsewhere in this review. In short, activation of TraR monomers via association with OOHL promotes dimerization of the transcriptional activator, thereby allowing it to bind to the promoter regions of the genes within its regulon (Luo & Farrand, 1999; Zhu & Winans, 1999, 2001; Qin *et al.*, 2000). It has been suggested that association with external OOHL molecules releases hydrophobic TraR monomers from their inner membrane location by causing them to dimerize, thereby allowing them to enter the cytoplasm and interact with DNA (Qin *et al.*, 2000). Localization of TraR monomers within the membrane may prevent precocious induction of this protein by signal molecules synthesized within the same cell. Alternatively, Zhu & Winans (2001) have proposed that monomeric TraR is susceptible to proteolytic degradation within *A. tumefaciens* cells and that association with OOHL and subsequent dimerization stabilizes the transcriptional regulator against such activity. Whether either of these models can be extended to other quorum-sensing systems remains to be seen.

Regulators of the TraRI system have been discovered and characterized. One of these is a LuxR homologue, designated TrlR (or TraS), encoded on octopine-type Ti plasmids, within an operon involved in opine uptake (Oger *et al.*, 1998; Zhu & Winans, 1998). A frameshift mutation within the gene encoding TrlR means that the resultant protein is truncated and lacks its carboxy-terminal (DNA-binding) domain. This defective protein is therefore capable of repressing the action of full-length TraR by forming inactive dimers with the LuxR homologue and preventing it from binding to DNA (Chai *et al.*, 2001). Another regulator, TraM, is encoded on both nopaline- and octopine-type Ti plasmids (Hwang *et al.*, 1995; Fuqua *et al.*, 1995). This 11 kDa repressor protein acts by binding to the carboxy-terminal domain of TraR, thereby preventing it from associating with target promoters (Hwang *et al.*, 1999; Luo *et al.*, 2000). The suggested physiological role of the protein has two parts. First, it may act to prevent TraR-mediated conjugal transfer at low cell densities (Piper & Farrand, 2000). Second, it may act to reduce the affinity of activated TraR dimers for DNA by binding to them and interfering with their activity (Luo *et al.*, 2000).

Quorum sensing and motility/dispersal

Quorum-sensing regulation of motility would theoretically allow a growing population of bacterial cells to disperse when the population has reached a threshold size, presumably with the beneficial consequence of lowering the competition for nutrients available to that population. Swarming motility is an organized bacterial behaviour in which cellular differentiation leads to the expression of hyper-flagellated, aseptate, multi-nucleoid and extremely motile cells which move co-ordinately to rapidly colonize surfaces (Harshey, 1994). The *swr* quorum-sensing system has been identified as responsible for initiation of swarming motility in *Serratia liquefaciens*. The *swrI* gene encodes a LuxI homologue responsible for the synthesis of two acyl HSLs, namely BHL and HHL (Eberl *et al.*, 1996). Disruption of *swrI* abolishes swarming motility; supplementing media with BHL and (to a lesser extent) HHL restores wild-type levels of swarming motility to a *swrI* mutant (Eberl *et al.*, 1996). The amount of biosurfactant, serrawettin W2, that is produced by the *swrI⁻* strain is concomitantly reduced with swarming (Lindum *et al.*, 1998). Serrawettin W2 was first identified as an exoproduct of *Serratia marcescens* and is required to condition surfaces prior to swarming through a reduction in surface tension (Matsuyama *et al.*, 1992). The *swrA* gene encodes a putative non-ribosomal peptide synthetase that presumably directs the biosynthesis of serrawettin W2 in *Serratia liquefaciens* (Lindum *et al.*, 1998). Transcription of *swrA* is up-regulated in the presence of exogenous BHL whilst the addition of serrawettin W2 to growth media restores swarming motility to *swrI⁻* strains (Lindum *et al.*, 1998). In *P.*

aeruginosa, swarming behaviour is also acyl-HSL-dependent as it relies upon cells to produce rhamnolipid biosurfactants (Köhler *et al.*, 2000). As previously mentioned, the synthesis of rhamnolipids is under the control of the *rhl* quorum-sensing system in this organism (Ochsner *et al.*, 1994; Brint & Ohman, 1995; Ochsner & Reiser, 1995).

Quorum sensing also plays a role in the control of other forms of bacterial dissemination. For instance, type IV pili dependent twitching motility in *P. aeruginosa* is controlled by quorum sensing (Glessner *et al.*, 1999). The YpsRI and YtbRI quorum-sensing systems of *Yersinia pseudotuberculosis* are also involved in the temperature-dependent regulation of motility and cellular aggregation (Atkinson *et al.*, 1999). The YpsR protein represses cellular clumping, presumably by down-regulating the production of a clumping factor. The protein also represses expression of the major flagellin subunit of the organism, as reflected by the temperature-dependent hypermotility of a *ypsR* mutant (Atkinson *et al.*, 1999). In *Rhodobacter sphaeroides*, inactivation of the *luxI* homologue *cerI* results in the formation of cellular aggregates, possibly due to the overproduction of an extracellular polysaccharide (Puskas *et al.*, 1997). Addition of exogenous N-(7-*cis*-tetradecenoyl)-L-homoserine lactone (tdeDHL), the acyl HSL synthesized by CerI, to cultures of the *cerI⁻* strain leads to the restoration of cellular dispersal (Puskas *et al.*, 1997).

Quorum sensing and pigmentation

Some bacteria use quorum-sensing systems for the regulation of pigment production. This has been exploited to use mutant strains of these micro-organisms as acyl HSL biosensors enabling facile detection of signalling molecule production. The most commonly used biosensor in this sense is *C. violaceum* CV026. This mutant strain can only synthesize a purple pigment, violacein, in the presence of exogenously supplied acyl HSLs (McClean *et al.*, 1997). This is because *C. violaceum* CV026 carries a mutation in a gene (*cviI*) which encodes a LuxI homologue that is responsible for synthesizing HHL, required for the induction of pigment synthesis. Violacein production is induced in the CV026 mutant in the presence of acyl HSLs containing acyl side chains of between four and eight carbon atoms. Acyl HSLs with longer acyl side chains can also be detected by CV026 through their ability to inhibit violacein production in the presence of an activating signal such as HHL (McClean *et al.*, 1997; Camara *et al.*, 1998). A *smaI* mutant of *Serratia* sp. ATCC 39006 can also be used in a similar manner to *C. violaceum* CV026 by virtue of this strain's dependence upon the presence of exogenous acyl HSLs for the up-regulation of the biosynthesis of a red pigment, 2-methyl-3-pentyl-6-methoxyprodigiosin (Thomson *et al.*, 2000).

Other widely used biosensors exist that are based on strains carrying monitor plasmids containing detectable reporter constructs fused to acyl-HSL-inducible promoters (Swift *et al.*, 1993; Shaw *et al.*, 1997; Winson *et al.*, 1998; Andersen *et al.*, 2001). Coupling any of the fore-mentioned biosensors with thin-layer chromatography enables putative identification of acyl HSLs extracted from spent culture supernatants. However, most of the detection systems are limited in the range of acyl HSLs to which they respond and there is no biosensor currently available which allows the detection of all known acyl HSL molecules (Ravn *et al.*, 2001). Hence the inability to detect a signal with a given sensor system does not exclude the possibility that acyl HSLs are being produced by an organism. A further caveat is that such biosensors can cross-react with structurally unrelated compounds such as DKPs that are produced by a wide range of bacteria (Holden *et al.*, 1999). Therefore, definitive identification of acyl HSLs does require further analysis via MS and NMR or chemical synthesis.

Other quorum-sensing micro-organisms

Acyl HSL production is very common among the genus *Rhizobium*. Species synthesize a wide array of signals and can often produce multiple short- and long-chained acyl HSLs. *Sinorhizobium meliloti*, *Rhizobium fredii* and *Rhizobium etli* have all been shown to synthesize acyl HSLs (Shaw *et al.*, 1997; Cha *et al.*, 1998; Rosemeyer *et al.*, 1998). The latter organism may use them to down-regulate nodulation (Rosemeyer *et al.*, 1998). Different strains of *Rhizobium leguminosarum* also control production of multiple compounds via a network of LuxI and LuxR homologues encoded both on Sym plasmids and on the chromosome (Gray *et al.*, 1996; Schripsema *et al.*, 1996; Rodelas *et al.*, 1999; Lithgow *et al.*, 2000, 2001). However, it is still unclear as to the precise roles of these quorum-sensing systems. In *Rhizobium leguminosarum* biovar *viciae*, disruption of one *luxI*-type gene (*rhiI*) causes a slight induction in nodule formation on plant hosts relative to the wild-type strain (Rodelas *et al.*, 1999). One of the other sets of LuxRI homologues (CinRI) may be involved in regulation of conjugation, although this is thought to be mediated via induction of an uncharacterized *luxRI* pair (Lithgow *et al.*, 2000). Interestingly, sequencing of the Sym plasmid of *Rhizobium* sp. NGR234 revealed that it contains genes sharing sequence identity with *traR*, *traI* and *traM* of *A. tumefaciens* (Freiberg *et al.*, 1997). Therefore, rhizobial regulation of conjugal transfer of Sym plasmids via a mechanism similar to that utilized by the latter organism seems likely.

Another genus in which acyl-HSL-based signalling is prevalent is that of *Yersinia*. *Yersinia frederiksenii*, *Yersinia kristensenii*, *Yersinia intermedia*, *Y. pseudotuberculosis*, *Yersinia enterocolitica*, *Yersinia pestis* and *Yersinia ruckeri* all produce acyl HSLs and *luxIR* homologues have been identified in the latter four organisms (Throup *et al.*, 1995b; Atkinson *et al.*, 1999; Swift *et al.*, 1999a; Temprano *et al.*, 2001). Apart from the role in control of motility in *Y. pseudotuberculosis*, the function of quorum sensing in this group of bacteria remains elusive. However, 2D PAGE analysis has revealed that acyl HSLs are required for the expression of unidentified proteins in *Y. enterocolitica* (Throup *et al.*, 1995b).

Many phytopathogenic pseudomonads have been demonstrated to produce acyl HSLs (Cha *et al.*, 1998; Dumenyo *et al.*, 1998; Elasri *et al.*, 2001). Surveys have revealed strains of *Pseudomonas corrugata*, *Pseudomonas savastanoi* and pathovars of *Pseudomonas syringae* all generate signals, probably via LuxI homologues (Dumenyo *et al.*, 1998). Although the functions of their quorum-sensing systems remain undetermined, it has been shown that acyl HSL production is more prevalent among plant-associated pseudomonad isolates than it is among pseudomonad isolates from soil (Elasri *et al.*, 2001). This indicates that acyl HSL production could be important for currently unrecognized plant–pseudomonad interactions. The closely related plant pathogen *Ralstonia solanacearum* also produces acyl HSLs that regulate unknown functions (Flavier *et al.*, 1997). In this case, the expression of the *luxIR* homologues (*solIR*) is subordinate to a regulatory system involving another diffusible signal, namely 3-hydroxypalmitic acid methyl ester (Fig. 1k) (reviewed by Schell, 2000; Whitehead *et al.*, 2001a).

Finally, the fish pathogen *V. anguillarum* produces acyl HSLs via two proteins, VanI and VanM (Milton *et al.*, 1997, 2001). The first of these LuxI-type enzymes catalyses the synthesis of N-(3-oxodecanoyl)-L-homoserine lactone (ODHL) while the latter LuxM-type enzyme is responsible for producing both HHL and N-(3-hydroxyhexa-noyl)-L-homoserine lactone (HHHL). Although expression of *vanI* is regulated by VanM, disruption of either acyl HSL synthase encoding gene does not affect production of proteases or virulence in a fish model (Milton *et al.*, 1997, 2001).

AI-2-TYPE SIGNALLING

As already discussed, the prototypic LuxRI-based quorum-sensing system was discovered in *V. fischeri*, where it is involved in the regulation of bioluminescence and, possibly, other phenotypes (Dunlap, 1999; Callahan & Dunlap, 2000). Later, it was shown that a second marine *Vibrio*, *V. harveyi*, also regulates light emission via quorum sensing (Cao & Meighen, 1989; Bassler *et al.*, 1993). However, the method and signals this micro-organism uses to do so differ from those used by *V. fischeri*. Genetic studies have revealed that in *V. harveyi*, two separate quorum-sensing systems operate in parallel to regulate bioluminescence (Fig. 3) (as reviewed by Schauder & Bassler, 2001). The first of these systems responds to an acyl HSL, N-(3-hydroxybutanoyl)-L-homoserine lactone (HBHL), which is synthesized by a non-LuxI-type enzyme called LuxM (Cao & Meighen, 1989; Bassler *et al.*, 1993; Milton *et al.*, 2001). *V. harveyi* cells detect this signal by use of a two-component hybrid sensor kinase termed LuxN (Bassler *et al.*, 1993; Freeman *et al.*, 2000). In the absence of sufficiently high concentrations of HBHL

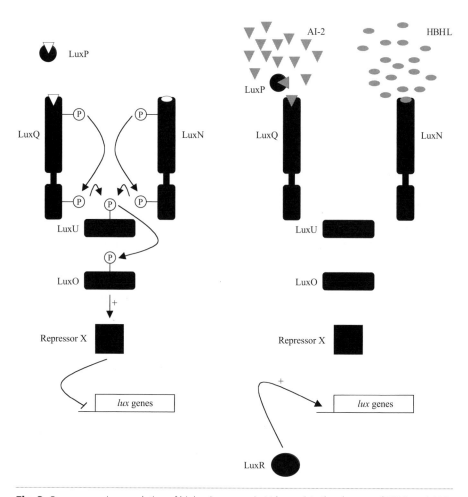

Fig. 3. Quorum-sensing regulation of bioluminescence in *V. harveyi*. In the absence of HBHL and AI-2, LuxN and LuxQ autophosphorylate and potentiate a phosphorelay through LuxU, ending in the phosphorylation of LuxO. Phosphorylated LuxO is thought to induce expression of a negative regulator of the *lux* genes (repressor X). In the presence of the two signalling molecules, LuxN and LuxQ gain phosphatase activity. Subsequent dephosphorylation of LuxO prevents up-regulation of repressor X, leaving LuxR to induce transcription of the *lux* genes.

(such as when *V. harveyi* population levels are low), LuxN autophosphorylates. This triggers a phosphorelay through a phosphotransfer protein, LuxU, eventually leading to phosphorylation of a response regulator, LuxO (Bassler *et al.*, 1994a; Freeman & Bassler, 1999a, b; Freeman *et al.*, 2000). When phosphorylated, LuxO, together with σ^{54}, is believed to activate the transcription of an unidentified regulator that acts by repressing expression of the bioluminescence (*lux*) structural genes (Lilley & Bassler, 2000). In the presence of inducing levels of HBHL, LuxN loses its kinase activity and gains phosphatase activity (Freeman *et al.*, 2000). Consequently, the dephosphorylation of LuxO results in its inactivation, thereby meaning that transcription of the unidentified repressor of the *lux* genes is dampened down and that bioluminescence can ensue unabated (Lilley & Bassler, 2000). A positive transcription factor, LuxR, is required for *lux* expression (Showalter *et al.*, 1990; Swartzman *et al.*, 1992). Despite its name, this DNA-binding protein is not closely related to LuxR homologues of acyl HSL quorum-sensing systems.

The second quorum-sensing system of *V. harveyi* is not acyl-HSL-based. Instead, the organism produces a further molecule called AI-2 via an enzyme termed LuxS (Surette *et al.*, 1999). It is believed that AI-2 primarily binds to a periplasmic protein, LuxP, which shares significant sequence identity with a ribose-binding protein of *Escherichia coli* (Bassler *et al.*, 1994b). The signal is then channelled to a two-component hybrid sensor kinase called LuxQ (Bassler *et al.*, 1994b). Similar to the first signalling pathway, at low concentrations of AI-2, LuxQ autophosphorylates and potentiates a phosphorelay through LuxU to the response regulator LuxO, thereby activating the latter protein. At high concentrations of AI-2, LuxQ gains phosphatase activity, causing eventual dephosphorylation and inactivation of LuxO (Lilley & Bassler, 2000).

It may at first seem wasteful for *V. harveyi* cells to possess parallel quorum-sensing circuits to control the same phenotype. However, there is evidence that the timing and intensity of response to the two signals is different (Freeman & Bassler, 1999a; Freeman *et al.*, 2000). Therefore, it has been suggested that *V. harveyi* utilizes two quorum-sensing systems to differentially respond to the presence of *V. harveyi* cells and non-*V. harveyi* cells (Bassler *et al.*, 1997). As such, Bassler *et al.* (1997) proposed that HBHL is recognized as an intraspecies signal, predominantly produced by *V. harveyi* cells, whereas AI-2 is sensed as an intra- and interspecies signal, produced by both *V. harveyi* and other micro-organisms. Corresponding to this theory, HBHL signalling does appear to be uncommon amongst the acyl-HSL-producing micro-organisms surveyed to date. Likewise, an initial screening programme (using a strain of *V. harveyi* mutated in *luxN* and therefore only able to emit light in the presence of AI-2 and not acyl HSLs) demonstrated that AI-2-like molecules were present in the supernatants of multiple micro-organisms. These included *Vibrio cholerae*, *Vibrio parahaemolyticus*, *V. anguillarum* and *Y. enterocolitica* (Bassler *et al.*, 1997). More recently, Surette & Bassler (1998) showed that certain strains of *E. coli* and *Salmonella typhimurium* could produce AI-2-like molecules when grown in LB media supplemented with certain carbohydrates. The genes necessary for AI-2 synthesis in these two micro-organisms were subsequently identified and both were termed *luxS* as both shared significant sequence identity with *luxS* of *V. harveyi* (Surette *et al.*, 1999). Interestingly, *E. coli* DH5α does not produce an AI-2-like molecule. This is because the *luxS* gene in this strain contains a frameshift mutation, resulting in it encoding a truncated LuxS protein (Surette *et al.*, 1999).

Database analysis has revealed that *luxS* genes are present in the genomes of over 30 species of bacteria, including both Gram-negative and Gram-positive micro-organisms (Surette *et al.*, 1999). Such species include *Borrelia burgdorferi*, *Haemophilus influenzae*, *Bacillus subtilis*, *Neisseria meningitidis*, *Neisseria gonorrhoeae*, *Enterococcus faecalis*, *Y. pestis*, *Helicobacter pylori*, *Streptococcus pneumoniae*, *Mycobacterium tuberculosis*, *Deinococcus radiodurans*, *Porphyromonas gingivalis* and *Campylobacter jejuni* (Surette *et al.*, 1999). Some of these bacteria have subsequently been shown to synthesize molecules with AI-2-like activity; disruption of the corresponding *luxS* gene invariably results in loss of AI-2 activity (Forsyth & Cover, 2000; Joyce *et al.*, 2000; Day & Maurelli, 2001; Chung *et al.*, 2001; Frias *et al.*, 2001).

In one of the bacteria just listed (*B. burgdorferi*), the *luxS* gene is positioned within a three-ORF operon along with two genes encoding proteins involved with the synthesis and utilization of SAM. This observation led Schauder *et al.* (2001) to hypothesize that LuxS might also be involved in this biochemical pathway. They therefore used purified LuxS proteins (isolated from several bacterial species) in biochemical and biosynthetic studies to show that AI-2 is produced from SAM via a three-step pathway, of which LuxS is the enzyme involved in the final step. The end product of this biochemical conversion is 4,5-dihydroxy-2,3-pentanedione, which is believed to spontaneously cyclize to form a furanone (Fig. 1l). The crystal structure of three LuxS proteins has been solved and these indicate that the proteins are metalloenzymes that are active as homodimers (Lewis *et al.*, 2001). This study also suggests that the substrate for LuxS is likely to be an amino acid derivative (predicted to be *S*-ribosylhomocysteine) as the authors observed binding of methionine to the putative substrate binding site of the enzyme.

While the identity of AI-2 is still not known, Schauder *et al.* (2001) have shown that various furanones do have AI-2-like activity. Furthermore, others have shown that LuxS drives the synthesis of 4-hydroxy-5-methyl-3(2*H*)-furanone (MHF) (K. Winzer & P. Williams, personal communication). However, although active, the precise relationship between this molecule and AI-2 has yet to be established. Early indications suggest that the structures of compounds with AI-2 activity are the same, therefore meaning that the non-species-specific nature of this signalling molecule makes it recognizable by all organisms with AI-2-sensing apparatus. This differs from the situation with acyl HSL signals, which can be species-specific depending upon the nature of respective LuxR homologues. Intriguingly, halogenated furanones (Fig. 1m) have been demonstrated to be inhibitors of LuxRI-type quorum-sensing systems (Givskov *et al.*, 1996; Manefield *et al.*, 1999; Rasmussen *et al.*, 2000). It is believed (but has, at the time of writing, yet to be conclusively shown) that these compounds interact with LuxR homologues, blocking them from binding their cognate acyl HSLs (Manefield *et al.*, 1999). Therefore, if AI-2 is eventually shown to be a furanone, how this molecule interacts with LuxR homologues and how halogenated furanones interact with AI-2-based quorum-sensing systems will be interesting lines of research to pursue.

Analysis of the production profile of AI-2 by a strain of *Salmonella typhimurium* shows that it is only made transiently in mid-exponential phase of growth and is degraded during stationary phase (Surette & Bassler, 1999). Other environmental factors impinge on AI-2 production, particularly the nutrient status of the growth media, pH, osmolarity and aerobiosis (Surette & Bassler, 1999; de Lisa *et al.*, 2001). Early characterization of an AI-2-degradative pathway suggests that the system is stimulated in conditions of low osmolarity and upon the depletion of certain nutrients in the growth media (Surette & Bassler, 1999).

The physiological role of AI-2 in most of the micro-organisms that produce it is yet to be elucidated. However, initial studies indicate that the signalling molecule is involved in the regulation of pathogenesis in some bacteria. For instance, in *Shigella flexneri*, the expression of a transcriptional regulator involved in co-ordinating expression of genes required to elicit an invasion phenotype is sensitive to the presence of AI-2 (Day & Maurelli, 2001). Differential display PCR analysis has also shown that disruption of *luxS* in *Porphyromonas gingivalis* results in the altered expression of several virulence-related gene products (Chung *et al.*, 2001). However, a *P. gingivalis luxS* mutant which exhibits reduced exoprotease and haemagglutinin activities is nevertheless as virulent as the parent strain in a soft tissue animal infection model (N. Burgess & P. Williams, personal communication). Mutation of *luxS* in *H. pylori* has no effect on any discernible phenotype and comparative 2D PAGE analysis of proteins extracted from wild-type and *luxS⁻* strains of the organism grown under identical conditions did not reveal any differences in resultant protein profiles (Forsyth & Cover, 2000; Joyce *et al.*, 2000). In contrast, the AI-2 molecules produced by enteropathogenic and enterohaemorrhagic *E. coli* O157:H7 (EHEC) have been demonstrated to modulate expression of a type III secretion system involved in virulence factor transport (Sperandio *et al.*, 1999). Moreover, a recent study has used a gene array approach to show that the expression of very many EHEC genes is modulated by the organism's AI-2 quorum-sensing system (Sperandio *et al.*, 2001). Some of the AI-2-sensitive genes identified encoded products involved in phenotypes such as cell division, chemotaxis, DNA replication and energy metabolism, thereby underlining the fact that LuxS of this organism is a pleiotropic regulator. Interestingly, one of the AI-2-induced genes highlighted in the previous study was *sdiA*. In *E. coli* and *Salmonella typhimurium*, this gene encodes a LuxR homologue involved in the regulation of genes that are important for cell division and pathogenesis

(Wang *et al.*, 1991; Ahmer *et al.*, 1998; Kanamaru *et al.*, 2000a; Wei *et al.*, 2001). Neither of the latter organisms produce acyl HSLs and it is not believed that SdiA interacts with AI-2, although it does seem to interact with an unidentified signal in *E. coli* supernatants (Surette & Bassler, 1999; Kanamaru *et al.*, 2000a, b). There is also evidence that this LuxR homologue is activated by acyl HSLs (Sitnikov *et al.*, 1996; Kanamaru *et al.*, 2000b) and so it is possible that the protein allows *E. coli* and *Salmonella typhimurium* cells to respond to such signals produced by competing micro-organisms.

To date, *V. harveyi* is the only organism from which the entire AI-2 sensing and response system has been characterized. However, as so many other organisms produce AI-2, one would predict that these also express proteins that function in an analogous manner to LuxP, LuxQ, LuxU and LuxO of *V. harveyi*. As such, it is interesting that *luxO* and *luxU* homologues have been uncovered in *V. fischeri* (Miyamoto *et al.*, 2000). Inactivation of *luxO* in this organism results in increased bioluminescence (Miyamoto *et al.*, 2000). It is therefore possible that, like *V. harveyi*, *V. fischeri* possesses two systems for the regulation of bioluminescence. It would be enlightening to discover how many other micro-organisms with LuxRI-type quorum-sensing systems also use AI-2-type systems to regulate the same phenotypes.

FINAL COMMENTS

Over the past 10 years, chemical communication systems like those regulating the esoteric phenomenon of bacterial bioluminescence have moved from being bizarre curiosities to centre stage in microbial physiology. Intercellular signalling in free-living bacteria is now known to be profoundly important in dictating and modulating diverse aspects of bacterial physiology. An important factor enabling the explosion of research activity in the field of quorum sensing was the development of various simple biosensor strains capable of detecting active biomolecules. Once an appropriate sensor strain became available then detection of activating molecules became facile and rapidly revealed the panorama of producing organisms. It would be foolish indeed to think that the families of molecules currently known to play key roles in quorum sensing are the only chemical communication systems operating in Gram-negative bacteria. No doubt, new intercellular signalling molecules remain to be discovered and, when appropriate biosensors become available, the distribution of such molecules will also become evident. We await such developments in excited anticipation!

Quorum-sensing systems are clearly of significance for our understanding of the fundamental biology of bacteria, but they may also be exploitable as genetic regulatory control systems for biotechnological processes (Robson *et al.*, 1997). Furthermore, the components of quorum-sensing pathways could be significant chemotherapeutic targets in human, animal and plant pathogens; inactivation by enzymes or inhibition by small molecules might render the respective pathogenic hosts avirulent (Whitehead *et al.*, 2001b). It seems likely, therefore, that the impressive recent advances in solving crystal structures of LuxS (Lewis *et al.*, 2001) and LuxI and LuxR homologues (M. Churchill & S. Winans, personal communication) could enable new developments in chemotherapeutic intervention in bacterial disease.

Despite rapid and fascinating progress in the past decade, there remain many structural, mechanistic and physiological questions about the phenomenon of quorum sensing. As a system for bacterial global gene regulation, quorum sensing affords a simple chemical messenger route linking bacterial cell density to the control of a spectrum of physiological processes. In a sense, therefore, this kind of signalling involves only the titration of one environmental cue – bacterial population size. Of course, bacteria are the evolutionary experts par excellence at adapting to diverse niches by constantly sensing a multitude of environmental cues and integrating these signals before responding appropriately. So, a key question to address is: in the 'real world' just how important is quorum sensing for bacteria? As already discussed within this chapter, there are already some clues that it is very important. Perhaps the answers to structural and physiological questions may contribute to our understanding of the true ecological and evolutionary advantages of intercellular communication in the bacterial world.

ACKNOWLEDGEMENTS

Work in the laboratories of G. P. C. S. and P. W. is funded by the BBSRC, MRC, Wellcome Trust and European Union, which is gratefully acknowledged. Thanks to Martin Welch and Holly Slater for many helpful discussions.

REFERENCES

Ahmer, B. M. M., van Reeuwijk, J., Timmers, C. D., Valentine, P. J. & Heffron, F. (1998). *Salmonella typhimurium* encodes an SdiA homolog, a putative quorum sensor of the LuxR family, that regulates genes on the virulence plasmid. *J Bacteriol* **180**, 1185–1193.

Albus, A. M., Pesci, E. C., Runyen-Janecky, L. J., West, S. E. H. & Iglewski, B. H. (1997). Vfr controls quorum sensing in *Pseudomonas aeruginosa*. *J Bacteriol* **179**, 3928–3935.

Andersen, J. B., Heydorn, A., Hentzer, M., Eberl, L., Geisenberger, O., Christensen, B. B., Molin, S. & Givskov, M. (2001). *gfp*-based *N*-acyl homoserine-lactone sensor systems for detection of bacterial communication. *Appl Environ Microbiol* **67**, 575–585.

Andersson, R. A., Eriksson, A. R. B., Heikinheimo, R., Mae, A., Pirhonen, M., Koiv, V., Hyytiainen, H., Tuikkala, A. & Palva, E. T. (2000). Quorum sensing in the plant pathogen *Erwinia carotovora* subsp. *carotovora*: the role of *expR*$_{Ecc}$. *Mol Plant–Microbe Interact* **13**, 384–393.

Atkinson, S., Throup, J. P., Stewart, G. S. A. B. & Williams, P. (1999). A hierarchical quorum-sensing system in *Yersinia pseudotuberculosis* is involved in the regulation of motility and clumping. *Mol Microbiol* **33**, 1267–1277.

Bainton, N. J., Stead, P., Chhabra, S. R., Bycroft, B. W., Salmond, G. P. C., Stewart, G. S. A. B. & Williams, P. (1992a). *N*-(3-Oxohexanoyl)-L-homoserine lactone regulates carbapenem antibiotic production in *Erwinia carotovora*. *Biochem J* **288**, 997–1004.

Bainton, N. J., Bycroft, B. W., Chhabra, S. R. & 8 other authors (1992b). A general role for the *lux* autoinducer in bacterial cell signalling: control of antibiotic biosynthesis in *Erwinia*. *Gene* **116**, 87–91.

Bassler, B. L., Wright, M., Showalter, R. E. & Silverman, M. R. (1993). Intercellular signalling in *Vibrio harveyi*: sequence and function of genes regulating expression of luminescence. *Mol Microbiol* **9**, 773–786.

Bassler, B. L., Wright, M. & Silverman, M. R. (1994a). Sequence and function of LuxO, a negative regulator of luminescence in *Vibrio harveyi*. *Mol Microbiol* **12**, 403–412.

Bassler, B. L., Wright, M. & Silverman, M. R. (1994b). Multiple signalling systems controlling expression of luminescence in *Vibrio harveyi*: sequence and function of genes encoding a second sensory pathway. *Mol Microbiol* **13**, 273–286.

Bassler, B. L., Greenberg, E. P. & Stevens, A. M. (1997). Cross-species induction of luminescence in the quorum-sensing bacterium *Vibrio harveyi*. *J Bacteriol* **179**, 4043–4045.

Beck von Bodman, S. & Farrand, S. K. (1995). Capsular polysaccharide biosynthesis and pathogenicity in *Erwinia stewartii* require induction by an *N*-acyl homoserine lactone autoinducer. *J Bacteriol* **177**, 5000–5008.

Beck von Bodman, S., Hayman, G. T. & Farrand, S. K. (1992). Opine catabolism and conjugal transfer of the nopaline Ti plasmid pTiC58 are coordinately regulated by a single repressor. *Proc Natl Acad Sci U S A* **89**, 643–647.

Beck von Bodman, S., Majerczak, D. R. & Coplin, D. L. (1998). A negative regulator mediates quorum-sensing control of exopolysaccharide production in *Pantoea stewartii* subsp. *stewartii*. *Proc Natl Acad Sci U S A* **95**, 7687–7692.

Brint, J. M. & Ohman, D. E. (1995). Synthesis of multiple exoproducts in *Pseudomonas aeruginosa* is under the control of RhlR-RhlI, another set of regulators in strain PAO1 with homology to the autoinducer-responsive LuxR-LuxI family. *J Bacteriol* **177**, 7155–7163.

Callahan, S. M. & Dunlap, P. V. (2000). LuxR- and acyl-homoserine-lactone-controlled non-*lux* genes define a quorum-sensing regulon in *Vibrio fischeri*. *J Bacteriol* **182**, 2811–2822.

Camara, M., Daykin, M. & Chhabra, S. R. (1998). Detection, purification, and synthesis of *N*-acylhomoserine lactone quorum sensing signal molecules. *Methods Microbiol* **27**, 319–330.

Cao, J. G. & Meighen, E. A. (1989). Purification and structural identification of an autoinducer for the luminescence system of *Vibrio harveyi*. *J Biol Chem* **264**, 21670–21676.

Cha, C., Gao, P., Chen, Y. C., Shaw, P. D. & Farrand, S. K. (1998). Production of acyl-homoserine lactone quorum-sensing signals by Gram-negative plant-associated bacteria. *Mol Plant–Microbe Interact* **11**, 1119–1129.

Chai, Y., Zhu, J. & Winans, S. C. (2001). TrlR, a defective TraR-like protein of *Agrobacterium tumefaciens*, blocks TraR function *in vitro* by forming inactive TrlR : TraR dimers. *Mol Microbiol* **40**, 414–421.

Chapon-Hervé, V., Akrim, M., Latifi, A., Williams, P., Lazdunski, A. & Bally, M. (1997).

Regulation of the *xcp* secretion pathway by multiple quorum-sensing modulons in *Pseudomonas aeruginosa. Mol Microbiol* **24**, 1169–1178.

Chatterjee, A., Cui, Y., Liu, Y., Dumenyo, C. K. & Chatterjee, A. K. (1995). Inactivation of *rsmA* leads to overproduction of extracellular pectinases, cellulases, and proteases in *Erwinia carotovora* subsp. *carotovora* in the absence of the starvation/cell density-sensing signal, *N*-(3-oxohexanoyl)-L-homoserine lactone. *Appl Environ Microbiol* **61**, 1959–1967.

Chernin, L. S., Winson, M. K., Thompson, J. M., Haran, S., Bycroft, B. W., Chet, I., Williams, P. & Stewart, G. S. A. B. (1998). Chitinolytic activity in *Chromobacterium violaceum*: substrate analysis and regulation by quorum sensing. *J Bacteriol* **180**, 4435–4441.

Chin-A-Woeng, T. F. C., van den Broek, D., de Voer, G., van der Drift, K. M. G. M., Tuinman, S., Thomas-Oates, J. E., Lugtenberg, B. J. J. & Bloemberg, G. V. (2001). Phenazine-1-carboxamide production in the biocontrol strain *Pseudomonas chlororaphis* PCL1391 is regulated by multiple factors secreted into the growth medium. *Mol Plant–Microbe Interact* **14**, 969–979.

Choi, S. H. & Greenberg, E. P. (1991). The C-terminal region of the *Vibrio fischeri* LuxR protein contains an autoinducer-independent *lux* gene activating domain. *Proc Natl Acad Sci U S A* **88**, 11115–11119.

Choi, S. H. & Greenberg, E. P. (1992a). Genetic dissection of DNA binding and luminescence gene activation by the *Vibrio fischeri* LuxR protein. *J Bacteriol* **174**, 4064–4069.

Choi, S. H. & Greenberg, E. P. (1992b). Genetic evidence for multimerization of LuxR, the transcriptional activator of *Vibrio fischeri* luminescence. *Mol Mar Biol Biotechnol* **1**, 408–413.

Chugani, S. A., Whiteley, M., Lee, K. M., D'Argenio, D., Nanoil, C. & Greenberg, E. P. (2001). QscR, a modulator of quorum-sensing signal synthesis and virulence in *Pseudomonas aeruginosa. Proc Natl Acad Sci U S A* **98**, 2752–2757.

Chung, W. O., Park, Y., Lamont, R. J., McNab, R., Barbieri, B. & Demuth, D. R. (2001). Signaling system in *Porphyromonas gingivalis* based on a LuxS protein. *J Bacteriol* **183**, 3903–3909.

Costa, J. M. & Loper, J. E. (1997). EcbI and EcbR: homologs of LuxI and LuxR affecting antibiotic and exoenzyme production by *Erwinia carotovora* subsp. *betavasculorum. Can J Microbiol* **43**, 1164–1171.

Cox, A. R. J., Thomson, N. R., Bycroft, B., Stewart, G. S. A. B., Williams, P. & Salmond, G. P. C. (1998). A pheromone-independent CarR protein controls carbapenem antibiotic synthesis in the opportunistic human pathogen *Serratia marcescens. Microbiology* **144**, 201–209.

Davies, D. G., Parsek, M. R., Pearson, J. P., Iglewski, B. H., Costerton, J. W. & Greenberg, E. P. (1998). The involvement of cell-to-cell signals in the development of a bacterial biofilm. *Science* **280**, 295–298.

Day, W. A. & Maurelli, A. T. (2001). *Shigella flexneri* LuxS quorum-sensing system modulates *virB* expression but is not essential for virulence. *Infect Immun* **69**, 15–23.

Dessaux, Y., Petit, A., Farrand, S. K. & Murphy, P. J. (1998). Opines and opine-like molecules involved in plant-Rhizobiaceae interactions. In *The Rhizobiaceae: Molecular Biology of Modern Plant Associated Bacteria*, pp. 199–233. Edited by H. P. Spaink, A. Kondrossi & P. J. J. Hooykaas. Dordrecht: Kluwer.

Dong, Y.-H., Xu, J.-L., Li, X. Z. & Zhang, L.-H. (2000). AiiA, an enzyme that inactivates the acylhomoserine lactone quorum-sensing signal and attenuates the virulence of *Erwinia carotovora. Proc Natl Acad Sci U S A* **97**, 3526–3531.

Dong, Y.-H., Wang, L.-H., Xu, J.-L., Zhang, H.-B., Zhang, X.-F. & Zhang, L.-H. (2001). Quenching quorum-sensing-dependent bacterial infection by an *N*-acyl homoserine lactonase. *Nature* **411**, 813–817.

Dumenyo, C. K., Mukherjee, A., Chum, W. & Chatterjee, A. K. (1998). Genetic and physiological evidence for the production of *N*-acyl homoserine lactones by *Pseudomonas syringae* pv. *syringae* and other fluorescent plant pathogenic *Pseudomonas* species. *Eur J Plant Pathol* **104**, 569–582.

Dunlap, P. V. (1999). Quorum regulation of luminescence in *Vibrio fischeri*. *J Mol Microbiol Biotechnol* **1**, 5–12.

Dunny, G. M. & Winans, S. C. (1999). *Cell-cell Signalling in Bacteria*. Washington, DC: American Society for Microbiology.

Dunphy, G., Miyamoto, C. & Meighen, E. (1997). A homoserine lactone autoinducer regulates virulence of an insect-pathogenic bacterium, *Xenorhabdus nematophilus* (*Enterobacteriaceae*). *J Bacteriol* **179**, 5288–5291.

Eberl, L., Winson, M. K., Sternberg, C., Stewart, G. S. A. B., Christiansen, G., Chhabra, S. R., Bycroft, B., Williams, P., Molin, S. & Givskov, M. (1996). Involvement of *N*-acyl-L-homoserine lactone autoinducers in controlling the multicellular behaviour of *Serratia liquefaciens*. *Mol Microbiol* **20**, 127–136.

Egland, K. A. & Greenberg, E. P. (1999). Quorum sensing in *Vibrio fischeri*: elements of the *luxI* promoter. *Mol Microbiol* **31**, 1197–1204.

Egland, K. A. & Greenberg, E. P. (2001). Quorum sensing in *Vibrio fischeri*: analysis of the LuxR DNA binding region by alanine-scanning mutagenesis. *J Bacteriol* **183**, 382–386.

Elasri, M., Delorme, S., Lemanceau, P., Stewart, G., Laue, B., Glickmann, E., Oger, P. M. & Dessaux, Y. (2001). Acyl-homoserine lactone production is more common among plant-associated *Pseudomonas* spp. than among soilborne *Pseudomonas* spp. *Appl Environ Microbiol* **67**, 1198–1209.

El-Sayed, A. K., Hothersall, J. & Thomas, C. M. (2001). Quorum-sensing-dependent regulation of biosynthesis of the polyketide antibiotic mupirocin in *Pseudomonas fluorescens* NCIMB 10586. *Microbiology* **147**, 2127–2139.

Farrand, S. K. (1998). Conjugation in Rhizobiaceae. In *The Rhizobiaceae: Molecular Biology of Modern Plant Associated Bacteria*, pp. 199–233. Edited by H. P. Spaink, A. Kondrossi & P. J. J. Hooykaas. Dordrecht: Kluwer.

Flavier, A. B., Ganova-Raeva, L. M., Schell, M. A. & Denny, T. P. (1997). Hierarchical autoinduction in *Ralstonia solanacearum*: control of acyl-homoserine lactone production by a novel autoregulatory system responsive to 3-hydroxypalmitic acid methyl ester. *J Bacteriol* **179**, 7089–7097.

Forsyth, M. H. & Cover, T. L. (2000). Intercellular communication in *Helicobacter pylori*: *luxS* is essential for the production of an extracellular signalling molecule. *Infect Immun* **68**, 3193–3199.

Fray, R. G., Throup, J. P., Daykin, M., Wallace, A., Williams, P., Stewart, G. S. A. B. & Grierson, D. (1999). Plants genetically modified to produce *N*-acylhomoserine lactones communicate with bacteria. *Nat Biotechnol* **17**, 1017–1020.

Freeman, J. A. & Bassler, B. L. (1999a). A genetic analysis of the function of LuxO, a two-component response regulator involved in quorum sensing in *Vibrio harveyi*. *Mol Microbiol* **31**, 665–677.

Freeman, J. A. & Bassler, B. L. (1999b). Sequence and function of LuxU: a two-component phosphorelay protein that regulates quorum sensing in *Vibrio harveyi*. *J Bacteriol* **181**, 899–906.

Freeman, J. A., Lilley, B. N. & Bassler, B. L. (2000). A genetic analysis of the functions of LuxN: a two-component hybrid sensor kinase that regulates quorum sensing in *Vibrio harveyi. Mol Microbiol* **35**, 139–149.

Freiberg, C., Fellay, R., Bairoch, A., Broughton, W. J., Rosenthal, A. & Perret, X. (1997). Molecular basis of symbiosis between *Rhizobium* and legumes. *Nature* **387**, 394–401.

Frias, J., Olle, E. & Alsina, M. (2001). Periodontal pathogens produce quorum sensing signal molecules. *Infect Immun* **69**, 3431–3434.

Fuqua, W. C. & Winans, S. C. (1994). A LuxR-LuxI type regulatory system activates *Agrobacterium* Ti plasmid conjugal transfer in the presence of a plant tumour metabolite. *J Bacteriol* **176**, 2796–3806.

Fuqua, C. & Winans, S. C. (1996a). Localization of OccR-activated and TraR-activated promoters that express two ABC-type permeases and the *traR* gene of Ti plasmid pTiR10. *Mol Microbiol* **20**, 1199–1210.

Fuqua, C. & Winans, S. C. (1996b). Conserved *cis*-acting promoter elements are required for density-dependent transcription of *Agrobacterium tumefaciens* conjugal transfer genes. *J Bacteriol* **178**, 435–440.

Fuqua, C., Burbea, M. & Winans, S. C. (1995). Activity of the *Agrobacterium* Ti plasmid conjugal transfer regulator TraR is inhibited by the product of the *traM* gene. *J Bacteriol* **177**, 1367–1373.

Gambello, M. J. & Iglewski, B. H. (1991). Cloning and characterization of the *Pseudomonas aeruginosa lasR* gene, a transcriptional activator of elastase expression. *J Bacteriol* **173**, 3000–3009.

Gambello, M. J., Kaye, S. & Iglewski, B. H. (1993). LasR of *Pseudomonas aeruginosa* is a transcriptional activator of the alkaline protease gene (*apr*) and an enhancer of exotoxin A expression. *Infect Immun* **61**, 1180–1184.

Gardiner, S. M., Chhabra, S. R., Harty, C., Williams, P., Pritchard, D. I., Bycroft, B. W. & Bennett, T. (2001). Haemodynamic effects of the bacterial quorum sensing signal molecule, *N*-(3-oxododecanoyl)-L-homoserine lactone in conscious normal and endotoxaemic rats. *Br J Pharmacol* **133**, 1047–1054.

Gilson, L., Kuo, A. & Dunlap, P. V. (1995). AinS and a new family of autoinducer synthesis proteins. *J Bacteriol* **177**, 6946–6951.

Givskov, M., de Nys, R., Manefield, M., Gram, L., Maximilien, R., Eberl, L., Molin, S., Steinberg, P. D. & Kjelleberg, S. (1996). Eukaryotic interference with homoserine lactone-mediated prokaryotic signalling. *J Bacteriol* **178**, 6618–6622.

Glessner, A., Smith, R. S., Iglewski, B. H. & Robinson, J. B. (1999). Roles of *Pseudomonas aeruginosa las* and *rhl* quorum-sensing systems in control of twitching motility. *J Bacteriol* **181**, 1623–1629.

Gotschlich, A., Huber, B., Geisenberger, O. & 11 other authors (2001). Synthesis of multiple N-acylhomoserine lactones is wide-spread among the members of the *Burkholderia cepacia* complex. *Syst Appl Microbiol* **24**, 1–14.

Govan, J. R. W. & Deretic, V. (1996). Microbial pathogenesis in cystic fibrosis: mucoid *Pseudomonas aeruginosa* and *Burkholderia cepacia. Microbiol Rev* **60**, 539–574.

Gray, K. M. & Garey, J. R. (2001). The evolution of bacterial LuxI and LuxR quorum sensing regulators. *Microbiology* **147**, 2379–2387.

Gray, K. M., Pearson, J. P., Downie, J. A., Boboye, B. E. A. & Greenberg, E. P. (1996). Cell-to-cell signaling in the symbiotic nitrogen-fixing bacterium *Rhizobium leguminosarum*: autoinduction of a stationary phase and rhizosphere-expressed genes. *J Bacteriol* **178**, 372–376.

Greenberg, E. P., Hastings, J. W. & Ulitzur, S. (1979). Induction of luciferase synthesis in *Beneckea harveyi* by other marine bacteria. *Arch Microbiol* **120**, 87–91.

Habeeb, L. F., Wang, L. & Farrand, S. C. (1991). Transcription of the octopine catabolism operon of the *Agrobacterium* tumour-inducing plasmid pTiA6 is activated by a LysR-type regulatory protein. *Mol Plant–Microbe Interact* **4**, 379–385.

Hanzelka, B. L. & Greenberg, E. P. (1995). Evidence that the N-terminal region of the *Vibrio fischeri* LuxR protein constitutes an autoinducer-binding domain. *J Bacteriol* **177**, 815–817.

Hanzelka, B. L., Parsek, M. R., Val, D. L., Dunlap, P. V., Cronan, J. E., Jr & Greenberg, E. P. (1999). Acylhomoserine lactone synthase activity of the *Vibrio fischeri* AinS protein. *J Bacteriol* **181**, 5766–5770.

Harshey, R. M. (1994). Bees aren't the only ones: swarming in Gram-negative bacteria. *Mol Microbiol* **13**, 389–394.

Hassett, D. J., Ma, J. F., Elkins, J. G. & 10 other authors (1999). Quorum sensing in *Pseudomonas aeruginosa* controls expression of catalase and superoxide dismutase genes and mediates biofilm susceptibility to hydrogen peroxide. *Mol Microbiol* **34**, 1082–1093.

Hoang, T. T., Yufang, M., Stern, R. J., McNeil, M. R. & Schweizer, H. P. (1999). Construction and use of low-copy number T7 expression vectors for purification of problem proteins: purification of *Mycobacterium tuberculosis* RmlD and *Pseudomonas aeruginosa* LasI and RhlI proteins, and functional analysis of purified RhlI. *Gene* **237**, 361–371.

Holden, M. T. G., Chhabra, S. R., de Nys, R. & 14 other authors (1999). Quorum-sensing cross talk: isolation and chemical characterization of cyclic dipeptides from *Pseudomonas aeruginosa* and other Gram-negative bacteria. *Mol Microbiol* **33**, 1254–1266.

Hwang, I., Li, P. L., Zhang, L., Piper, K. R., Cook, D. M., Tate, M. E. & Farrand, S. K. (1994). TraI, a LuxI homologue, is responsible for production of conjugation factor, the Ti plasmid *N*-acylhomoserine lactone autoinducer. *Proc Natl Acad Sci U S A* **91**, 4639–4643.

Hwang, I., Cook, D. M. & Farrand, S. K. (1995). A new regulatory element modulates homoserine lactone-mediated autoinduction of Ti plasmid conjugal transfer. *J Bacteriol* **177**, 449–458.

Hwang, I., Smyth, A. J., Luo, Z. Q. & Farrand, S. K. (1999). Modulating quorum sensing by antiactivation: TraM interacts with TraR to inhibit activation of Ti plasmid conjugal transfer genes. *Mol Microbiol* **34**, 282–294.

Jander, G., Rahme, L. G. & Ausubel, F. M. (2000). Positive correlation between virulence of *Pseudomonas aeruginosa* mutants in mice and insects. *J Bacteriol* **182**, 3843–3845.

Jiang, Y., Camara, M., Chhabra, S. R., Hardie, K. R., Bycroft, B. W., Lazdunski, A., Salmond, G. P. C., Stewart, G. S. A. B. & Williams, P. (1998). In vitro biosynthesis of the *Pseudomonas aeruginosa* quorum-sensing signal molecule *N*-butanoyl-L-homoserine lactone. *Mol Microbiol* **28**, 193–203.

Jones, S., Yu, B., Bainton, N. J. & 11 other authors (1993). The *lux* autoinducer regulates the production of exoenzyme virulence determinants in *Erwinia carotovora* and *Pseudomonas aeruginosa*. *EMBO J* **12**, 2477–2482.

Joyce, E. A., Bassler, B. L. & Wright, A. (2000). Evidence for a signalling system in *Helicobacter pylori*: detection of a *luxS*-encoded autoinducer. *J Bacteriol* **182**, 3638–3643.

Kanamaru, K., Kanamaru, K., Tatsuno, I., Tobe, T. & Sasakawa, C. (2000a). SdiA, an

Escherichia coli homologue of quorum-sensing regulators, controls the expression of virulence factors in enterohaemorrhagic *Escherichia coli* O157 : H7. *Mol Microbiol* **38**, 805–816.

Kanamaru, K., Kanamaru, K., Tatsuno, I., Tobe, T. & Sasakawa, C. (2000b). Regulation of virulence factors of enterohemorrhagic *Escherichia coli* O157 : H7 by self-produced extracellular factors. *Biosci Biotechnol Biochem* **64**, 2508–2511.

Kaplan, H. B. & Greenberg, E. P. (1985). Diffusion of autoinducer is involved in regulation of the *Vibrio fischeri* luminescence system. *J Bacteriol* **163**, 1210–1214.

de Kievit, T., Seed, P. C., Nezezon, J., Passador, L. & Iglewski, B. H. (1999). RsaL, a novel repressor of virulence gene expression in *Pseudomonas aeruginosa*. *J Bacteriol* **181**, 2175–2184.

de Kievit, T. R., Gillis, R., Marx, S., Brown, C. & Iglewski, B. H. (2001). Quorum-sensing genes in *Pseudomonas aeruginosa* biofilms: their role and expression patterns. *Appl Environ Microbiol* **67**, 1865–1873.

Kim, H. & Farrand, S. K. (1997). Characterization of the *acc* operon from the nopaline-type Ti plasmid pTiC58, which encodes utilization of agrocinopines A and B and suscepti-bility to agrocin 84. *J Bacteriol* **179**, 7559–7572.

Kleerebezem, M., Quadri, L. E. N., Kuipers, O. P. & de Vos, W. M. (1997). Quorum sensing by peptide pheromones and two-component signal-transduction systems in Gram-positive bacteria. *Mol Microbiol* **24**, 895–904.

Köhler, T., Curty, L. K., Barja, F., van Delden, C. & Pechere, J. C. (2000). Swarming of *Pseudomonas aeruginosa* is dependent on cell-to-cell signaling and requires flagella and pili. *J Bacteriol* **182**, 5990–5996.

Latifi, A., Winson, M. K., Foglino, M., Bycroft, B. W., Stewart, G. S. A. B., Lazdunski, A. & Williams, P. (1995). Multiple homologues of LuxR and LuxI control expression of virulence determinants and secondary metabolites through quorum sensing in *Pseudomonas aeruginosa* PAO1. *Mol Microbiol* **17**, 333–343.

Latifi, A., Foglino, M., Tanaka, K., Williams, P. & Lazdunski, A. (1996). A hierarchical quorum-sensing cascade in *Pseudomonas aeruginosa* links the transcriptional activa-tors LasR and RhlR (VsmR) to expression of the stationary-phase sigma factor RpoS. *Mol Microbiol* **21**, 1137–1146.

Laue, B. E., Jiang, Y., Chhabra, S. R., Jacob, S., Stewart, G. S. A. B., Hardman, A., Downie, J. A., O'Gara, F. & Williams, P. (2000). The biocontrol strain *Pseudomonas fluorescens* F113 produces the *Rhizobium small* bacteriocin, *N*-(3-hydroxy-7-*cis*-tetradecenoyl) homoserine lactone, via HdtS, a putative novel *N*-acylhomoserine lactone synthase. *Microbiology* **146**, 2469–2480.

Lawrence, R. N., Dunn, W. R., Bycroft, B., Camara, M., Chhabra, S. R., Williams, P. & Wilson, V. G. (1999). The *Pseudomonas aeruginosa* quorum-sensing signal mol-ecule, *N*-(3-oxododecanoyl)-L-homoserine lactone, inhibits porcine arterial smooth muscle contraction. *Br J Pharmacol* **128**, 845–848.

Leadbetter, J. D. & Greenberg, E. P. (2000). Metabolism of acyl-homoserine lactone quorum-sensing signals by *Variovorax paradoxus*. *J Bacteriol* **182**, 6921–6926.

Lewenza, S. & Sokol, P. A. (2001). Regulation of ornibactin biosynthesis and *N*-acyl-L-homoserine lactone production by CepR in *Burkholderia cepacia*. *J Bacteriol* **183**, 2212–2218.

Lewenza, S., Conway, B., Greenberg, E. P. & Sokol, P. A. (1999). Quorum sensing in *Burkholderia cepacia*: identification of the LuxRI homologs CepRI. *J Bacteriol* **181**, 748–756.

Lewis, H. A., Furlong, E. B., Laubert, B. & 9 other authors (2001). A structural genomics approach to the study of quorum sensing: crystal structures of three LuxS orthologs. *Structure* **9**, 527–537.

Lilley, B. N. & Bassler, B. L. (2000). Regulation of quorum sensing in *Vibrio harveyi* by LuxO and sigma-54. *Mol Microbiol* **36**, 940–954.

Lindum, P. W., Anthoni, U., Christophersen, C., Eberl, L., Molin, S. & Givskov, M. (1998). *N*-Acyl-L-homoserine lactone autoinducers control production of an extracellular lipopeptide biosurfactant required for swarming motility of *Serratia liquefaciens* MG1. *J Bacteriol* **180**, 6384–6388.

de Lisa, M. P., Valdes, J. J. & Bentley, W. E. (2001). Mapping stress-induced changes in autoinducer AI-2 production in chemostat-cultivated *Escherichia coli* K-12. *J Bacteriol* **183**, 2918–2928.

Lithgow, J. K., Wilkinson, A., Hardman, A., Rodelas, B., Wisniewski-Dye, F., Williams, P. & Downie, J. A. (2000). The regulatory locus *cinRI* in *Rhizobium leguminosarum* controls a network of quorum-sensing loci. *Mol Microbiol* **37**, 81–97.

Lithgow, J. K., Danino, V. E., Jones, J. & Downie, J. A. (2001). Analysis of *N*-acyl homoserine-lactone quorum-sensing molecules made by different strains and biovars of *Rhizobium leguminosarum* containing different symbiotic plasmids. *Plant Soil* **232**, 3–12.

Luo, Z. Q. & Farrand, S. K. (1999). Signal-dependent DNA binding and functional domains of the quorum-sensing activator TraR as identified by repressor activity. *Proc Natl Acad Sci U S A* **96**, 9009–9014.

Luo, Z. Q., Qin, Y. & Farrand, S. K. (2000). The antiactivator TraM interferes with the autoinducer-dependent binding of TraR to DNA by interfering with the C-terminal region of the quorum-sensing activator. *J Biol Chem* **275**, 7713–7722.

Lutter, E., Lewenza, S., Dennis, J. J., Visser, M. B. & Sokol, P. A. (2001). Distribution of quorum-sensing genes in the *Burkholderia cepacia* complex. *Infect Immun* **69**, 4661–4666.

McClean, K. H., Winson, M. K., Fish, L. & 9 other authors (1997). Quorum sensing and *Chromobacterium violaceum*: exploitation of violacein production and inhibition for the detection of *N*-acylhomoserine lactones. *Microbiology* **143**, 3703–3711.

McGowan, S., Sebaiha, M., Jones, S., Yu, B., Bainton, N., Chan, P. F., Bycroft, B., Stewart, G. S. A. B., Williams, P. & Salmond, G. P. C. (1995). Carbapenem antibiotic production in *Erwinia carotovora* is regulated by CarR, a homologue of the LuxR transcriptional activator. *Microbiology* **141**, 541–550.

McGowan, S. J., Sebaihia, M., Porter, L. E., Stewart, G. S. A. B., Williams, P., Bycroft, B. W. & Salmond, G. P. C. (1996). Analysis of bacterial carbapenem antibiotic production genes reveals a novel β-lactam biosynthesis pathway. *Mol Microbiol* **22**, 415–426.

McGowan, S. J., Sebaihia, M., O'Leary, S., Hardie, K. R., Williams, P., Stewart, G. S. A. B., Bycroft, B. W. & Salmond, G. P. C. (1997). Analysis of the carbapenem gene cluster of *Erwinia carotovora*: definition of the antibiotic biosynthetic genes and evidence for a novel β-lactam resistance mechanism. *Mol Microbiol* **26**, 545–556.

McGowan, S. J., Holden, M. T. G., Bycroft, B. W. & Salmond, G. P. C. (1999). Molecular genetics of carbapenem antibiotic biosynthesis. *Antonie Leeuwenhoek Int J Gen Mol Microbiol* **75**, 135–141.

McKnight, S. L., Iglewski, B. H. & Pesci, E. C. (2000). The *Pseudomonas* quinolone signal regulates *rhl* quorum sensing in *Pseudomonas aeruginosa*. *J Bacteriol* **182**, 2702–2708.

Manefield, M., de Nys, R., Kumar, N., Read, R., Givskov, M., Steinberg, P. & Kjelleberg,

S. (1999). Evidence that halogenated furanones from *Delisea pulchra* inhibit acylated homoserine lactone (AHL)-mediated gene expression by displacing the AHL signal from its receptor protein. *Microbiology* **145**, 283–291.

Matsuyama, T., Kaneda, K., Nakagawa, Y., Isa, K., Hara-Hotta, H. & Yano, I. (1992). A novel extracellular cyclic lipopeptide which promotes flagellum-dependent and independent spreading growth of *Serratia marcescens*. *J Bacteriol* **174**, 1769–1776.

Mavrodi, D. V., Ksenzenko, V. N., Bonsall, R. F., Cook, R. J., Boronin, A. M. & Thomashow, L. S. (1998). A seven-gene locus for synthesis of phenazine-1-carboxylic acid by *Pseudomonas fluorescens* 2-79. *J Bacteriol* **180**, 2541–2548.

Milton, D. L., Hardman, A., Camara, M., Chhabra, S. R., Bycroft, B. W., Stewart, G. S. A. B. & Williams, P. (1997). Quorum sensing in *Vibrio anguillarum*: characterization of the *vanI/vanR* locus and identification of the autoinducer *N*-(3-oxodecanoyl)-L-homoserine lactone. *J Bacteriol* **179**, 3004–3012.

Milton, D. L., Chalker, V. J., Kirke, D., Hardman, A., Camara, M. & Williams, P. (2001). The LuxM homologue VanM from *Vibrio anguillarum* directs the synthesis of *N*-(3-hydroxyhexanoyl) homoserine lactone and *N*-hexanoyl homoserine lactone. *J Bacteriol* **183**, 3537–3547.

Miyamoto, C. M., Lin, Y. H. & Meighen, E. A. (2000). Control of bioluminescence in *Vibrio fischeri* by the LuxO signal response regulator. *Mol Microbiol* **36**, 594–607.

Moré, M. I., Finger, D., Stryker, J. L., Fuqua, C., Eberhard, A. & Winans, S. C. (1996). Enzymatic synthesis of a quorum-sensing autoinducer through use of defined substrates. *Science* **272**, 1655–1658.

Nasser, W., Bouillant, M. L., Salmond, G. & Reverchon, S. (1998). Characterization of the *Erwinia chrysanthemi expI-expR* locus directing the synthesis of two *N*-acyl-homoserine lactone signal molecules. *Mol Microbiol* **29**, 1391–1405.

Ochsner, U. A. & Reiser, J. (1995). Autoinducer-mediated regulation of rhamnolipid biosurfactant synthesis in *Pseudomonas aeruginosa*. *Proc Natl Acad Sci U S A* **92**, 6424–6428.

Ochsner, U. A., Koch, A. K., Fiechter, A. & Reiser, J. (1994). Isolation and characterization of a regulatory gene affecting rhamnolipid biosurfactant synthesis in *Pseudomonas aeruginosa*. *J Bacteriol* **176**, 2044–2054.

Oger, P., Kim, K. S., Sackett, R. L., Piper, K. R. & Farrand, S. K. (1998). Octopine-type Ti plasmids code for a mannopine-inducible dominant-negative allele of *traR*, the quorum-sensing activator that regulates Ti plasmid conjugal transfer. *Mol Microbiol* **27**, 277–288.

O'Toole, G. A. & Kolter, R. (1998). Flagellar and twitching motility are necessary for *Pseudomonas aeruginosa* biofilm development. *Mol Microbiol* **30**, 295–304.

Parker, W. L., Rathnum, M. L., Wells, J. S., Trejo, W. H., Principe, P. A. & Sykes, R. B. (1982). SQ27850, a simple carbapenem produced by species of *Serratia* and *Erwinia*. *J Antibiot* **35**, 653–660.

Parsek, M. R., Val, D. L., Hanzelka, B. L., Cronan, J. E., Jr & Greenberg, E. P. (1999). Acyl homoserine-lactone quorum-sensing signal generation. *Proc Natl Acad Sci U S A* **96**, 4360–4365.

Passador, L., Cook, J. M., Gambello, M. J., Rust, L. & Iglewski, B. H. (1993). Expression of *Pseudomonas aeruginosa* virulence genes requires cell-to-cell communication. *Science* **260**, 1127–1130.

Pearson, J. P., Gray, K. M., Passador, L., Tucker, K. D., Eberhard, A., Iglewski, B. H. & Greenberg, E. P. (1994). Structure of the autoinducer required for expression of *Pseudomonas aeruginosa* virulence genes. *Proc Natl Acad Sci U S A* **91**, 197–201.

Pearson, J. P., Pesci, E. C. & Iglewski, B. H. (1997). Roles of *Pseudomonas aeruginosa las* and *rhl* quorum-sensing systems in control of elastase and rhamnolipid biosynthesis genes. *J Bacteriol* **179**, 5756–5767.

Pearson, J. P., van Delden, C. & Iglewski, B. H. (1999). Active efflux and diffusion are involved in transport of *Pseudomonas aeruginosa* cell-to-cell signals. *J Bacteriol* **181**, 1203–1210.

Pearson, J. P., Feldman, M., Iglewski, B. H. & Prince, A. (2000). *Pseudomonas aeruginosa* cell-to-cell signaling is required for virulence in a model of acute pulmonary infection. *Infect Immun* **68**, 4331–4334.

Perombelon, M. C. M. & Kelman, A. (1980). Ecology of the soft rot erwinias. *Annu Rev Phytopathol* **18**, 361–387.

Pesci, E. C., Pearson, J. P., Seed, P. C. & Iglewski, B. H. (1997). Regulation of *las* and *rhl* quorum sensing in *Pseudomonas aeruginosa*. *J Bacteriol* **179**, 3127–3132.

Pesci, E. C., Milbank, J. B. J., Pearson, J. P., McKnight, S., Kende, A. S., Greenberg, E. P. & Iglewski, B. H. (1999). Quinolone signaling in the cell-to-cell communication system of *Pseudomonas aeruginosa*. *Proc Natl Acad Sci U S A* **96**, 11229–11234.

Pessi, G. & Haas, D. (2000). Transcriptional control of the hydrogen cyanide biosynthetic genes *hcnABC* by the anaerobic regulator ANR and the quorum-sensing regulators LasR and RhlR in *Pseudomonas aeruginosa*. *J Bacteriol* **182**, 6940–6949.

Pierson, L. S., III & Pierson, E. A. (1996). Phenazine antibiotic production in *Pseudomonas aureofaciens*: role in rhizosphere ecology and pathogen suppression. *FEMS Microbiol Lett* **136**, 101–108.

Pierson, L. S., III, Keppenne, V. D. & Wood, D. W. (1994). Phenazine antibiotic biosynthesis in *Pseudomonas aureofaciens* 30-84 is regulated by PhzR in response to cell density. *J Bacteriol* **176**, 3966–3974.

Piper, K. R. & Farrand, S. K. (2000). Quorum sensing but not autoinduction of Ti plasmid conjugal transfer requires control by the opine regulon and the antiactivator TraM. *J Bacteriol* **182**, 1080–1088.

Piper, K. R., Beck von Bodman, S. & Farrand, S. K. (1993). Conjugation factor of *Agrobacterium tumefaciens* regulates Ti plasmid transfer by autoinduction. *Nature* **362**, 448–450.

Piper, K. R., Beck von Bodman, S., Hwang, I. & Farrand, S. K. (1999). Hierarchical gene regulatory systems arising from fortuitous gene associations: controlling quorum sensing by the opine regulon in *Agrobacterium*. *Mol Microbiol* **32**, 1077–1089.

Pirhonen, M., Flego, D., Heikinheimo, R. & Palva, E. T. (1993). A small diffusible signal molecule is responsible for the global control of virulence and exoenzyme production in *Erwinia carotovora*. *EMBO J* **12**, 2467–2476.

Pollack, M. (1990). *Pseudomonas aeruginosa*. In *Principles and Practice of Infectious Disease*, pp. 1673–1691. Edited by G. L. Mandell, R. G. Douglas & J. E. Bennett. New York: Churchill Livingstone.

Puskas, A., Greenberg, E. P., Kaplan, S. & Schaefer, A. L. (1997). A quorum-sensing system in the free-living photosynthetic bacterium *Rhodobacter sphaeroides*. *J Bacteriol* **179**, 7530–7537.

Py, B., Barras, F., Harris, S., Robson, N. & Salmond, G. P. C. (1998). Extracellular enzymes and their role in *Erwinia* virulence. *Methods Microbiol* **27**, 158–168.

Qin, Y., Luo, Z. Q., Smyth, A. J., Gao, P., Beck von Bodman, S. & Farrand, S. K. (2000). Quorum-sensing signal binding results in dimerization of TraR and its release from membranes into the cytoplasm. *EMBO J* **19**, 5212–5221.

Rashid, M. H., Rumbaugh, K., Passador, L., Davies, D. G., Hamood, A. N., Iglewski,

B. H. & Kornberg, A. (2000). Polyphosphate kinase is essential for biofilm development, quorum sensing, and virulence of *Pseudomonas aeruginosa*. *Proc Natl Acad Sci U S A* **97**, 9636–9641.

Rasmussen, T. B., Manefield, M., Andersen, J. B., Eberl, L., Anthoni, U., Christophersen, C., Steinberg, P., Kjelleberg, S. & Givskov, M. (2000). How *Delisea pulchra* furanones affect quorum sensing and swarming motility in *Serratia liquefaciens* MG1. *Microbiology* **146**, 3237–3244.

Ravn, L., Beck Christensen, A., Molin, S., Givskov, M. & Gram, L. (2001). Methods for detecting acylated homoserine lactones produced by Gram-negative bacteria and their application in studies of AHL-production kinetics. *J Microbiol Methods* **44**, 239–251.

Reimmann, C., Beyeler, M., Latifi, A., Winteler, H., Foglino, M., Lazdunski, A. & Haas, D. (1997). The global activator GacA of *Pseudomonas aeruginosa* PAO positively controls the production of the autoinducer *N*-butyryl-homoserine lactone and the formation of the virulence factors pyocyanin, cyanide, and lipase. *Mol Microbiol* **24**, 309–319.

Reverchon, S., Bouillant, M. L., Salmond, G. & Nasser, W. (1998). Integration of the quorum-sensing system in the regulatory networks controlling virulence factor synthesis in *Erwinia chrysanthemi*. *Mol Microbiol* **29**, 1407–1418.

Rhodius, V. A. & Busby, S. J. W. (1998). Positive activation of gene expression. *Curr Opin Microbiol* **1**, 152–159.

Riedel, K., Ohnesorg, T., Krogfelt, K. A., Hansen, T. S., Omori, K., Givskov, M. & Eberl, L. (2001). *N*-Acyl-homoserine lactone-mediated regulation of the Lip secretion system in *Serratia liquefaciens* MG1. *J Bacteriol* **183**, 1805–1809.

Robson, N. D., Cox, A. R. J., McGowan, S. J., Bycroft, B. W. & Salmond, G. P. C. (1997). Bacterial *N*-acyl-homoserine-lactone-dependent signalling and its potential biotechnological applications. *Trends Biotechnol* **15**, 458–464.

Rodelas, B., Lithgow, J. K., Wisniewski-Dye, F., Hardman, A., Wilkinson, A., Economou, A., Williams, P. & Downie, J. A. (1999). Analysis of quorum-sensing-dependent control of rhizosphere-expressed (*rhi*) genes in *Rhizobium leguminosarum* bv. *viciae*. *J Bacteriol* **181**, 3816–3823.

Rosemeyer, V., Michiels, J., Verreth, C. & Vanderleyden, J. (1998). *luxI*- and *luxR*-homologous genes of *Rhizobium etli* CNPAF512 contribute to synthesis of autoinducer molecules and nodulation of *Phaseolus vulgaris*. *J Bacteriol* **180**, 815–821.

Rumbaugh, K. P., Griswold, J. A., Iglewski, B. H. & Hamood, A. N. (1999). Contribution of quorum sensing to the virulence of *Pseudomonas aeruginosa* in burn wound infections. *Infect Immun* **67**, 5854–5862.

Salmond, G. P. C., Bycroft, B. W., Stewart, G. S. A. B. & Williams, P. (1995). The bacterial 'enigma': cracking the code of cell-cell communication. *Mol Microbiol* **16**, 615–624.

Schaefer, A. L., Val, D. L., Hanzelka, B. L., Cronan, J. E., Jr & Greenberg, E. P. (1996). Generation of cell-to-cell signals in quorum sensing: acyl homoserine lactone synthase activity of a purified *Vibrio fischeri* LuxI protein. *Proc Natl Acad Sci U S A* **93**, 9505–9509.

Schauder, S. & Bassler, B. L. (2001). The languages of bacteria. *Genes Dev* **15**, 1468–1480.

Schauder, S., Shokat, K., Surette, M. G. & Bassler, B. L. (2001). The LuxS family of bacterial autoinducers: biosynthesis of a novel quorum-sensing signal molecule. *Mol Microbiol* **41**, 463–476.

Schell, M. A. (2000). Control of virulence and pathogenicity genes of *Ralstonia solanacearum* by an elaborate sensory network. *Annu Rev Phytopathol* **38**, 263–292.

Schripsema, J., de Rudder, K. E. E., van Vliet, T. B., Lankhorst, P. P., de Vroom, E., Kijne, J. W. & van Brussel, A. A. N. (1996). Bacteriocin *small* of *Rhizobium leguminosarum* belongs to the class of *N*-acyl-L-homoserine lactone molecules, known as autoinducers and as quorum sensing co-transcription factors. *J Bacteriol* **178**, 366–371.

Shadel, G. S., Young, R. & Baldwin, T. O. (1990). Use of regulated cell lysis in a lethal genetic selection in *Escherichia coli*: identification of the autoinducer-binding region of the LuxR protein from *Vibrio fischeri* ATCC7744. *J Bacteriol* **172**, 3980–3987.

Shapiro, J. A. (1998). Thinking about bacterial populations as multicellular organisms. *Annu Rev Microbiol* **52**, 81–104.

Shaw, P. D., Ping, G., Daly, S. L., Cha, C., Cronan, J. E., Jr, Rinehart, K. L. & Farrand, S. K. (1997). Detecting and characterizing *N*-acyl-homoserine lactone signal molecules by thin-layer chromatography. *Proc Natl Acad Sci U S A* **94**, 6036–6041.

Showalter, R. E., Martin, M. O. & Silverman, M. R. (1990). Cloning and nucleotide sequence of *luxR*, a regulatory gene controlling bioluminescence in *Vibrio harveyi*. *J Bacteriol* **172**, 2946–2954.

Sitnikov, D. M., Schineller, J. B. & Baldwin, T. O. (1996). Control of cell division in *Escherichia coli*: regulation of transcription of *ftsQA* involves both *rpoS* and SdiA-mediated autoinduction. *Proc Natl Acad Sci U S A* **93**, 336–341.

Slock, J., van Riet, D., Kolibachuk, D. & Greenberg, E. P. (1990). Critical regions of the *Vibrio fischeri* LuxR protein defined by mutational analysis. *J Bacteriol* **172**, 3974–3979.

Sperandio, V., Mellies, J. L., Nguyen, W., Shin, S. & Kaper, J. B. (1999). Quorum sensing controls expression of the type III secretion gene transcription and protein secretion in enterohemorrhagic and enteropathogenic *Escherichia coli*. *Proc Natl Acad Sci U S A* **96**, 15196–15201.

Sperandio, V., Torres, A. G., Giron, J. A. & Kaper, J. B. (2001). Quorum sensing is a global regulatory mechanism in enterohemorrhagic *Escherichia coli* O157 : H7. *J Bacteriol* **183**, 5187–5197.

Stevens, A. M. & Greenberg, E. P. (1997). Quorum sensing in *Vibrio fischeri*: essential elements for activation of the luminescence genes. *J Bacteriol* **179**, 557–562.

Stevens, A. M., Dolan, K. M. & Greenberg, E. P. (1994). Synergistic binding of the *Vibrio fischeri* LuxR transcriptional activator domain and RNA polymerase to the *lux* promoter region. *Proc Natl Acad Sci U S A* **91**, 12619–12623.

Stevens, A. M., Fujita, N., Ishihama, A. & Greenberg, E. P. (1999). Involvement of the RNA polymerase α-subunit C-terminal domain in LuxR-dependent activation of the *Vibrio fischeri* luminescence genes. *J Bacteriol* **181**, 4704–4707.

Stintzi, A., Evans, K., Meyer, J. M. & Poole, K. (1998). Quorum-sensing and siderophore biosynthesis in *Pseudomonas aeruginosa*: *lasR/lasI* mutants exhibit reduced pyoverdine biosynthesis. *FEMS Microbiol Lett* **166**, 341–345.

Suh, S. J., Silo-Suh, L., Woods, D. E., Hassett, D. J., West, S. E. H. & Ohman, D. E. (1999). Effect of *rpoS* mutation on the stress response and expression of virulence factors in *Pseudomonas aeruginosa*. *J Bacteriol* **181**, 3890–3897.

Surette, M. G. & Bassler, B. L. (1998). Quorum sensing in *Escherichia coli* and *Salmonella typhimurium*. *Proc Natl Acad Sci U S A* **95**, 7046–7050.

Surette, M. G. & Bassler, B. L. (1999). Regulation of autoinducer production in *Salmonella typhimurium*. *Mol Microbiol* **31**, 585–595.

Surette, M. G., Miller, M. B. & Bassler, B. L. (1999). Quorum sensing in *Escherichia coli*, *Salmonella typhimurium*, and *Vibrio harveyi*: a new family of genes responsible for autoinducer production. *Proc Natl Acad Sci U S A* **96**, 1639–1644.

Swartzman, E., Silverman, M. & Meighen, E. A. (1992). The *luxR* gene product is a transcriptional activator of the *lux* promoter. *J Bacteriol* **174**, 7490–7493.

Swift, S., Winson, M. K., Chan, P. F. & 11 other authors (1993). A novel strategy for the isolation of *luxI* homologues: evidence for the widespread distribution of a LuxR : LuxI superfamily in enteric bacteria. *Mol Microbiol* **10**, 511–520.

Swift, S., Karlyshev, A. V., Fish, L., Durant, E. L., Winson, M. K., Chhabra, S. R., Williams, P., Macintyre, S. & Stewart, G. S. A. B. (1997). Quorum sensing in *Aeromonas hydrophila* and *Aeromonas salmonicida*: identification of the LuxRI homologs AhyRI and AsaRI and their cognate *N*-acylhomoserine lactone signal molecules. *J Bacteriol* **179**, 5271–5281.

Swift, S., Isherwood, K. E., Atkinson, S., Oyston, P. C. F. & Stewart, G. S. A. B. (1999a). Quorum sensing in *Aeromonas* and *Yersinia*. In *Microbial Signalling and Communication*, pp. 85–104. Society for General Microbiology Symposium 57. Edited by R. England, G. Hobbs, N. Bainton & D. Roberts. Cambridge: Cambridge University Press.

Swift, S., Lynch, M. J., Fish, L., Kirke, D. F., Tomas, J. M., Stewart, G. S. A. B. & Williams, P. (1999b). Quorum sensing-dependent regulation and blockade of exoprotease production in *Aeromonas hydrophila*. *Infect Immun* **67**, 5192–5199.

Swift, S., Downie, J. A., Whitehead, N. A., Barnard, A. M. L., Salmond, G. P. C. & Williams, P. (2001). Quorum sensing as a population-density-dependent determinant of bacterial physiology. *Adv Microb Physiol* **45**, 199–270.

Tan, M. W., Rahme, L. G., Sternberg, J. A., Tompkins, R. G. & Ausubel, F. M. (1999). *Pseudomonas aeruginosa* killing of *Caenorhabditis elegans* to identify *P. aeruginosa* virulence factors. *Proc Natl Acad Sci U S A* **96**, 2408–2413.

Tang, H. B., Dimango, E., Bryan, R., Gambello, M. J., Iglewski, B. H., Goldberg, J. B. & Prince, A. (1996). Contribution of specific *Pseudomonas aeruginosa* virulence factors to pathogenesis of pneumonia in a neonatal mouse model of infection. *Infect Immun* **64**, 37–43.

Telford, G., Wheeler, D., Williams, P., Tomkins, P. T., Appleby, P., Sewell, H., Stewart, G. S. A. B., Bycroft, B. W. & Pritchard, D. I. (1998). The *Pseudomonas aeruginosa* quorum-sensing signal molecule *N*-(3-oxododecanoyl)-L-homoserine lactone has immunomodulatory activity. *Infect Immun* **66**, 36–42.

Temprano, A., Yugueros, J., Hernanz, C., Sanchez, M., Berzal, B., Luengo, J. M. & Naharro, G. (2001). Rapid identification of *Yersinia ruckeri* by PCR amplification of *yruI-yruR* quorum sensing. *J Fish Dis* **24**, 253–261.

Teplitski, M., Robinson, J. B. & Bauer, W. D. (2000). Plants secrete substances that mimic bacterial *N*-acyl homoserine lactone signal activities and affect population density-dependent behaviors in associated bacteria. *Mol Plant–Microbe Interact* **13**, 637–648.

Thomashow, L. S. & Weller, D. M. (1988). Role of a phenazine antibiotic from *Pseudomonas fluorescens* in biological control of *Gaeumannomyces graminis* var. *tritici. J Bacteriol* **170**, 3499–3508.

Thomson, N. R., Thomas, J. D. & Salmond, G. P. C. (1999). Virulence determinants in the bacterial phytopathogen *Erwinia*. *Methods Microbiol* **29**, 347–426.

Thomson, N. R., Crow, M. A., McGowan, S. J., Cox, A. & Salmond, G. P. C. (2000). Biosynthesis of carbapenem antibiotic and prodigiosin pigment in *Serratia* is under quorum sensing control. *Mol Microbiol* **36**, 539–556.

Throup, J., Winson, M. K., Bainton, N. J., Bycroft, B. W., Williams, P. & Stewart, G. S. A. B. (1995a). Signalling in bacteria beyond bioluminescence. In *Bioluminescence and*

Chemiluminescence: Fundamentals and Applied Aspects, pp. 89–92. Edited by A. Campbell, L. Kricka & P. Stanley. Chichester: Wiley.

Throup, J. P., Camara, M., Briggs, G. S., Winson, M. K., Chhabra, S. R., Bycroft, B. W., Williams, P. & Stewart, G. S. A. B. (1995b). Characterisation of the *yenI/yenR* locus from *Yersinia enterocolitica* mediating the synthesis of two *N*-acyl homoserine lactone signal molecules. *Mol Microbiol* **17**, 345–356.

Toder, D. S., Gambello, M. J. & Iglewski, B. H. (1991). *Pseudomonas aeruginosa* LasA: a second elastase under the transcriptional control of *lasR*. *Mol Microbiol* **5**, 2003–2010.

Wang, X. D., de Boer, P. A. J. & Rothfield, L. I. (1991). A factor that positively regulates cell division by activating transcription of the major cluster of essential cell division genes of *Escherichia coli*. *EMBO J* **10**, 3363–3372.

Wang, L., Helmann, J. D. & Winans, S. C. (1992). The *A. tumefaciens* transcriptional activator OccR causes a bend at a target promoter, which is partially relaxed by a plant tumour metabolite. *Cell* **69**, 659–667.

Wei, Y., Lee, J. M., Smulski, D. R. & La Rossa, R. A. (2001). Global impact of *sdiA* amplification revealed by comprehensive gene expression profiling of *Escherichia coli*. *J Bacteriol* **183**, 2265–2272.

Welch, M., Todd, D. E., Whitehead, N. A., McGowan, S. J., Bycroft, B. W. & Salmond, G. P. C. (2000). *N*-Acyl homoserine lactone binding to the CarR receptor determines quorum-sensing specificity in *Erwinia*. *EMBO J* **19**, 631–641.

Weller, D. M. & Cook, R. J. (1983). Suppression of take-all of wheat by seed treatments with fluorescent pseudomonads. *Phytopathology* **73**, 463–469.

Whitehead, N. A. & Salmond, G. P. C. (2000). Quorum sensing and the role of diffusible signalling molecules in plant-microbe interactions. In *Plant-Microbe Interactions*, vol. 5, pp. 43–92. Edited by G. Stacey & N. T. Keen. Minnesota: APS Press.

Whitehead, N. A., Barnard, A. M. L., Slater, H., Simpson, N. J. L. & Salmond, G. P. C. (2001a). Quorum sensing in Gram-negative bacteria. *FEMS Microbiol Rev* **25**, 365–404.

Whitehead, N. A., Welch, M. & Salmond, G. P. C. (2001b). Silencing the majority. *Nat Biotechnol* **19**, 735–736.

Whiteley, M., Lee, K. M. & Greenberg, E. P. (1999). Identification of genes controlled by quorum sensing in *Pseudomonas aeruginosa*. *Proc Natl Acad Sci U S A* **96**, 13904–13909.

Whiteley, M., Parsek, M. R. & Greenberg, E. P. (2000). Regulation of quorum sensing by RpoS in *Pseudomonas aeruginosa*. *J Bacteriol* **182**, 4356–4360.

Williams, P., Bainton, N. J., Swift, S., Chhabra, S. R., Winson, M. K., Stewart, G. S. A. B., Salmond, G. P. C. & Bycroft, B. W. (1992). Small-molecule mediated density-dependent control of gene expression in prokaryotes: bioluminescence and the biosynthesis of carbapenem antibiotics. *FEMS Microbiol Lett* **100**, 161–168.

Williams, P., Camara, M., Hardman, A., Swift, S., Milton, D., Hope, V. J., Winzer, K., Middleton, B., Pritchard, D. I. & Bycroft, B. W. (2000). Quorum sensing and the population-dependent control of virulence. *Philos Trans R Soc Lond B* **355**, 667–680.

Winson, M. K., Camara, M., Latifi, A. & 10 other authors (1995). Multiple *N*-acyl-L-homoserine lactone signal molecules regulate production of virulence determinants and secondary metabolites in *Pseudomonas aeruginosa*. *Proc Natl Acad Sci U S A* **92**, 9427–9431.

Winson, M. K., Swift, S., Fish, L., Throup, J. P., Jorgensen, F., Chhabra, S. R., Bycroft, B. W., Williams, P. & Stewart, G. S. A. B. (1998). Construction and analysis of *luxCDABE*-based plasmid sensors for investigating *N*-acyl homoserine lactone-mediated quorum sensing. *FEMS Microbiol Lett* **163**, 185–192.

Winzer, K. & Williams, P. (2001). Quorum sensing and the regulation of virulence gene expression in pathogenic bacteria. *Int J Med Microbiol* **291**, 131–143.

Winzer, K., Falconer, C., Garber, N. C., Diggle, S. P., Camara, M. & Williams, P. (2000). The *Pseudomonas aeruginosa* lectins PA-IL and PA-IIL are controlled by quorum sensing and by RpoS. *J Bacteriol* **182**, 6401–6411.

Withers, H., Swift, S. & Williams, P. (2001). Quorum sensing as an integral component of gene regulatory networks in Gram-negative bacteria. *Curr Opin Microbiol* **4**, 186–193.

Wood, D. W. & Pierson, L. S., III (1996). The *phzI* gene of *Pseudomonas aureofaciens* 30-84 is responsible for the production of a diffusible signal required for phenazine antibiotic production. *Gene* **168**, 49–53.

Wood, D. W., Gong, F., Daykin, M. M., Williams, P. & Pierson, L. S., III (1997). *N*-Acyl-homoserine lactone-mediated regulation of phenazine gene expression by *Pseudomonas aureofaciens* 30-84 in the wheat rhizosphere. *J Bacteriol* **179**, 7663–7670.

Wu, H., Song, Z., Givskov, M., Doring, G., Worlitzsch, D., Mathee, K., Rygaard, J. & Høiby, N. (2001). *Pseudomonas aeruginosa* mutations in *lasI* and *rhlI* quorum sensing systems result in milder chronic lung infection. *Microbiology* **147**, 1105–1113.

Zhang, L., Murphy, P. J., Kerr, A. & Tate, M. E. (1993). *Agrobacterium* conjugation and gene regulation by *N*-acyl-L-homoserine lactones. *Nature* **362**, 446–448.

Zhu, J. & Winans, C. (1998). Activity of the quorum-sensing regulator TraR of *Agrobacterium tumefaciens* is inhibited by a truncated, dominant defective TraR-like protein. *Mol Microbiol* **27**, 289–297.

Zhu, J. & Winans, S. C. (1999). Autoinducer binding by the quorum-sensing regulator TraR increases affinity for target promoters *in vitro* and decreases TraR turnover rates in whole cells. *Proc Natl Acad Sci U S A* **96**, 4832–4837.

Zhu, J. & Winans, S. C. (2001). The quorum-sensing transcriptional regulator TraR requires its cognate signaling ligand for protein folding, protease resistance, and dimerization. *Proc Natl Acad Sci U S A* **98**, 1507–1512.

Zhu, J., Oger, P. M., Schrammeijer, B., Hooykaas, P. J. J., Farrand, S. K. & Winans, S. C. (2000). The bases of crown gall tumorigenesis. *J Bacteriol* **182**, 3885–3895.

INDEX

References to tables/figures are shown in italics